Redirecting Science

Niels Bohr, 1885–1962.

Redirecting Science

NIELS BOHR, PHILANTHROPY,
AND THE RISE OF NUCLEAR PHYSICS

Finn Aaserud

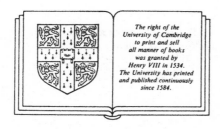

The right of the
University of Cambridge
to print and sell
all manner of books
was granted by
Henry VIII in 1534.
The University has printed
and published continuously
since 1584.

CAMBRIDGE UNIVERSITY PRESS

CAMBRIDGE
NEW YORK PORT CHESTER MELBOURNE SYDNEY

PUBLISHED BY THE PRESS SYNDICATE OF THE UNIVERSITY OF CAMBRIDGE
The Pitt Building, Trumpington Street, Cambridge, United Kingdom

CAMBRIDGE UNIVERSITY PRESS
The Edinburgh Building, Cambridge CB2 2RU, UK
40 West 20th Street, New York NY 10011–4211, USA
477 Williamstown Road, Port Melbourne, VIC 3207, Australia
Ruiz de Alarcón 13, 28014 Madrid, Spain
Dock House, The Waterfront, Cape Town 8001, South Africa

http://www.cambridge.org

© Cambridge University Press 1990

First published 1990
First paperback edition 2002

A catalogue record for this book is available from the British Library

Library of Congress Cataloging-in-Publication Data
Aaserud, Finn.
Redirecting science : Niels Bohr, philanthropy, and the rise of nuclear
physics / Finn Aaserud.
p.   cm.
Includes bibliographical references.
ISBN 0 521 35366 1 hardback
1. Nuclear physics – Research – Denmark – History.   2. Science and state –
Denmark – History.   3. Research – Denmark – Finance – History.
4. Bohr, Niels Henrik David, 1885–1962.   I. Title.
QC789.2.D4A27   1990
539.7′0720489 – dc20   89-48317
CIP

ISBN 0 521 35366 1 hardback
ISBN 0 521 53067 9 paperback

To Hilde and Knud

Contents

Acknowledgements

Many people have taken part in the long journey to the completion of this book, and it is impossible to thank them all. The project started as a PhD dissertation in the History of Science Department of the Johns Hopkins University, where I worked from 1980 to 1984. I am particularly grateful to my two dissertation advisers, Russell McCormmach, who guided me until he left Hopkins in 1983, and Robert Kargon, who saw me through to completion. I could hardly have made it without either the long, quiet, and intense discussions with Russ or Bob's benevolent impatience to have me finish. Many students at Hopkins also contributed importantly to my thinking. I am particularly grateful to Michael Freedman for spending innumerable hours to penetrate the structure and content of my dissertation. His labors made me see more clearly both what my work was all about and how I could communicate it to others.[1]

When I moved to Copenhagen in 1980, I soon came to know Hilde Levi, who had come to Bohr's institute as a refugee from Germany in 1934. She was now working on a biography of George Hevesy, with whom she had worked as an assistant from the mid-1930s until the Second World War. Never hesitating to provide unrestrained criticism of my work, she soon became, in effect, my on-site adviser and close friend. Through her, I came to know Knud Max Møller, another amateur devotee to the history of science, with unmatched knowledge of local sources in any form. Like so many others who have known them, I benefited from their services without being given a sense that anything was expected in return; I have counted on their help throughout my work on this historical project. In thankfulness, I dedicate this book to them.

I am equally thankful to Erik Rüdinger, who, as director of the

Niels Bohr Archive (NBA) and general editor of Niels Bohr's Collected Works, was always ready to provide help and advice, including a close reading of an early version of the manuscript. His knowledge of historical material on Bohr is unmatched. Like Hilde and Knud, he continued to help through extensive letter writing after I left Copenhagen in 1985.

I would like to extend my special thanks to Niels Bohr's son, Aage Bohr, who carried on the tradition of excellence in physics. Despite a heavy schedule, Aage always took time to express his confidence in my work. At various times, he also shared his experiences with me, and provided detailed criticism of an early manuscript for this book. I hope I have lived up to Aage's confidence at least to some extent.

My thanks go also to Erik's staff at the NBA and to my many friends at the Niels Bohr Institute who added to my life both personally and intellectually. I am particularly grateful to the physicists – in residence or visiting – who shared their experience of the institute between the world wars and at several stages provided constructive criticism.

For personal reasons, I spent the academic year 1981–82 in Oslo. I am grateful to Hans Skoie of the research division of the Norwegian Research Council for Science and the Humanities for providing me with an office and working facilities during this crucial phase of my work. It was a pleasure to work there.

In 1985, I moved to New York City to take up work at the Center for History of Physics of the American Institute of Physics (AIP). Working next to the Center's Niels Bohr Library, with its extensive holdings – especially of oral history interviews – I was never far from the basic sources for my book. Although both my historical and administrative tasks at the Center had nothing to do with my book project, my employers Spencer Weart and Joan Warnow were always encouraging me in my labors to complete my book. I am grateful to them for that, as well as for Spencer's comments on a late version of the manuscript. I would also like to thank the Center's staff for all the different kinds of help it has provided.

This book is based to a great extent on archival work, also outside the NBA. In that connection, I would like to express special gratitude to the late Warren Hovious – and also to Thomas Rosenbaum – at the Rockefeller Archive Center in Pocantico Hills, New York, to Niels Petri at the Carlsberg Foundation Archive in Copenhagen, and, again, to Spencer Weart and Joan Warnow at the AIP. Their help has been inestimable.

I want to direct my special thanks to everybody who allowed me to correspond with and interview them about their experiences of Niels Bohr and his institute between the world wars. Their contributions were invaluable both in bringing history to life and in providing encouragement for the completion of my work. A partial list of such contributors is given in the "Notes on sources" at the end of the book.

Toward the end of the work, this book has profited immeasurably from the critical eye of two people whom I want to thank. First, Tom Cornell, former fellow student and now historian of science at the Rochester Institute of Technology, read the manuscript and annotated it almost beyond recognition. I have taken most of his suggestions into account. Subsequently, my manuscript editor Ronald Cohen introduced an equal number of suggestions, changes, and corrections into the manuscript, improving the end result even further. I despair at the thought of how the book would have read without the help of Tom and Ronald.

Helle Bonaparte, secretary at the NBA, provided crucial criticism at the proofing stage.

In addition to the people already mentioned, many others have commented on some or all parts of my developing manuscript, and at different length. I would like to extend special thanks to Pnina Abir-Am, Stephen Cross, Shannon Davies, David Favrholdt, Paul Forman, Robert Friedman, Carsten Jensen, Jørgen Kalckar, Robert Kohler, Sharon Kingsland, Helge Kragh, Jens Lindhard, Abraham Pais, Philip Pauly, Mogens Pihl, Roy Porter, Nils Roll-Hansen, Stefan Rozental, Jan Teuber, Jan Vaagen, Charles Weiner, Sheila Weiss, Victor Weisskopf, and Aage Winther. Three anonymous readers of the manuscript for Cambridge University Press made several useful comments. I would also like to thank Helen Wheeler, my editor at Cambridge, for her positive attitude toward my work. It has been a pleasure to work with her.

In this evolving computer age, the manuscript has been through as many wordprocessing programs as revisions. Many people have contributed to helping me with the programs and transfers between programs. I would like to extend special thanks to Bitten Brøndum and Björn Nilsson at the Niels Bohr Institute; to Carol Weinreich, formerly at the Johns Hopkins Computing Center; and to Roman Czujko of the AIP.

I am thankful to several people and institutions for permission to use a variety of material. Above all, I want to thank Aage Bohr, who allowed me to quote from his father's papers. Spencer Weart, Thomas Rosenbaum, A.E.B. Owen, Niels Petri, Saundra Taylor, Ehud Benamy, and Anita Kerkmann have permitted me to quote from material respectively from the American Institute of Physics, the Rockefeller Archive Center, the Cambridge University Library, the Carlsberg Foundation Archive, the Lilly Library of Indiana University, the Hebrew University of Jerusalem, and the Staatsbibliothek preussischer Kulturbesitz in West Berlin. Hilde Levi and Victor Weisskopf have allowed me to quote from unpublished interviews with them and to use photographs from the collections of the NBA. The widows of Felix Bloch, Max Delbrück, and Samuel Goudsmit have allowed me to quote from their husbands' unpublished papers, while Jenny Arrhenius, Elisabeth Lisco, Bodil Schmidt-Nielsen, and Louise Slater Huntington – daughters respectively of George Hevesy, James Franck, August Krogh, and John Slater – have permitted me to quote from unpublished material of their fathers. D.K Hill has allowed me to quote his father, the physiologist A.V. Hill.

Concerning reproduction of photographs, I would like to thank Hilde Levi for permission to use photographs from the collections at the NBA; when no credit is given with the caption of a photograph, it comes from the NBA. I would like to mention, among other sources for photographs, John Wheeler, who not only allowed me to reproduce two pictures from a series of previously unused photographs that he took in the 1930s, but also personally saw to it that I could have prints in time for publication of this book.

In its first stages, my work was supported by the Norwegian Research Council for Science and the Humanities. It has also been supported by the Danish Ministry of Education and the U.S. '76 Foundation – a temporary organization established in Denmark from the proceeds of the Danish involvement in the bicentennial of the American Declaration of Independence. Grants from the Rockefeller Archive Center facilitated my research there. I am grateful to these sources for making my work financially possible. Professor Aage Winther at the Niels Bohr Institute helped in providing working facilities and in securing my financial existence there.

I have reserved my last expression of gratitude to Gro Synnøve Næs, my life's companion and best friend, and our little son, Andreas. Gro has contributed importantly to this work in all places and at all levels, from structuring the general argument to typing the individual words. My thanks are due to her for that great effort, as well as for her encouragement all along this long journey. Andreas's exemplary behavior and cheerful disposition during his first two years of life made the completion of this work both possible and rewarding.

Introduction

It is generally agreed that natural science today is an integral part of our politics, our economy, and our culture. It is seen as the major force behind the enormous technological advances of the last few decades, and has become an important part of the budget of any developed society. We are in the era of "big science," in which scientific practice is planned on a large scale nationally and internationally. More than ever, natural science is part of the general cultural debate, posing important economic, social, and ethical questions.[1]

This perception of the importance of natural science is relatively new. It grew rapidly after the Second World War, with its unparalleled massive efforts on projects such as the atomic bomb and radar. For scientists and historians alike, the Second World War was the turning point, when science lost its innocence. But the historical roots of the way science works today must be sought in an earlier time. I would like this book to be a contribution to the search for these roots.[2]

Long before the twentieth century, disciplines such as physics, astronomy, and chemistry had become esoteric enterprises, understood and pursued by a select few. Yet, like other human enterprises, they were part of human history. How does the historian integrate them into the larger historical picture without losing sight of all the complex scientific detail?[3]

Many observers have pointed to scientific change as a way to detect these historical links. During periods of change, scientific preconceptions are themselves in flux, and interact more readily with general historical trends. Even so, the discussion between historians, sociologists, and philosophers of science continues about whether general historical developments are responsible

1

for the rate of growth, the direction, or even the content of science.[4]

Funding provides one obvious means of influencing basic science research. Before the Second World War, science was severely underfunded by today's standards. Yet it had sources of support. The period between the two world wars, which this study will focus on, was the heyday of organized private philanthropy's support of basic science. The Rockefellers, in particular, spent substantial sums to develop institutional and individual science research. Although much has been written about science funding in the interwar period, its effect on actual scientific work has received much less attention.[5]

With hindsight, we see that one particular change in physics research between the wars was of special importance for subsequent historical developments. This was the transition to a concerted effort in the 1930s to theoretical and experimental research on the atomic nucleus, which would become the scientific basis for building the atomic bomb a decade later. Like the general features of science funding during this period, internal and technical aspects of the rise of nuclear physics in the 1930s have received substantial historical attention. Considerably less is known, however, about the relationship between the emergence of nuclear physics and external developments, including its sources of funding.[6]

This book attempts to bridge that gap. It brings together scholarship about the internal origins and development of nuclear physics in the 1930s and concurrent changes in private support for international basic science – in particular, the Rockefeller philanthropies. In doing so, the book places the emergence of nuclear physics into a larger historical context.

I have not attempted to study how all of nuclear physics interacted with all science funding. Instead, I have focused on one carefully chosen scientific institution. The resulting detail is essential to understanding the complex interrelations between basic science and its sources of financial support.

The typical scientific institute between the world wars was large enough to require administrative attention, yet small enough for the director to be personally responsible for both science and policy. In fact, all the activities at such an institute went through the director. Hence, the interconnections between science and funding in an institutional context can be reduced to the scientific and policy endeavors of its director.

Concentration on a single institution has limitations. At any time, the development of physics depends on the joint effort of several individuals at many institutions. This was particularly true of the development of quantum mechanics between the world wars. Indeed, Niels Bohr, the major figure in this account, repeatedly played down his own and his institute's contributions to these developments, emphasizing instead the collective nature of the scientific enterprise.[7]

The institution I have chosen is the Institute for Theoretical Physics at Copenhagen University, directed by Niels Bohr, and in 1965 renamed the Niels Bohr Institute. It is generally agreed that Bohr's institute was a mecca for theoretical physicists between the wars. Under Bohr's direction, the institute was a focal point for the developments that culminated in the formulation of quantum mechanics in the mid-1920s and revolutionized physics. In the following years, Bohr, together with his collaborators, developed the "Copenhagen interpretation" of the new physics, a position that came to be accepted by the majority of physicists.

Then, in the mid-1930s, after the so-called miraculous year of 1932 had created a watershed in experimental discoveries about the atomic nucleus, interest and research at the institute turned to nuclear physics. In 1936, Bohr proposed his highly influential "compound nucleus" model of the atomic nucleus. During the next few years, he oversaw both an increasing theoretical attention to the atomic nucleus and the installation of new and expensive equipment devoted to its investigation. By the outbreak of the Second World War, Otto Robert Frisch, working at the institute, and his aunt Lise Meitner, then in Sweden, explained the process of nuclear fission. At about the same time, the new equipment was put to successful use. The institute had become as central to the new field of nuclear physics as it had been to the development of quantum mechanics in the previous decade. For this reason, it is the natural choice for our study. This book puts the transition to theoretical and experimental nuclear physics at the institute into the broader context of its director's role as policymaker and fund-raiser.[8]

Although little has been written about the relationship between physics and funding at the institute and other places, interesting historical questions arise from even a cursory look at the basic historical documentation. In the next few paragraphs, I will use the publication record at the institute to pose some of these questions. This will provide a general idea of the issues considered in the book.

After the number of publications reached a peak of forty-seven in 1927, publication activity decreased gradually, reaching a low of seventeen papers in 1933. At that time, the publications were yet to reflect concerted research on the atomic nucleus. In 1934 the total number of publications rose to twenty-four as relative emphasis on the nucleus increased. By 1936, when twenty-six papers were published, nuclear physics was the main area of research at the institute. The emphasis on nuclear physics continued, the number of publications reaching a new peak of forty in 1937. Activity then fell slowly, until it almost ceased after the outbreak of the Second World War and the German occupation of Denmark.[9]

If these numbers signify trends, are they to be accounted for by developments outside or inside physics? Specifically, was the decrease after 1927 due to a decline in relevant scientific questions or to a decrease in funding? What was the main research interest at the institute during this period of low activity? What triggered the sudden theoretical and experimental turn to the atomic nucleus after a seeming lack of interest in such questions immediately after the miraculous year? How did Bohr obtain the funds necessary to make the transition? Were his fund-raising efforts motivated by a clearcut policy decision to introduce nuclear physics, or did the turn to a new field emerge as a byproduct of fund-raising efforts motivated by other factors?

From the list of research publications at the institute it is evident that there was substantial activity in biology during this period. This activity was led by the physical chemist George Hevesy and was represented by such articles as "Atomic Dynamics of Plant Growth" and "The Circulation of Phosphorus in the Body." From 1936 until the Second World War, biological work comprised as much as a quarter of the publication output from the institute, the rest being devoted to nuclear problems. Although it is well known that the institute became a major center for nuclear research in the 1930s, historians and reminiscing scientists have paid little attention to the concurrent activity in biology there. How and why did this activity begin, and what was its relationship, if any, to the contemporaneous transition to nuclear physics? Whatever the final answer to these questions, the existence of a biological research program at the institute complicates the picture of the activities there and further motivates a broad approach to the study of the institute's history in the 1930s.[10]

The prologue following this introduction presents a picture of the special atmosphere at the institute as subsequently remembered by the people working there. The prologue has two complementary functions. On the one hand, it provides a general introduction to life at the institute before we turn to the particular case of nuclear physics research. On the other hand, however, the physicists' accounts reflect not established facts, but rather the perception of the research participants. Contrary to the general argument of this book, these physicists remember few or no extrascientific developments affecting their work. Nevertheless, as I will argue, this difference does not refute my own interpretation. In fact, a side effect of the book is to give the reminiscences of those physicists a new perspective. It is therefore appropriate to turn now to their recollections of the atmosphere at the institute.

Prologue: The Copenhagen spirit

In their accounts of life at the institute between the wars, physicists have often portrayed their experience in terms of what they refer to as the "Copenhagen spirit." This term, or rather its German equivalent *der Kopenhagener Geist,* was first used in print by Werner Heisenberg. In his 1930 textbook on quantum theory, which was based on a lecture series given at the University of Chicago in the spring of 1929, Heisenberg referred to "the Copenhagen spirit of quantum theory," implying a special approach to the conceptual problems of physics. Subsequently, the term has come to refer to the atmosphere and style of work, rather than to specific ideas. It is this later meaning that I will consider here.[1]

By general agreement, the Copenhagen spirit signifies a broad approach to problems and the freedom to pursue one's own research. Victor Weisskopf, an Austrian physicist who spent several years at the institute during the 1930s, points to Bohr's unique ability to develop independent minds rather than assigning specific pieces of work. Weisskopf suggests that Ernest Rutherford, under whom Bohr worked in Manchester after obtaining his PhD in Copenhagen, is Bohr's only possible equal in this respect. Stefan Rozental from Poland – who arrived at the institute in 1938 and, except for the last part of the Second World War, has stayed there ever since – sees the resulting freedom to pursue one's own independent research as the dynamic necessary for the success of any physics research institute. For Rozental, physics research cannot be guided, but develops by circumstance.[2]

According to the physicists who worked at the institute, the cultivation of this intellectual independence included encouraging an informal intellectual and social atmosphere to both work

6

and play. The German experimental physicist Otto Robert Frisch, who spent several years at the institute in the 1930s, tells of his stupefaction when shortly after his arrival he went to a colloquium at which he found Bohr in eager discussion with the Russian physicist Lev Davidovich Landau. Landau was lying on his back on a table, while Bohr seemed to take no notice of this unusual position. Frisch found such unconventional behavior radically different from what he had experienced during his stays in Hamburg and London. Referring to the Landau incident, Frisch wrote later that "it took me a while to get used to the informal habits at the Institute for Theoretical Physics in Copenhagen, where a man was judged purely by his ability to think clearly and straight." The thoroughly informal atmosphere at the institute is recalled by many as a central ingredient in the Copenhagen spirit.[3]

This informal atmosphere extended beyond the discussion of physics. Several physicists report, for example, that Bohr enjoyed American Western movies, and they frequently joined him. The following is a favorite story of at least two visitors. After seeing a Western with a few visiting physicists, Bohr offered a theory as to why the hero always won the gunfight duels provoked by the villain. Bohr reasoned that reaching a decision by free will always takes longer than reacting mechanically; therefore the villain who sought to kill in cold blood acted more slowly than the hero who reacted spontaneously. To test Bohr's theory in a scientific manner, the group bought two toy guns. The theory was duly tested, and verified, with the Russian George Gamow in the role of the villain against the hero Niels Bohr.[4]

However, Bohr did not agree to all the younger physicists' pranks. The Danish physicist Christian Møller reports, for example, that Bohr did not object to a game of Ping-Pong in the institute's library; it was Gamow's insistence on using books as paddles that he objected to. Likewise Frisch, reflecting upon Hendrik Brugt Casimir's swim across a Copenhagen lake fully clothed, considered such behavior inconsistent with Bohr's personality and his running of the institute. Rather apologetically, Frisch suggests that such "childish" behavior by the physicists should be explained as part of their special qualities as human beings: "A scientist *has* to be curious like a child; perhaps one

Animated discussion at the blackboard at the Institute for Theoretical Physics, c. 1930. Left to right: Bohr, Pauli, Lothar Nordheim, and others. [Courtesy of AIP Niels Bohr Library: Landé Collection]

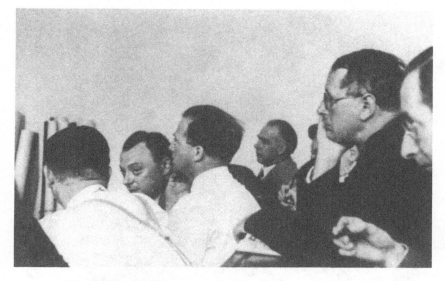

The audience gives rapt attention to a lecture at the institute, c. 1930.
Left to right: Jordan, Pauli, Heisenberg, Bohr, and others.

can understand that there are other childish features he hasn't grown out of." Bohr's special ability to deal with younger physicists, Frisch seems to imply, was not that he encouraged such childish behavior. Rather, realizing that this behavior was a necessary expression of the physicist's questioning mind, Bohr was wise enough not to discourage it. In all of these accounts Bohr appears as a benevolent father watching his children in amusement from a distance, politely informing his younger colleagues of his disapproval when the play went too far.[5]

The image of Bohr as a father figure is expressed more fully in the many accounts of how he communicated intellectually with his younger colleagues. For many physicists this was the most valued aspect of the Copenhagen spirit. The most common and efficient context for communicating scientific (and nonscientific) ideas in Copenhagen was not formal lectures and seminars. The encounters with Bohr remembered most vividly consisted of personal discussions, which could take place at the institute, at Bohr's summer house, or during sailing trips along the Danish coast.[6]

Margrethe and Niels Bohr outside the Carlsberg mansion.

Among such encounters, the informal gatherings in Bohr's
home, especially from 1932 when Bohr moved from his residence at the institute to the Carlsberg mansion, are remembered
with particular fondness. This nineteenth-century house had
been bequeathed "for life to a man or woman appreciated by
society for his or her activity in science, literature, art, or in
other respects" by its owner Jacob Christian Jacobsen, director
of the Carlsberg Breweries, on his death in 1914. Frisch, in fact,
resorted to nothing less than the image of classical Greece to
evoke his experience of Carlsberg. "Here," Frisch wrote, "I
felt was Socrates come to life again lifting each argument onto a
higher plane, drawing wisdom out of us which we didn't know
was in us (and which, of course, wasn't) . . . and when I cycled
home through the streets of Copenhagen, fragrant with lilac or
wet with rain, I felt intoxicated with the heady spirit of Platonic
dialogue." In a similar statement of youthful impression, the
Belgian Léon Rosenfeld, Bohr's close collaborator in the 1930s,
wrote: ". . . when [the young physicist] joins the group of disciples around the kindly smiling master, truly he feels himself received into a spiritual family, strongly united under Niels Bohr's
paternal aegis." The association of Bohr with Socrates and Je-
sus suggests an uncommon reverence for him and underscores
the closeness within the group of younger physicists surrounding
Bohr.[7]

Bohr's special way of communicating, deemed by his collaborators as conducive to creative scientific work, was not just a
strategy on Bohr's part to attract and inspire students, but reflected a fundamental trait in his personality. Frequently described as a less than perfect public speaker with a barely audi-
ble voice, Bohr found it difficult to express his physics reasoning
in the form of a polished monologue. According to many of his
co-workers, he always needed a human sounding board in order
to work out his intellectual problems – in physics or otherwise.
Rosenfeld acted in this capacity during his stay at the institute
from 1931 to 1940. Before that time, the Dutchman Hendrik Anthony Kramers, the Swede Oskar Klein, the German Werner
Heisenberg, and the Austrian Wolfgang Pauli had likewise
worked as Bohr's "helpers." Weisskopf served in this capacity
from the fall of 1936 to early 1937, when Rosenfeld was away,

while Stefan Rozental continued the tradition after Rosenfeld's departure in February 1940.[8]

Rosenfeld, Weisskopf, and Rozental all report that the main task assigned to them consisted of helping Bohr ready his manuscripts for publication. For this work they had to be available virtually any time. In addition to the long and irregular hours, the nature of the work was exhausting; writing down Bohr's thoughts was the least demanding part of it. Always expected to respond meaningfully to any of his suggestions, the helpers had to be alert to Bohr's thinking at any moment. Because Bohr's capacity for work was tremendous, this was no small task, even for his much younger collaborators. That Bohr could never be expected to be happy the next morning with a manuscript completed the day before only added to the strain. After having "slept on the problem" – one of Bohr's favorite expressions – Bohr could demand that the topic in question be attacked from an altogether different angle. An article usually went through a great number of drafts before Bohr would even consider it for publication. This was true of papers in physics as well as with articles of a more popular or philosophical bent.

Many physicists recall that this reluctance to publish was not so much a reflection of Bohr's perfectionism in formulating a finished train of thought as of his belief in a complete unity between the form and content of the written word. Every new formulation was not just an improvement of the presentation, but a continuation of the very thinking process. In poring over every sentence, Bohr and his helper thus laid themselves intellectually bare to one another. Close collaboration with Bohr demanded complete honesty and openness. It is a testament to the strength of the relationships between Bohr and his helpers that the vulnerabilities thus exposed very rarely, if ever, led to mistrust and hampered relations.[9]

Collaboration with Bohr could be painful in a more direct, physical way. One of the many anecdotes coming out of the Copenhagen milieu relates how Rosenfeld's wife obtained a doctor's certificate for her husband recommending him to restrict his working hours. This incident, which produced many a humorous comment at the institute, shows that work for the intellectual father could easily compete with family relationships as more conventionally understood. Indeed, Bohr's unflinching energy, even in physical endeavors, is reported by many of his collaborators who, although usually younger than Bohr, often found it difficult to keep up. Bohr's energy carried over into his enthusiasm for his work, and it appeared that he could discuss the most pressing questions in physics endlessly.[10]

As a result, the pressure on Bohr's helpers made it difficult for them to pursue independent scientific work, a fact they readily admitted. Yet nobody would even dream of criticizing Bohr for being too demanding. On the contrary, it was considered a great privilege to be asked to help Bohr, and the work was regarded with hindsight as one of life's greatest experiences. Although the problems under study were generally defined by Bohr, his helpers never felt dictated to. Instead, they saw themselves as participating in, and sometimes even contributing to, the master's very thinking process. Their complete acceptance of the chores involved shows once more the strong bonds between Bohr and his younger collaborators.

As in most families, there were limits to the children's capitulation to the father, however great the respect for him. At times, the fear of losing one's self-reliance might prevail over the exhilaration of working with Bohr. In an interview with me, Weisskopf expressed doubt that such work was always a "purely positive experience." Half in jest, he employed the term "victim," or even "slave," in place of the more gratifying term "helper," and suggested that working with Bohr for a long period could have the effect of extinguishing the younger physicist as an independent thinker.[11]

I have been able to find only two exceptions to the uniquely positive experience at the institute in the 1920s and 1930s. The first and most well known is the recollection of the American physicist John Clarke Slater, who arrived in Copenhagen in December 1923 to work on his ideas on the wave–particle duality of light. Earlier in the year he had received his PhD from Harvard University. Before going to Copenhagen he had spent the fall semester at Cambridge University.

The outcome of Slater's visit was a famous paper coauthored with Bohr and his assistant, Kramers. The refutation of the concept of energy conservation for microscopical processes contained in this paper was soon abandoned because of contrary experimental evidence. Furthermore, at the time of the article's publication Bohr had still not accepted the concept of light as particles, or photons. The Bohr–Kramers–Slater paper, in other words, soon became an anachronism in the history of physics.

According to his own recollection, Slater never agreed to give up the notion of energy conservation. In fact, he recalls that Bohr and Kramers, after having taken up one of Slater's ideas, sat down to write the paper without letting him in on the process. They only invited him to sign it after it had been completed. As for the relationship between Bohr and his disciples, Slater notes:

> [Kramers] was trying to act like the wise papa who was telling the little boy [that] he has to know how to handle the great man, or how to behave towards the great man. Oh it was very much the case of the great man and the little boy (in the corner). I wasn't used to this. Nobody at Harvard had ever acted that way.

Slater describes his time in Copenhagen as "horrible." He even relates that sometime after the paper had been submitted for publication, his understanding landlady offered him an escape to her summer house. He stayed there for the rest of his time in Denmark, working by himself. He did not see Bohr again during his stay.[12]

The other negative experience at the institute is recalled by the noted American molecular chemist and two-time Nobel Prize winner Linus Pauling. Pauling began his career as a physicist, and in that capacity spent a month at the institute in the spring of 1927. Pauling reports that during his visit Bohr showed no interest in his work, and he can recall no seminars at the institute. By contrast, he found Arnold Sommerfeld, with whom he spent the year in Munich before going to Copenhagen, an excellent teacher who was always responsive to the work of his students. As Pauling notes, one month is too short a period to make a final judgement on the Copenhagen milieu. Indeed, that Pauling and Slater were not able to establish themselves with the group around Bohr may even be evidence of the closeness of the group already there. Weisskopf, for example, admits to such elitism by the Copenhagen physicists. He also remembers a certain anti-American sentiment, which he insists cannot be traced back to Bohr.[13]

With the singular exceptions of Slater's and Pauling's recollections, then, the physicists agree remarkably well in their assessment of the institute. In my correspondence and interviews with people connected with Bohr and the institute between the world wars, most have enthusiastically recounted experiences like those described in the literature. Completely unhampered by external pressures of teaching or administration, the physicists could devote their entire time and energy to physics. In this endeavor, Bohr, intellectual stimulator and center, served as the guide and rallying point.[14]

In the narrative that follows, the experiences recalled by the Copenhagen physicists will be discussed in relation to Bohr's more mundane activities as policymaker and fund-raiser for the institute. From the perspective of the Copenhagen spirit it is difficult to understand how the concerted redirection of scientific research to nuclear physics in the 1930s, involving both theory and experiment, could even have occurred. The reminiscing physicists give the impression that there was a continuous, self-propagating discussion, rather than deliberate plans for changing an experimental and theoretical research program. However, Bohr, as director, had other concerns in addition to the theoretical discussion reported by the physicists. As we will see, there may have been times when Bohr's enthusiasm for the problems of physics impeded his efforts to direct the institute. But he was always ready for such chores when he had to be. We will see that the redirection to nuclear physics can only be fully understood in the context of concurrent changes in funding opportunities for basic science and in Bohr's response to them.

Yet this book is by no means a simplistic rejection of the physicists' perspective. It will be argued that the broader range of Bohr's activities, and the context in which he pursued them, provide a framework for appreciating the physicists' recollections of the Copenhagen spirit. Equally, the existence of a Copenhagen spirit will be corroborated to a significant extent.

1

Science policy and fund-raising up to 1934

It is symptomatic of the general lack of attention to the economic plight of basic science before the Second World War that the many accounts by physicists who worked at the institute rarely mention Bohr's activities as a policymaker and fund-raiser, and never suggest a relationship between the economic realities of the institute and the science pursued there. Indeed, these scientists go beyond pleading ignorance of such a relationship; they often imply that it did not exist, and that a fusion of pure science and economic considerations would be detrimental to productive disinterested research. To them, the unique stature of the institute in the 1920s and 1930s can be ascribed in part to the *lack* of such a connection. The institute was the ultimate place where science could be pursued for its own sake, without the distracting concerns of policy and funding.

Despite these physicists' reminiscences, Bohr, like the director of any physics institute, had to involve himself in securing funds, facilities, and personnel; in short, he had to work within the bounds of existing economic conditions. In fact, Bohr devoted substantial time and energy to defining and carrying out policy for the institute. The physicists' accounts, then, should not be interpreted as implying Bohr's noninvolvement in such matters; on the contrary – though without so intending – these accounts are evidence of Bohr's skills in shielding his collaborators from these concerns. A visitor's involvement in policy matters would have been of no use to Bohr nor for the effectiveness of the work of the institute.

This chapter not only establishes Bohr's strong involvement in policy and funding; it goes a long way toward explaining why this aspect of Bohr's activities has not been touched upon in the reminiscences of the physicists who worked there.

EMPHASIS ON EXPERIMENT

In his application to Copenhagen University's science faculty in April 1917 for the construction of a new institute in connection with his recently established professorship, Bohr presented the need for experimental facilities as an important justification for the creation of the institute. He noted that the validity of classical physics, which had previously been trusted to provide a sound theoretical basis for physical phenomena, was being severely challenged. Consequently, theoreticians would have to rely more than ever before on empirical data in their considerations. Bohr concluded that it was necessary that theoretical physicists "have the opportunity to carry out and guide scientific experiments in direct connection with the theoretical investigations."[1]

In his inaugural speech for the new institute nearly four years later, Bohr reiterated that a close cooperation between theoreticians and experimentalists was absolutely necessary in physics. He called for the construction of experimental facilities at the institute that could answer the most urgent questions of the theoreticians. At the time that meant apparatus for investigating the electrons orbiting the atomic nucleus. Specifically, Bohr needed spectroscopical equipment to measure the electromagnetic radiation emitted or absorbed by electrons when an atom moved from one energy state to another.[2]

To obtain this experimental apparatus, Bohr worked hard to secure funding. His elaborate applications – which like his academic writing went through several drafts – appealed to the Danish government and the Carlsberg Foundation for experimental equipment costing more than half the total amount for constructing the institute itself. The Carlsberg Foundation – a private Danish agency established in 1876 from the fortune of the Danish brewer Jacob Christian Jacobsen "for the advancement of scientific purposes" – had already supported Bohr during his education.[3]

In addition to securing funds for apparatus, Bohr involved himself in securing salaries for a permanent staff, which consisted of a scientific assistant, a secretary, and a trained mechanic. He also succeeded in persuading the Carlsberg Foundation to quadruple his annual allowance, which covered apparatus as

The Institute for Theoretical Physics in 1921.

well as scientific assistance. Bohr's efforts to secure funds for the establishment of the institute were aggravated by the economic situation in Denmark. From 1915 to 1920, wholesale prices went up more than 50 percent, including a 22 percent increase in a single year – 1919 to 1920.[4]

From the outset, the opportunity for experimental research was Bohr's basic rationale for the establishment of the institute. In the process, he gained experience in obtaining the necessary funds to build up the institute according to his vision.

RISING PRESTIGE

In addition to his diligence in approaching potential sources of support, Bohr's rising scientific prestige helped enormously in establishing his success as a policymaker and fund-raiser. By the 1920s, Bohr had achieved worldwide stature. He was regarded by his colleagues as the pioneer in the construction of a quantum physical basis for the atom. Not surprisingly, he had received several tempting job offers from foreign countries. As early as the First World War, for example, the renowned British experimental physicist Ernest Rutherford, under whom Bohr had worked for two years, tried to lure him to Manchester. In 1920, Max Planck, a leader in German physics, offered him a salaried membership in the Berlin Academy of Science, similar to the one accepted by Albert Einstein in 1914; this position would have enabled him to devote his entire time to research. In both instances Bohr declined, declaring his attachment to Denmark and his responsibilities toward Danish science. In 1922, at the early age of thirty-seven, Bohr was awarded the most coveted and prestigious scientific award – the Nobel Prize.[5]

The attempts to lure Bohr from Copenhagen continued. In the summer of 1923, he was offered one of the new professorships created by the Royal Society of London. These positions were designed to provide the most able scientists with ample personal means as well as complete freedom from all administrative duties. The salary of £1,400 ($6,400) annually – almost three times the amount paid him by Copenhagen University – was particularly tempting. Moreover, as a Royal Society professor Bohr would be free to affiliate himself with the universities or research organizations in Britain of his choice.[6]

In mid-July the physicist James Jeans, a prominent member of the committee appointed to propose candidates for the new professorships, informed Bohr that he was the committee's first choice. The offer included the prospect of a personal laboratory in Cambridge where he could work closely with Rutherford, who had been appointed a professor there. Bohr was further tempted by a personal letter from Rutherford expressing hope for a close collaboration.[7]

Bohr considered the offer seriously, but he wanted to maintain his connection with Copenhagen. Learning that this would be difficult, Bohr offered to give up his Copenhagen professorship while continuing as director of the institute. Only when told toward the end of August 1923 that the new position in all likelihood would require a complete break with Copenhagen did Bohr withdraw his application.[8]

The prospect of the Royal Society professorship led directly to a substantial improvement of Bohr's economic situation at home. In early September, two fellow Danish academics and members of the Carlsberg Foundation's board of directors sent a letter to the foundation. The two were the physiologist Valdemar Henriques, who had taken over the professorship of Niels Bohr's father, Christian Bohr, at the latter's death in 1911, and the mathematician Johannes Hjelmslev. In their letter they referred explicitly to the tempting offer from the Royal Society. They proposed that Bohr be freed from all administrative duties at Copenhagen University, including the obligation to teach, and that the foundation provide 10,000 Danish kroner ($1,800) annually, virtually doubling his salary. The idea of providing an individual with complete freedom to pursue research, the application read,

> . . . could hardly find any more perfect or satisfactory application than in the present case, in which the C[arlsberg] F[oundation] join to create in cooperation [with the Danish government] a truly ideal position as a free scientist for one of the most distinguished sons of our country, who is in the full power of his youth and carries one of the greatest names in the world. In this way the C.F. would also acquire great merit by supplying its considerable contribution to vindicate our country, in this nationally quite important matter, in competition with the grandiose endeavors of a rich and powerful country to improve and strengthen its scientific work by calling the best talents of the world to the colors.

Henriques and Hjelmslev appealed to a nationalistic sentiment in order to prevent a powerful rival country from stealing their nation's most prominent scientist. In the process they also pointed effectively to Bohr's unique talents. They succeeded in their venture. At the next opportunity the Carlsberg Foundation increased its contribution to Bohr's salary by a substantial amount.[9]

The flattering offer from the Royal Society seems also to have affected the Danish Education Ministry's attitude toward Bohr's application in early 1923 for new appointments at the institute. Thus, while expressing unwillingness in July to connect lectureship privileges to the permanent assistantship held by Bohr's close collaborator Hendrik Anthony Kramers, the ministry accepted Bohr's application in late September. As a result, Kramers formally took over Bohr's teaching duties. Bohr's simultaneous application for an annually renewable assistantship to help with the experimental work at the institute was also accepted, and the Dane Sven Werner was appointed in May 1924. Clearly, Bohr's growing prestige helped greatly in increasing the resources for the institute.[10]

In February 1924, Bohr received an offer from the Franklin Institute in Philadelphia, described to Bohr by its secretary, Robert Bowie Owens, as "what I believe to be the most desirable research opportunity in this country." In addition to ample laboratory facilities, assistants, and complete freedom to pursue his own research, Bohr was offered a salary of $10,000 annually, free housing, free travel, and a generous retirement allowance.[11]

Bohr did not even consider this or any later offers. After the offer from the Royal Society, his sense of attachment and responsibility toward his homeland always prevailed over the temptation of higher income or better working conditions. Having decided to stay in Copenhagen for good, he could devote his prestige entirely to improving the situation at his own institution. As we will see, the funding situation at the time was ideal for pursuing this strategy.

THE INTERNATIONAL EDUCATION BOARD

The International Education Board (IEB), based in the United States and backed by the Rockefeller fortune, was particularly responsive to the prestige of Bohr and others in deciding which scientists and scientific institutions to support. From its establishment in 1923 the IEB was a major contributor to Bohr and his institute. That economic support provides a splendid example of the importance of scientific prestige in obtaining funding for basic science in the 1920s.

By 1923 the Danish government and private funding agencies supported the institute in distinctly different ways. This division of labor continued throughout the interwar period. Whereas the government provided salaries for the permanent personnel and day-to-day operating expenses, the foundations supported expansion of facilities, acquisition of new equipment, and some of the operating expenses for research resulting from the expansion. They also paid for young foreign scientists who visited the institute for a year or two. Hence the private foundations were particularly crucial when Bohr sought to expand or redefine research at the institute. Because of the extent of its funding, and because it had the most clearly defined funding policy among private foundations, the IEB played a unique role in supporting the institute.[12]

The involvement of Rockefeller philanthropies in support of international basic science began in 1923, the very year of Bohr's first approach to the IEB. In that year the IEB was established under the leadership of Wickliffe Rose. Born in 1862, Rose was a professor of philosophy until 1907, when he began administering philanthropic work. He joined the Rockefeller philanthropies' work in 1910, taking responsibility for the successful eradication of hookworm disease in the southern United States. As director of the International Health Board of the newly established Rockefeller Foundation, Rose was later able to expand his health work to the international scene. Having been a member of the General Education Board, another Rockefeller philanthropy, since 1910, Rose was asked in 1922 to become its director. According to its charter, the work of the General Education Board was limited to the United States, and Rose made his directorship conditional upon the creation of an international education board. It is a testament to Rose's influence that his condition was accepted, and that his brainchild came into being with him as director.[13]

Rose limited the educational activities supported by the IEB to the natural sciences and agriculture. Strongly believing that the development of science in a country "affects the entire system of education and carries with it the remaking of civilization," Rose's strategy for supporting higher education consisted of finding the very best international centers for natural scientific research – which he considered to be situated in Europe – and exposing only the most promising students to them for a limited period. This was the rationale for the fellowship system, which Rose viewed as the pinnacle of his support of higher education. He may have been guided in his effort to create a fellowship system for European students by a similar kind of program established in 1919 for promising young American scientists. This program was paid for by the Rockefeller Foundation, and administered by the National Research Council (NRC). In Rose's plan, however, the fellowships were construed as part of a comprehensive system of higher education. The fellowship program was at its peak during the five-year period from 1924 to 1929, and was terminated in 1933. In all, it provided 509 fellowships in the natural sciences, with physics receiving 163, more than any other field.[14]

Although the fellowships were the pinnacle of Rose's policy, they accounted for only a small part of the IEB's total expenditure. To make the best centers that offered fellowships even better, the IEB supplied funds for the expansion, and sometimes even the creation, of basic research centers – reflecting Rose's explicit dictum "to make the peaks higher." In its European office in Paris, the IEB developed its own expertise in evaluating the abilities of potential fellows and institutions applying for support. In practice, as well as in theory, the funding of higher education was in this way reduced to a problem of promoting the best basic science research at the best institutions, which Rose usually found at universities.[15]

The IEB's emphasis on quality alone was fundamentally different from the approach of the Rockefeller Foundation, established ten years earlier "to promote the well-being of mankind throughout the world." The Rockefeller Foundation's dedication to the solution of specific health problems was a problem-oriented, managerial approach. In contrast, the IEB supported a spectrum of meritorious efforts in mathematics, physics, chemistry, and biology, preferring the research at individual institutions to be as broadly based as possible. Bohr took advantage of this preference when he formulated his request to the IEB for new equipment in terms of the general need for a close connection between theory and experiment.[16]

Support for equipment

Unable to obtain Danish support to expand the institute further, Bohr first decided to seek American funding in early 1923. By then, the number of visitors had increased beyond the capacity of the small institute. Furthermore, Bohr wanted equipment to improve his spectroscopical investigations and to be able to expand his measurements of electromagnetic radiation emitted by atoms to shorter and longer wavelengths than had previously been possible. Finally, although prices in Denmark had decreased somewhat since the record highs at the establishment of the institute, the financial situation was still uncertain.[17]

Bohr was encouraged to approach American sources by influential Danish friends, who subsequently established the necessary contact with Rockefeller-supported philanthropies. Thus Hans Marius Hansen, a physicist colleague of Bohr's at the institute, met with Christen Lundsgaard, a Danish physician then working at the Rockefeller Institute of Medical Research in New York; this institute, established in 1901, was the first to be financed with Rockefeller money. Aage Berlème, a financier who had gone to high school with Bohr, and Knud Faber, a professor of medicine at Copenhagen University, had supported Bohr in the establishment of the institute. Now they provided important help and advice once more. Lundsgaard brought the matter of obtaining American funding to the attention of Abraham Flexner, long-time secretary of the General Education Board. Flexner, who was director of the IEB's educational studies, took the case to Rose. Receiving positive signals, Bohr prepared an application, and sent it to Lundsgaard. Characteristically prudent, Bohr suggested that the institute would need $20,000. Lundsgaard doubled this amount, then sent the application on to Rose.[18]

After an interview at the IEB's headquarters in November 1923 in connection with his visit to Yale University to give the prestigious Silliman Lectures, Bohr was granted the full amount requested – $40,000. It is a tribute to Bohr's prominent position that this support was the first provided by the IEB to any physics research institute; his application was approved even before Rose set out on a five-month tour of Europe in December 1923 to survey some fifty universities and other institutions of education and research.[19]

It was the IEB's general policy that support for building and equipment could be provided only if additional funding were obtained from other sources. In fulfillment of this requirement, the land needed for the planned expansion was granted as a gift from the City of Copenhagen, while the Danish government provided the increased annual maintenance costs, as well as a second annually renewable assistantship, which was occupied by the Danish experimentalist Jacob Christian Georg Jacobsen from January 1925.[20]

There is no doubt that the economic situation at the institute was helped by favorable changes in both the wholesale and cost-of-living indexes in Denmark; after having been unusually high at the time of the founding of the institute, from 1925 to 1926 wholesale prices fell by as much as 22 percent. Yet the newly acquired funds proved to be insufficient. In January 1925, Bohr turned successfully to the Carlsberg Foundation, and in the fall of the same year, by submitting an entirely new application, he was able to obtain an additional amount from the IEB as well. Toward the end of the following year he applied for and obtained a final amount from the Carlsberg Foundation to complete the expansion of the institute. Bohr's fruitful approach to the IEB had enabled him to secure Danish support that would otherwise have been difficult to obtain. As a result of his successful fund-raising efforts, the expansion of the institute was completed by the end of 1926.[21]

Fellowships

In contrast to his activities to secure experimental equipment, which stemmed from his belief in the unity of theory and experiment, Bohr's efforts to obtain young physicists from abroad stemmed from his belief in international cooperation. From its inception, the institute had received support from the Danish Rask–Ørsted Foundation for young foreign visitors who were completing or had just completed their advanced university education. This foundation had been established in late 1919 "by the Danish State for the support of Danish science in connection with international [*mellemfolkelig*] research." During the 1920s it provided support for an average of three such visitors per year to the institute. Beginning in 1929, the Rask–Ørsted Foundation also supported the annual informal physics conferences at the institute. These conferences played an important role in theoretical physics in the 1930s, and we will encounter them on several occasions in later chapters.[22]

Yet a significant rise in the number of visitors to the institute occurred only after the IEB began its fellowship program in

1924. This number reached a peak of twenty-four in 1927, and fell rapidly to a low of seven in 1930, when the IEB fellowship system was in the process of being discontinued. By then, of the more than sixty young visitors staying for a substantial period, the IEB had supported fifteen and the Rask–Ørsted Foundation thirteen. Together, these two agencies paid for more than four times as many fellowships as any other agency. Among the IEB fellows were Werner Heisenberg and Pascual Jordan from Germany, Samuel Goudsmit from Holland, and George Gamow from the Soviet Union. The Rask–Ørsted Foundation supported Wolfgang Pauli from Austria, Yoshio Nishina from Japan, and others.[23]

Bohr's helpers, paid for primarily by the IEB and the Rask–Ørsted Foundation, arrived in Copenhagen prepared to spend all their energies working under the acknowledged master in their field, Niels Bohr. These newcomers had no prospects or ambitions for careers in Denmark, and hence no motivation for becoming involved in questions of administration and funding. For his part, Bohr got more out of his young co-workers by not involving them in such concerns. No wonder, then, that the physicists remember the collaboration with Bohr as devoted solely to basic physics. The true implication of their recollections, however, is not that Bohr was uninvolved in administrative concerns. On the contrary, they reflect Bohr's success as a policymaker; his astute use of funding opportunities contributed importantly to the cultivation of the Copenhagen spirit.

Other basic physics research institutes also profited from IEB's generous funding in the 1920s. In that decade, Göttingen University, another major center for theoretical physics and the development of quantum mechanics, received more than $80,000 for research in physics and $275,000 for building and equipping a mathematical institute. Amounts close to $100,000 were also provided to Heike Kamerlingh Onnes's low-temperature laboratory at the University of Leiden in Holland, and for the establishment of an Institute for Cosmic Physics in Tromsø, Norway. By far the largest institutional grant for physics – $430,000 – went to the Council for the Extension of Research and Scientific Investigations in Spain, which was destroyed during the Spanish Civil War. The IEB provided the largest grant for any field – $6 million – for the construction of the 200-inch telescope at Mount Palomar, California. Although the grant was made in 1928, the telescope was not put to use until over twenty years later.[24]

Of all the institutions supported by the IEB, only Bohr's institute has been identified unequivocally with a certain "spirit" of research. Therefore, the funding policy of the IEB alone can hardly be claimed to be responsible for the Copenhagen spirit. But it did help Bohr substantially in creating it.

Scientific accomplishments

By the mid-1920s, Bohr had largely achieved his goal of unifying theory and experiment. During this period, many experimental physicists joined the theoreticians at the institute to make use of the growing collection of apparatus. Among the foreign visitors, James Franck from Germany and Dirk Coster from Holland made particularly significant contributions to spectroscopical work during the first few years. Bohr's Danish colleagues Hans Marius Hansen, who moved into biophysics after the mid-1920s, and Sven Werner also did important experiments in spectroscopy. Jacobsen, who remained an experimentalist at the institute throughout his career, contributed experimentally both in spectroscopy and in radioactivity, the work in radioactivity untypically having little direct impact on the theoretical work at the institute. Although the discovery of the chemical element hafnium in 1922, which dramatically corroborated Bohr's atomic theory just in time for him to present this finding in his Nobel Prize lecture, is probably the most publicized case of the close interrelationship between theory and experiment at the institute, experimental activity complementing theoretical work went far beyond a single incident.[25]

Nevertheless, the most outstanding work was theoretical. During the early years of the institute, Bohr himself was able to explain the table of chemical elements in terms of his quantum theory of the atom. Then, in the mid-1920s, perhaps the most important breakthrough in physics in this century occurred. Bohr and his scientific collaborators played a crucial role in laying the groundwork for this revolutionary development. In 1925, Werner Heisenberg, by then a close collaborator of Bohr's and a regular visitor to the institute, was able to formulate a full-fledged theory of the atom – quantum mechanics. The new theory broke thoroughly with the concepts of classical physics, at the same time showing strong predictive power. With the search for a new theory complete, there was time for consolidation. During the next few years, the institute was a center for interpreting and fully understanding the new theory and applying it to the fullest extent.[26]

THE INTERNATIONAL EDUCATION BOARD AND OTHER INSTITUTES AT COPENHAGEN UNIVERSITY

By funding the best science with no strings attached, and by supporting both personnel and scientific equipment at the institute, the IEB enhanced the institute's role in developing and consolidating the new physics. In the 1920s the IEB accepted two more applications for building and equipment from Copenhagen University – new facilities for physiological research, which included funding for the prominent physiologist August Krogh, and Johannes Brønsted's institute for physical chemistry. As we will see in later chapters, Krogh would play a significant role when Bohr responded to changes in the policy of Rockefeller-supported philanthropies toward basic science in the early 1930s. For its part, Brønsted's new institute was built near to Bohr's, and the prospect of collaboration between the two institutes was an important consideration in the IEB's decision to support Brønsted. Furthermore, as we will also see in later chapters, the Rockefeller Foundation would be crucial in the redirection of the institute to nuclear physics in the mid-1930s. For this reason, a digression from the immediate activities of Bohr and the institute that addresses the IEB's funding of other segments of Copenhagen University will be useful here. It will also bring further perspective to the importance of Bohr's prestige and how and why it secured him a unique status with the Rockefeller-supported IEB.

Physiology

Schack August Steenberg Krogh, the Danish physiologist, has been described as "something of a scientific twin to Niels

August Krogh in his new laboratory, c. 1930.

Bohr.'' The older Krogh's slower rise to prominence virtually made up for their age difference of eleven years. In 1908, after serving as a student of and assistant to Niels Bohr's father Christian Bohr, Krogh obtained a permanent position in animal physiology at Copenhagen University's science faculty. Two years

Building constructed in 1928 at Copenhagen University for physiological research and financed by the Rockefeller Foundation and the IEB.

TO COMMEMORATE
THE GIFT OF HOUSING AND EQUIPMENT
BY
THE INTERNATIONAL EDUCATION BOARD
TO
THE LABORATORY OF ZOOPHYSIOLOGY
MCMXXVIII
OUR GRATITUDE SHALL BE SHOWN BY WORK

Plaque at the entrance to Krogh's laboratory thanking the Rockefeller
Foundation for its support.

later he was given his own modest laboratory. Only in 1916,
however – the year in which Bohr obtained his professorship –
did Krogh also become full professor. He won the Nobel Prize
in physiology or medicine in 1920, two years before Bohr won
his in physics, and gave the prestigious Silliman Lectures at Yale
University in 1922, the year before Bohr. By that time, the two
had become international leaders in their respective fields, both
receiving foreign visitors who wanted to pursue research at their
laboratories. Yet there seems to have been little contact between
the two men.[27]

In the spring of 1923, only days after the first approach had
been made to the IEB on behalf of Bohr, Krogh wrote to the
Rockefeller Foundation asking about the possibility of obtaining
funds to expand his institute. Unlike Bohr's application,
Krogh's request for $100,000 was not forwarded to the IEB, but
was handed to the Rockefeller Foundation's own newly established Division of Medical Education under the directorship of
Richard M. Pearce. By October, the IEB too had become involved in Krogh's application, and in March 1924, months after
Bohr had received his grant, Pearce told Rose that he did not
think it advisable to fund Krogh alone. Instead, Pearce presented a plan in which all the scattered activities in physiology

at Copenhagen University would be brought together under one roof. The plan was consummated later that year when the Rockefeller Foundation and the IEB provided $300,000 and $100,000, respectively, to move five laboratories for physiological research into one building constructed specifically for that purpose. Krogh's institute and the new biophysics laboratory established for Bohr's collaborator Hans Marius Hansen were two of the three laboratories belonging to the natural science faculty; the other two were in the medical faculty.[28]

The Rockefeller philanthropies went through a much more complex procedure in handling Krogh's application than Bohr's. Indeed, the support of Krogh's institute involved the collaboration of two funding agencies with contrasting policy traditions: the managerial hands-on approach used by the Rockefeller Foundation in defining the organization of physiology at Copenhagen University, and the IEB's nonintrusive "best science" policy based exclusively on scientific excellence. In later chapters we will see that remnants of this difference still existed when the Rockefeller Foundation took over the IEB's responsibilities in the 1930s for funding basic science, including the funding of Bohr's institute.

Physical chemistry

The IEB's third large-scale funding venture for natural science in Copenhagen in the 1920s was the establishment of a new institute for physical chemistry. Johannes Nicolaus Brønsted, who had become the first professor at Copenhagen University in that discipline in 1908, began inquiring about the possibilities for funding in early 1925. When Brønsted visited the United States for the academic year 1926–27, his application was still pending, and after meeting Rose in January Brønsted wrote to Bohr asking about the possibilities for supplementary funding from Danish sources. It is a testament to Bohr's influence in prominent Danish circles that he was immediately able to assure Brønsted that the Carlsberg Foundation would almost certainly grant Dkr 30,000 ($8,000) for instruments. Government funding, Bohr continued in his letter to Brønsted, might be provided if the IEB deal required it.[29]

In February 1925, after twenty years as a physics professor

at Princeton University, Augustus Trowbridge became the first director of the IEB's Paris office. In March 1927, Rose asked him for a "list of outstanding scientists in Copenhagen . . . to interpret Brønsted's proposal in terms of its scientific setting." Trowbridge formulated his reply in ten typewritten pages, beginning with the question of whether Brønsted's case should be treated in terms of support to a prominent international center (Copenhagen) or to an extraordinary scientist (Brønsted). Having concluded that Copenhagen could not hope to maintain its current position at the forefront of international natural science, Trowbridge could recommend only the extraordinary-scientist option. This he did, but not without expressing some doubt. Brønsted, he wrote to Rose,

> . . . is not a great personality and not one who will arouse enthusiasm in his field, as Niels Bohr has done so notably in his own, but he certainly should be counted among the first half dozen men in his line in the world.[30]

In a follow-up letter to Rose, Trowbridge retreated further from his position that Brønsted's request "be treated as aid to an individual." As justification he stated that "the matter should be considered with its bearing on our policy as regards outstanding men at centers where adequate support has been for long delayed from local sources." In an interview with Rose in April 1927, Simon Flexner, Abraham's brother and the influential director of the Rockefeller Institute, implied that Brønsted's abilities were inferior to Bohr's, and presented the planned proximity between Bohr's and Brønsted's institutes as his main argument for funding Brønsted. The closeness between the two institutes, the plans for collaboration between them, and Bohr's recommendation of Brønsted's case – all figured prominently in the long deliberations within the IEB. In June 1927, two and a half years after the first contact had been made, the IEB decided to provide $100,000 for the establishment of the new institute, which was put to full use in the fall of 1930.[31]

The IEB's considerable efforts to practice its elitist funding policy are demonstrated with unique clarity in Brønsted's case. After years of postponement because of what was regarded as Brønsted's somewhat inadequate scientific standing, Bohr's reputation came to play a crucial role in the final decision to provide

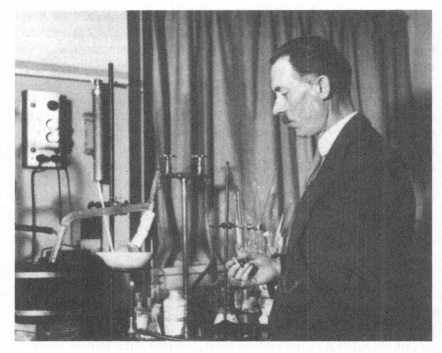

J. N. Brønsted in his new laboratory, 1930. [Courtesy of Carlsberg Foundation Archives]

funding. Brønsted's case provides especially convincing evidence of the importance of Bohr's renown in the IEB's practice of its funding policy.

ACTIVITIES UP TO 1934

From 1925 to the mid-1930s, there was no change in funding policy for international basic science, as defined primarily by the IEB. At Bohr's institute, in particular, the IEB continued to support young scientists from other countries, and Danish foundations did not change their demands on the kind of science they supported. Nevertheless, the boom in the funding of Bohr's institute, in which the IEB had taken the lead, did not continue. After 1925, Bohr did not seek substantial new funding, and financial support for the institute stagnated.

Funding sought by Bohr in 1932 does not change the picture. In 1932, Bohr and the mathematics teachers at Copenhagen University submitted a proposal to the Carlsberg Foundation, suggesting that a new institute for mathematics be built next to Bohr's. A gift from the Carlsberg Foundation for building such an institute had been pending since the University's 450th anniversary in 1929.[32]

The proposal was quickly endorsed by the Carlsberg Foundation. The new Institute of Mathematics was constructed in 1933, and the opening ceremony took place on 8 February 1934. The institute was headed by Niels Bohr's brother, Harald. From now on the two institutes would share the budget for nonscientific operating expenses paid by the Danish government, and the close connection between them would permit a greater flexibility in the use of space and facilities. Yet the construction of a new building for the mathematicians at Copenhagen University did not add substantially to the economic well-being of Bohr's institute.[33]

In April 1933, Bohr applied to the Carlsberg Foundation for a small increase in the annual appropriations to the institute. He argued that the "domain of atomic theory" now required new mathematical tools, and that he needed increased support to obtain "collaborators with special education in the relevant mathematical–physical domain." With this formulation, Bohr showed continued interest in pursuing a research program devoted to atomic theory. Unlike previous applications for funding, this one used the changing role of mathematics, not the need for new experimental equipment, as the main argument. This argument conformed with the simultaneous plans to establish a new institute of mathematics adjacent to Bohr's institute. Bohr's application was accepted, and the annual support rose from Dkr 18,000 ($3,900) to Dkr 24,000 ($5,200), beginning in June 1934.[34]

Between 1926 and 1932, Danish wholesale prices fell another 30 percent, after which they began to rise slowly. Although this development undoubtedly helped the finances of the institute, during this period Bohr was notably passive in his fund-raising efforts. Compared with the early years of the institute's development, the economic situation there from the mid-1920s to the mid-1930s was stagnant.[35]

A similar trend is reflected in the changing number of long-term visitors at the institute, as well as in the number of publications stemming from research there. As already noted, the number of visitors rose steadily during the first few years of the institute's history, reaching a peak of twenty-four in 1927. It then dropped off sharply to ten already the following year. The number reached a low of seven in 1930, and did not exceed fourteen in any year before 1936. The number of publications also rose at a relatively constant rate through 1927, when it reached a high of forty-seven. By the following year the number had dropped to twenty-one, staying at about that level through 1933. As measured in personnel and publications, scientific activity at the institute dropped after economic support leveled off in the mid-1920s. In the next chapter we will investigate how this trend related to the scientific activity itself during this period.[36]

CONCLUSION

For the first fifteen years, Bohr involved himself in setting policy and raising funds for the institute. He was particularly active during the first five years. This involvement, which has been ignored by physicists reminiscing about their work there, was decisive for the institute's work and success.

In the mid-1920s, in particular, Bohr applied successfully for substantial support from the Rockefeller-funded International Education Board (IEB). This agency had an elitist funding policy toward basic science, providing funds for the best institutions and scientists and setting no specific conditions on the kind of work done. As one of the most prestigious physicists of his time, Bohr profited substantially from this policy. By obtaining funding both to expand the institute and to acquire new experimental equipment, he was able to maintain the unity between theoretical and experimental work in atomic physics that provided the rationale for establishing the institute in the first place. After the IEB had demonstrated its interest, other funding agencies in Denmark followed suit, permitting Bohr to complete the expansion of the institute.

In addition, IEB funding was crucial to the realization of Bohr's goal of international cooperation among physicists at the institute – it enabled the finest young physicists from various countries to visit for periods of one to two years. In this, the IEB's objectives matched Bohr's exactly. Although other agencies – notably the Danish Rask–Ørsted Foundation – gave similar support, the IEB soon took a leading role, providing others with an incentive to follow its example.

Having established Bohr as a policymaker and fund-raiser, we have completed the first step in understanding the redirection of his institute in the mid-1930s. An equally necessary step is to account for the scientific activities of Bohr and the institute prior to the redirection, particularly as they relate to matters involved in the subsequent transition. This is the goal of the next chapter.

2

The Copenhagen spirit at work, late 1920s to mid-1930s

In 1927, theoretical and experimental research at the institute, aimed at understanding the outer part of the atom, reached a climax both scientifically and economically. In that year, Bohr presented his "complementarity argument," which has come to be seen as securing the conceptual framework of quantum mechanics in terms of the generally accepted "Copenhagen interpretation." At that time, too, as seen in Chapter 1, the number of visitors and publications reached a peak after Bohr's successful funding efforts. In the next few years, Bohr submitted no new applications for substantial expansion and equipment for the institute. At the same time, the scientific activity there, as measured by the number of scientific personnel and publications, fell dramatically.[1]

What were the scientific concerns of Bohr and the institute after the consolidation of quantum mechanics, and how were they pursued? How do these developments relate to the economic stagnation and the slowdown in activity? Even though Bohr did not appeal for substantial new *funding* for the institute into the year 1934, was there anything in the *science* pursued there that anticipated the scientific redirection that would begin soon after? Specifically, was there any discussion or work relating to the atomic nucleus or biological matters, which would constitute the main elements in the transition? If so, to what extent can these concerns be said to lead to the subsequent redirection?

This chapter will consider these questions. As we will see, Bohr showed interest both in the physics of the nucleus and in aspects of biology during this period; as a consequence, I have divided the chapter into two main sections – one for each interest. I will argue, however, that none of these concerns led Bohr to contemplate a concerted redirection of research. On the contrary, Bohr and his younger colleagues pursued these interests in an atmosphere of open and free-wheeling discussion that was the exact antithesis of a concerted plan for a well-defined theoretical and experimental research program. Indeed, these discussions provide an excellent case study of the Copenhagen spirit at work. Although discussion doubtless led to profound theoretical insight, it did not produce concerted scientific change. During this period, in fact, the unity between theory and experiment preached and practiced by Bohr during the early years of the institute was replaced by an open-ended approach to theoretical questions for their own sake.

INTEREST IN THE ATOMIC NUCLEUS UP TO 1934

The first half of the 1930s is generally known among historians of science as the period of emergence of nuclear physics as a subfield of theoretical physics; because of the flood of new discoveries in 1932, that year, in particular, has been termed the "miraculous year" of nuclear physics. In the first section of this chapter, we will see that Bohr followed these discoveries closely. However, by early 1934 they had not provided a justification for redirecting research at the institute. On the contrary, there was a continuity both in the kind and content of discussion that did not portend a transition to a concerted theoretical and experimental study of the atomic nucleus.[2]

Background

A separation between the study of the inner and outer constituents of the atom had already begun in the early years of the century. In 1911, Ernest Rutherford, whom Bohr was later to call "the founder of nuclear science," postulated on experimental grounds that the atom consisted of a small massive nucleus surrounded by light electrons at relatively large distances. Soon after, Bohr came to study at Rutherford's Manchester laboratory. Profiting from Rutherford's experimental insight, Bohr began to formulate a quantum theory valid for the system of electrons orbiting the nucleus, without addressing the question of the

Bohr and Rutherford after a walk in the countryside, c. 1912.

structure of the nucleus itself. From the very outset, Bohr was able to distinguish between phenomena arising from the nucleus, such as radioactivity, and phenomena arising from his atomic model. Indeed, in 1913 Bohr was the first to conclude in print that the radioactive phenomenon of beta decay was of nuclear origin.[3]

Bohr's development of a quantum theory for the outer part of the atom proved to be such a successful research effort that it soon became the main preoccupation of theoretical physicists. By 1920, for example, Bohr was able to explain the elaborate but little understood classification of the chemical elements on the basis of his atomic theory, thus making a significant step toward establishing physics as the fundamental natural science. Moreover, his efforts toward constructing a unity between theory and experiment at the new institute in the early 1920s, discussed briefly in Chapter 1, amounted to a continuation of his research program to understand the outer part of the atom. In the mid-1920s, this research reached a climax with the formulation of quantum mechanics by Bohr's young collaborator Werner Heisenberg, Erwin Schrödinger, and others, which amounted to a true revolution in physics.[4]

An important corollary to the new quantum mechanics was the so-called uncertainty principle, which Heisenberg developed in part as a result of close scientific discussion and collaboration with Bohr. This principle stated that Planck's constant, a fundamental value in quantum physics, set a limit to how precisely two so-called conjugate quantities, such as position and linear momentum, could be measured in a single experiment. For example, if one wanted to measure exactly the position of a particle, one would have to give up all information about its momentum. After Heisenberg's formulation of his principle, Bohr characteristically continued to use his younger colleague as a sounding board for his own thinking. It was in the course of these sometimes tense discussions that Bohr's revolutionary interpretation of quantum mechanics in terms of his "complementarity" argument came to fruition.[5]

Bohr first presented his complementarity argument in a lecture in Como, Italy, in September 1927. He concluded from Heisenberg's uncertainty principle that in any physical observation the quantities measured and the apparatus employed in measuring them constitute an inseparable whole. From this perspective, the observation of light, sometimes as particles and sometimes as waves, was not contradictory. Because the different manifestations of light could be correlated with the different experimental arrangements under which it was observed, a full description of light would consist of a description of two such complementary manifestations, the observation of one excluding the simultaneous observation of the other. Similarly, a precise measurement at any time of the positions of the individual electrons in an atom would preclude simultaneous knowledge of the energy states of the atom; complete knowledge of the coordinates of an atom's constituent particles was complementary to a dynamic description of the atom in terms of the conservation of energy principle. This complementarity implied a renunciation of traditional causality. Bohr considered this renunciation to be a necessary consequence of the existence of Planck's constant. As such, he saw it as eliminating the meaningless consequence of classical theory that atoms were not stable entities but would collapse into themselves. As we will see, it was a typical case of Bohr's explanation by renunciation.[6]

From the very beginning, Bohr considered his complementarity argument as going beyond quantum mechanics. As such, it provided what Bohr considered a "lesson" to be used in other branches of science as well as in human affairs more generally. Bohr expressed this attitude clearly in one of the several printed versions of the Como lecture, which was first published in early 1928. He predicted that in order to adapt relativity theory to quantum theory, one would have to

> . . . be prepared to meet with a renunciation as to visualization in the ordinary sense going still further than in the formulation of the quantum laws considered here.

Thus the complementarity argument might have a different bearing on the *generalization* of quantum physics.[7]

Referring these difficulties to "the fact that, so to say, every word in the language refers to our ordinary perception," he expressed the hope

> . . . that the idea of complementarity is suited to characterise the situation, which bears a deep-going analogy to the general difficulty in the formation of human ideas, inherent in the distinction between subject and object.

Already at this stage Bohr viewed his complementarity argument within a substantially broader framework than quantum mechanics, and even physics.[8]

After the successful formulation of quantum mechanics in the mid-1920s, theoretical physicists did not turn directly to the atomic nucleus as a confined area of research. Instead, they sought to generalize quantum mechanics to areas that had already been treated, at least approximately, by classical physics. Theoretical physicists at Bohr's institute and elsewhere sought in this way to formulate quantum theories for particles with relativistic velocities, for the electromagnetic field, and for the interaction between field and matter. As we will see, the interest in the nucleus at the institute in the late 1920s and early 1930s can only be understood in this general theoretical context.[9]

In 1927, for example, Paul Adrien Maurice Dirac, a young English theoretical physicist who in the course of work at the institute had contributed significantly to the formulation of quantum mechanics, proposed an equation describing the behavior of one electron with arbitrary velocity in a classical electromagnetic field. The solution of the Dirac equation, as it came to be called, did not only include particles with energy above a certain positive value; the equation also predicted an infinite number of particles with negative mass and energy. This double set of solutions was also obtained from the corresponding classical equation. In the classical case, however, the set of negative energy solutions – which had no obvious physical interpretation – could be disregarded, because classically energy could not change spontaneously from a positive to a negative value. Quantum physically, such transitions *could* take place; hence, all solutions to Dirac's equation were interdependent, and none could be disregarded without damaging the whole theoretical structure. The Dirac equation thus proved to be difficult to interpret physically.[10]

In 1929, Dirac proposed an interpretation that incorporated positively charged protons, as well as electrons, into his scheme. Wolfgang Pauli, another younger colleague of Bohr's, had previously formulated the so-called exclusion principle, which stated that no two electrons could occupy the same energy state. In accordance with this principle, Dirac suggested that all negative energy states predicted by his equation were filled up by electrons, which thus constituted a uniform "sea" of unobservable entities. It could happen, however, that one of the unobservable particles underwent a quantum transition to become an observable electron with positive energy. When that happened, a "hole" would appear in the unobservable sea of negative energy electrons. The principle of the conservation of charge asserted that this hole would be observed as a particle with positive elementary charge, which Dirac in his paper identified with the proton, the only empirically known particle with the required property.

As a tentative answer to the question of whether his theory could account for the large difference between the masses of the electron and the proton, Dirac noted that interaction between the electrons introduced a dissymmetry in the description of the negative and positive particles; he expressed the hope that this dissymmetry might explain the difference in the masses. According to Dirac's interpretation, the electron and proton "are connected in such a way that actually there is only one fundamental kind of particle in nature." Like other physicists, Dirac saw it as a goal to reduce the number of nature's fundamental constituents.[11]

In 1927, before arriving at his electron theory, Dirac also contributed to the quantum theoretical treatment of the electromagnetic radiation field. However, the description of the interaction of this field with matter soon caused problems. Even in classical electrodynamics the electron had to be assigned a minimum radius in order not to obtain an infinite self-energy that would be physically meaningless. The resulting departure from the concept of a point charge made a proper relativistic treatment impossible. These problems were now carried over into the quantum case. Heisenberg and Pauli were particularly active in formulating an elaborate quantum field theory incorporating the interaction with matter. Yet their work – and that of others – encountered serious difficulties.[12]

As a result of these difficulties, the elation among theoretical physicists at having developed a quantum mechanics applicable to nonrelativistic particle velocities was soon replaced by a sense of profound crisis. As early as the summer of 1928, Heisenberg reported to Bohr that he was turning to other work in physics "almost in despair"; two years later he did in fact give up this work for a period of eighteen months. In early 1931, Pauli was so unhappy with the situation that he considered it futile to participate in the annual informal physics conference at the institute, a custom established in 1929.[13]

The problems of establishing a relativistic quantum physics provided the context for Bohr's speculation in 1928, referred to earlier, that a solution might be based on a generalization of his complementarity argument, which in turn would require further renunciations of classical theory. Unlike most of his colleagues, however, Bohr did not despair of the situation; on the contrary, his views represented what the historian of science John Heilbron has aptly called his sentiment of "enthusiastic resignation."[14]

Despite the general reluctance of theoretical physicists to attack the structure of the nucleus directly, in the late 1920s there was one particularly successful attempt to apply quantum mechanics to the nucleus. In 1928, Ronald Gurney and Edward Condon at Princeton, and independently George Gamow at Göttingen, were able to explain quantum mechanically how radioactive nuclei could emit alpha particles – that is, helium nuclei – with less than their nuclear binding energy. Gamow, a Russian physicist traveling in Europe on fellowships, subsequently spent the academic year 1928–29 at the institute, where he continued to contribute importantly with quantum mechanical treatments of nuclear alpha decay and gamma decay, the third manifestation (in addition to alpha and beta decay) of radioactivity, which consisted of high-frequency electromagnetic radiation. Before leaving Copenhagen, Gamow also conceived of the liquid drop model of the atomic nucleus, which would play a crucial role in the development of nuclear physics generally and in Bohr's contributions to it specifically. Even before the 1930s, Gamow's contributions had brought hope that theoretical concepts originating from the study of the outer part of the atom could finally be applied to the nucleus.[15]

However, as Gamow was well aware, the progress was severely limited; in particular, it did not contribute to the fundamental problem of including electrons in the atomic nucleus. Both of the then known fundamental entities of physics – protons and electrons – were required in order to explain the basic properties of nuclei: Whereas heavy protons accounted for the nuclear mass, the relatively weightless and negatively charged electrons had to be included to account for the observed net charge. In the late 1920s, the inclusion of these nuclear electrons posed ever graver theoretical problems.[16]

One problem concerned beta decay. In beta decay, an atom emitted electrons with a continuous range of energies up to a certain maximum value. This phenomenon had been known since 1914, but it was not until 1927 that Charles Drummond Ellis, a British experimental physicist, and his colleague William Alfred Wooster were able to establish conclusively that the energies were distributed continuously at the electrons' emission from the nucleus. Before this result, Ellis had engaged in a long dispute with Lise Meitner in Germany, who held that the electrons were slowed down unevenly only after being emitted. Combined with the fundamental principle of energy conservation, Ellis and Wooster's result implied that the radioactive nuclei themselves had a continuous range of energies, and hence were not identical to each other either before or after decaying. This conclusion, however, was generally seen as untenable, because it contradicted the elementary and well-established empirical law that for a given species of radioactive nuclei, the radioactive half-life – the time required for half of the nuclei in a sample to disintegrate – was well-defined. The problem of the continuous energy spectrum was peculiar to beta decay, and did not apply to the other two manifestations of radioactivity – alpha and gamma decay – which Gamow had explained using traditional quantum mechanics.[17]

A second problem arose from Heisenberg's uncertainty principle, which implied that if an electron were confined to nuclear dimensions, it would possess an energy much greater than typical nuclear binding energies. It therefore seemed paradoxical that electrons could be contained in the nucleus at all.[18]

A third problem of incorporating electrons in the nucleus was that it led to a value for the nuclear spin predicted by quantum mechanics that did not agree with experiment. Accurate analysis of experiments on the nitrogen 14 nucleus published in 1928 by Ralph de Laer Kronig, and subsequent experiments on other nuclei, showed that electrons, which were known to have spin outside the nucleus, did not contribute to nuclear spin at all. These spin measurements, in fact, indicated that nuclei did not contain electrons. Similar problems arose in situations involving nuclear magnetic moment and nuclear statistics.[19]

Gamow continued to struggle with these problems. From Copenhagen he went to Rutherford's Cavendish Laboratory in

Cambridge for a year's stay, after which he returned to Copenhagen to complete the first monograph by a theoretical physicist entirely devoted to the atomic nucleus. This book shows with particular clarity the problems posed by the nuclear electrons. In the manuscript the jocose Gamow illustrated several sections with a drawing of a skull and crossbones to indicate the danger involved in accepting statements about nuclear electrons at face value. The English publisher replaced Gamow's drawings with less colorful S's, which indicated "the more speculative passages about nuclear electrons." The German publisher was even more inflexible, refusing to identify the more uncertain parts of Gamow's book at all. Clearly, Gamow was completely aware of the limitations of his book and hence of the uncertain state of theoretical nuclear physics. Indeed, within a year the book would become almost obsolete. When it was published in 1931, however, it was evidence of the continuing gap between the successful physics of the outer and the insufficient understanding of the inner part of the atom.[20]

As the problems of the nuclear electron became increasingly evident, theoretical physicists came more and more to see these problems as bound up with the task of establishing a relativistic quantum physics – which was also encountering serious difficulties, and also involved the problematic electron. Thus, in the first of their two joint articles on quantum field theory, Heisenberg and Pauli lamented that the jumps between positive and negative energies in Dirac's theory seemed "to thwart a more detailed treatment of the constitution of the nucleus." Somewhat later, Bohr was able to convince Heisenberg that his most recent approach toward formulating a relativistic quantum physics was mistaken. Heisenberg replied in exasperation: "Perhaps we must first await the complete development of nuclear physics before we can go any further with this."[21]

Although increasingly aware of the problems posed by the atomic nucleus, Bohr and other theoretical physicists continued to consider their overarching task to be the generalization of quantum mechanics to the relativistic domain. They did not see nuclear physics in isolation, but viewed the particular problem of the nuclear electron as part of a more general task of establishing a generalized quantum physics. Hence their renewed efforts around 1930 did not constitute the beginning of an autonomous nuclear physics. Instead, they expected that the problem of the nuclear electrons would be just one of several questions to be resolved by a sweeping relativistic quantum theory that was yet to be formulated. This was the broad theoretical context for Bohr's early tentative statements about the atomic nucleus around 1930.[22]

Nuclear physics concerns up to 1932

In the summer of 1929, Bohr wrote his first comment directly addressing the problem of the atomic nucleus. It was made in a letter to the editor of *Nature* proposing a theoretical approach to the problem of beta decay. Bohr suggested that the principle of energy conservation was violated in "the close interaction of the constituent particles of the nucleus." This was the second time Bohr had suggested such a violation; the first time had been for atomic processes in 1924, shortly before quantum mechanics made the suggestion superfluous. Although Bohr's proposal had precedents, it was truly radical; for more than a century, energy conservation had been viewed as one of the securest pinnacles of physics. Bohr's proposal typified the Copenhagen spirit's call for an open attitude. At the end of his letter, Bohr even suggested an application of his proposed solution, speculating that the violation of energy conservation might also help explain astrophysical phenomena, notably the energy source in the sun.[23]

As he often did, Bohr sent his manuscript to his younger colleagues for criticism. Typically, they did not hesitate to state their opinions unequivocally. Pauli, for one, did not find Bohr's argument sufficiently constructive to merit publication. He advised Bohr to lay his note aside and to "let the stars shine in peace." Whether or not because of Pauli's criticism, Bohr never submitted his letter to *Nature*.[24]

Dirac was equally negative toward Bohr's proposal of energy nonconservation, writing that he

> . . . should prefer to keep rigorous conservation of energy at all costs and would rather abandon even the concept of matter consisting of separate atoms and electrons than the conservation of energy.

He made this statement in the course of an extensive correspondence, in which Bohr confronted Dirac's newly developed "hole" interpretation with his own ideas. In their animated correspondence, Bohr argued that Dirac's findings necessitated a wholly new approach to quantum physics in the relativistic domain. Thus he wrote:

> In the difficulties of your old theory I still feel inclined to see a limit of the fundamental concepts on which atomic theory hitherto rests rather than a problem of interpreting the experimental evidence in a proper way by means of these concepts. Indeed according to my view the fatal transition from positive to negative energy should not be regarded as an indication of what may happen under certain conditions but rather as a limitation in the applicability of the energy concept.

As in his lecture in Como in 1927, Bohr saw the renunciation of a traditional physical concept as the basis for resolving a fundamental paradox.[25]

This correspondence with Dirac provides a fascinating glimpse into the discussion of general conceptual problems represented by the Copenhagen spirit. While disagreeing on detail, both men sought to exploit the difficulties involved in gaining a theoretical understanding of the nucleus in order to highlight the problems of quantum physics more generally. At the same time, their discussion combining nuclear physics with relativistic quantum theory was typical of contemporary discussions of theoretical physics.

In subsequent public lectures, Bohr continued to discuss the atomic nucleus exclusively in connection with more general problems of theoretical physics. As a rule, these lectures contained an overview of the contemporary state of quantum physics, and concluded with a brief discussion of beta decay as an example of the problems involved. Typical are his Scott Lectures at Cambridge University, followed by a Faraday Lecture for the Chemical Society of London in May 1930.[26]

Statements made by Bohr at a conference on nuclear physics in 1931 are particularly interesting. In October of that year, the prominent Italian physicist Enrico Fermi hosted a pioneering conference in Rome devoted specifically to "nuclear physics"; this event is regarded by historians as the first conference of theoretical physicists devoted exclusively to the problems of the atomic nucleus. Even there, however, Bohr's presentation had the same structure as previous lectures. Referring to his propos-

Rome conference on nuclear physics, October 1931. The five people to the left are Heisenberg, Otto Stern, Bohr, Peter Debye, and Fermi.

al of energy nonconservation in nuclear processes, he concluded:

> Just as we have been forced to renounce the ideal of causality in the atomistic interpretation of the ordinary physical and chemical properties of matter, we may be led to further renunciations in order to account for the stability of the atomic constituents themselves.

In this way, Bohr clearly related his proposal to his complementarity argument: Just as he had shown that an explanation of atomic stability required the abandonment of traditional causality, so he speculated that the explanation of nuclear stability would require the abandonment of traditional energy conservation. In other words, his disagreement with Dirac, as well as his views on the atomic nucleus, was based on his belief in the complementarity argument as a guide to resolving problems posed by relativistic quantum physics. For Bohr, complementarity had a wider range of applicability than the quantum mechanics from which it originated.[27]

The historian John Heilbron has linked the reluctance of British physicists, including Dirac, to accept Bohr's complementarity argument to their pragmatic opposition to any philosophical outlook in physics. In the second part of this chapter we will see how Bohr's interest even in biological questions, which arose at the same time, was motivated by a desire to expand his complementarity argument to new areas. Notwithstanding Dirac's opposition, such general reasoning suited perfectly the style of work represented by the Copenhagen spirit.[28]

Reaction to the "miraculous year," 1932

The first reference to 1932 as an *annus mirabilis*, or "miraculous year," of physics was probably made by the British physicist Arthur Eddington in a brochure he wrote in October 1934 in order to obtain funds for a major extension of Rutherford's Cavendish Laboratory. For Eddington, the year signified a belated victory for experiment over theory, rather than for nuclear physics over atomic physics. Thus he wrote:

> For some years previously, the centre of advance had been in theoretical physics whilst experimental physics plodded patiently on. Then in rapid succession came a series of experimental achievements, not only startling in themselves but presenting immense possibilities for further advance. The laboratories of the world are now pressing forward in an orgy of experiment which has left the theoretical physicists gasping – though not entirely mute.

Historians writing at a later date, however, saw these developments as constituting the miraculous year of *nuclear physics*, and as providing the basis for the study of the atomic nucleus as an autonomous subfield of theoretical physics.[29]

As we will see, Bohr, while paying close attention to these developments, did not see them as portending a far-reaching reorientation of research; rather, he sought their implications for quantum physics generally and continued to pursue questions of theoretical physics through open discussions with his younger colleagues. In this respect his reaction was far from unique among theoretical physicists, who viewed the new experimental developments as contributing to the resolution of problems they had been grappling with for some time. Thus the miraculous year did not immediately set the stage, at Bohr's institute and most other places, for nuclear physics as an independent subfield of theoretical physics.

The neutron

Physicists and historians alike have considered the first event of the miraculous year – the discovery of the neutron – to be especially important. They have pointed out that the existence of a new fundamental particle, in addition to the electron and the proton, made it possible to envision an atomic nucleus that did not contain the problematic electron. As a consequence, they have concluded, the atomic nucleus could be studied independently of the seemingly insurmountable problems of relativistic quantum physics. As some recent historical studies have made clear, however, this conclusion was not drawn immediately, either by Bohr or by other prominent physicists, who for some time considered the neutron not as a fundamental particle, but rather as a combination of an electron and a proton. As we have already seen, Dirac, in working out his hole theory, endeavored to formulate a theoretical approach uniting the two known elementary entities – the electron and the proton – into one conceptual scheme. This endeavor represented a common desire among physicists to keep the number of elementary entities to a minimum. It is not surprising, therefore, that the nuclear electron continued to be seen as problematical even after the discovery of the neutron. For some time still, Bohr and other theoretical physicists expected that the nucleus could only be understood through a resolution of the more general problems of relativistic quantum physics.[30]

As early as 1920, in his Bakerian Lecture to the Royal Society of London, Rutherford had speculated that nuclei might contain a composite neutral component, consisting of a collapsed "atom" of a proton and an electron. Subsequently, Rutherford's students J.L. Glasson and J.K. Roberts carried out experiments in order to observe this neutral entity, which Glasson called the "neutron"; however, they did not succeed. More than ten years would elapse before James Chadwick announced the experimental discovery of the neutron in late February 1932. As Chadwick has indicated himself, it was no coincidence that the discovery was made by one of the collaborators of the original proponent of the new entity. By the time of Chadwick's discovery, in fact, Walther Bothe and Herbert Becker in Berlin, and subsequently Irène Curie and Frédéric Joliot in Paris, had already made similar experiments, without being able to observe the new particle. Moreover, Chadwick, like Rutherford, saw the new particle not as an elementary entity, but as a combination of the two fundamental charged particles.[31]

It is a tribute to the close connection between Rutherford's Cavendish Laboratory and the institute that Chadwick informed Bohr about his discovery before it was published. Less than a week later, Ralph Fowler, the senior theoretician at the Cavendish, wrote to Bohr that "Chadwick's neutrons have rather overwhelmed everything for the moment." Bohr, who shared Fowler's sense of the discovery's importance, told Heisenberg that the neutron would be the foremost item for discussion at the annual informal physics conference at the institute, and invited Chadwick to give an account of his discovery at the conference.[32]

Chadwick was unable to come to the Copenhagen conference, which took place in the second week of April 1932. Bohr was sufficiently enthusiastic, however, to give his own lecture on the neutron. In this lecture, he sought to demonstrate why the neutron had eluded the attention of experimentalists for so long. As Bohr saw it, this elusiveness was due to the low rate of reaction between neutrons and the orbiting atomic electrons of the matter they were penetrating. He now set out to explain this low reaction rate. Thus the framework for Bohr's lecture was the established subject of the penetration of particles through matter, and not the internal constitution of the neutron itself or its role in nuclear structure. Indeed, Bohr expressed his continued bewilderment about the internal constitution of the neutron. "Of course," he said,

> . . . it[s] mass and charge suggest that the neutron is formed
> by a combination of a proton and an electron, but we can not
> explain why these particles combine in this way, as little as we
> can explain why 4 protons and 2 electrons should combine to
> form a helium nucleus or an α-particle.

Clearly, Bohr shared Rutherford's original view that the neutron, like the alpha particle or any atomic nucleus, was a composite entity built up of electrons and protons. Hence, the problem of the nuclear electron remained, and the understanding of the atomic nucleus was still closely bound up with its resolution.[33]

In contrast to Bohr, Heisenberg immediately sought to incorporate the neutron into a new theory of nuclear constitution. In the second half of June 1932, he sent Bohr the proofs of his first article of a trilogy that was to break new ground in the development of an autonomous field of theoretical nuclear physics. But the accompanying letter reveals that he too viewed the neutron as essentially composite: "The basic idea is: to shove all principal difficulties onto the neutron, and to apply quantum mechanics in the nucleus." In the last of his three papers, published in February 1933, Heisenberg noted carefully the limitation of his approach. Like Bohr, he presented beta decay as evidence *against* conservation of energy and *for* the need for an entirely new relativistic quantum physics. Even in Heisenberg's classic article, then, the most basic problems of the nucleus were still part of the more general quest for a new relativistic quantum physics. Theoretical nuclear physics was not yet a field of investigation in its own right.[34]

Even outside Bohr's immediate circle, the view of the neutron as a composite prevailed for some time. At the annual meeting in York of the British Association for the Advancement of Science in early September 1932, for example, physicists discussed both "the conservation of energy and nuclear phenomena" and "the neutron." In the course of the latter discussion, Owen Willans Richardson, professor at the Royal Society of London and director of the University of London, said that he "was glad to hear that Dr. Chadwick regarded the neutron as some kind of a combination of a proton and an electron." Although aware that if the neutron were regarded "as some entirely new kind of ultimate structure," it could account for the otherwise inexplicable nuclear spin of nitrogen 14, Richardson felt "that this is too small a matter to invoke an entirely new material entity to account for." New fundamental entities were not accepted easily among physicists. At the conference, Richardson also expressed sympathy toward nonconservation of energy as an explanation of beta decay.[35]

In early November 1932, another exchange of letters between Bohr and a younger colleague shows that Bohr's view of the neutron had not changed. Samuel Goudsmit, who had recently moved from Holland to the United States, wrote to Bohr referring to Heisenberg's lectures on his own theory of the nucleus at the University of Michigan the previous summer:

> It is strange and regrettable that the discovery of the neutron did not give some more fertile clues for progress. In many respects the situation has not changed much from what it was at the Rome meeting, a year ago, except that the difficulties can now be formulated more sharply.

After having reported the concern at the institute about the problems of the nucleus, in particular in connection with "the fundamental difficulties of relativistic quantum mechanics," Bohr replied:

> Still, I quite agree with you as regards the very preliminary character of any attempt hitherto made to attack the problem on such lines [i.e., accounting for the nuclear problems by incorporating the neutron].

Bohr and other theoretical physicists still considered the neutron as a composite. For them, only a satisfactory relativistic quantum physics of the electron could provide hope for a complete understanding of the atomic nucleus as well as other physical phenomena. As we will see, the problem of the nuclear electron would remain at least until 1934, when Fermi secured the theoretical basis for another new particle – the neutrino. In the meantime, Bohr and others continued their struggle to understand how the electron could be incorporated into the atomic nucleus.[36]

Accelerator-induced artificial disintegration
In a letter to Bohr dated 21 April 1932, Rutherford exclaimed: "It never rains but it pours. . . ." Rutherford was referring to another event of the miraculous year. In recent experiments at his High-Tension Laboratory, Rutherford's two collaborators

John Douglas Cockcroft and Ernest Thomas Sinton Walton had obtained the first nuclear reactions produced by man-made accelerators. Cockcroft and Walton used the high voltages produced by their newly developed voltage-multiplier circuit (subsequently called a Cockcroft–Walton generator) to accelerate protons, which in turn disintegrated light nuclei through bombardment. Characteristically understating the case, Rutherford wrote to Bohr: "You can easily appreciate that these results may open up a wide line of research in transmutation generally."[37] Bohr was also elated. He replied:

> Progress in the field of nuclear constitution is at the moment really so rapid, that one wonders what the next post will bring, and the enthusiasm of which every line in your letter tells will surely be common to all physicists.

He added that he wished to be closer to Rutherford's laboratory, particularly during an exciting period like the present.[38]

The background for his interest in the new developments, however, had not changed. He enclosed with the letter a reprint of his Faraday Lecture, which had just been published, and promised to send the text of his lecture in Rome as soon as it was printed. He expressed the hope that the problem of the nuclear electrons, discussed in these publications, would be illuminated as a consequence of the new experiments. He thus wrote:

> If it should be possible to excite electron emission from nuclei by means of the recently discovered powerful agencies, it would perhaps be possible to settle this fundamental problem.

Bohr, in other words, was interested in the new discovery not primarily because of its implications for nuclear processes as such, but because of the light it would shed on nuclear electrons, and the resulting more general contribution to the establishment of a relativistic quantum physics.[39]

The positron

A third momentous event commonly associated with the miraculous year was the discovery of yet another fundamental particle – the positron. With hindsight, this particle, with opposite charge and the same mass as the electron, has come to substantiate the phenomenon of particle creation. By implication, the new particle made it possible to consider the electron in beta decay as created only at its departure from the nucleus. As a consequence, there was no longer any need to retain the electron within the nucleus itself. The nuclear electron having thus disappeared from the scene, nuclear physics was no longer subsumed under the elusive relativistic quantum physics, and the road to an autonomous theoretical nuclear physics independent of relativistic quantum theory lay open.[40]

Again, however, we are dealing with an oversimplification of the new particle's historical impact. Despite the traditional account of the positron's discovery, Bohr and other physicists did not immediately draw the eventual conclusion; instead, they continued to see a close relationship between nuclear problems and relativistic quantum physics.

The American physicist Carl David Anderson, working in Robert Andrews Millikan's laboratory at the California Institute of Technology (Caltech) in Pasadena, announced the discovery of the new particle in the late summer of 1932. He made the discovery fortuitously during investigations of cosmic rays exposed to a magnetic field in a Wilson cloud chamber. Just as had happened before the discovery of the neutron, other scientists had made observations similar to Anderson's, without ascribing them to the existence of a new particle. Again, the Joliot-Curies in Paris were notably among them.[41]

In early 1933, Patrick Maynard Stewart Blackett and Giuseppe Paulo S. Occhialini, working in the Cavendish Laboratory, confirmed Anderson's result with an improved experimental technique, and provided an elaborate theoretical framework for the new particle. Identifying it with the hole in Dirac's sea of negative energy electrons, Blackett and Occhialini considered the discovery to be an important verification of Dirac's electron theory. Dirac, who had become convinced in the meantime that the hole must have the same mass as the electron, and hence could not be identified with the proton, quickly endorsed Blackett and Occhialini's argument.[42]

The findings of Anderson and Blackett and Occhialini did not immediately convince Bohr of the existence of the new particle. Nor did they alter his skeptical opinion of Dirac's electron theory. By early 1933, Bohr's Swedish colleague and close friend

Oskar Klein was established in Stockholm. When Klein expressed his enthusiasm for Blackett and Occhialini's conclusions, Bohr replied:

> Regarding the positive electrons I cannot, however, quite share your enthusiasm. I am at least as yet very skeptical as regards the interpretation of Blackett's photographs, and am afraid that it will take a long time before we can have any certain knowledge about the existence or non-existence of the positive electrons. Nor as regards the applicability of Dirac's theory to this problem I feel certain, or, more correctly, I doubt it, at least for the moment.[43]

A month and a half later, during an extensive tour of the United States, Bohr was the main speaker at a two-week symposium at Caltech. According to a wire service report written for the American Science Service by the resident theoretical physicist Rudolph Meyer Langer, Bohr "said that after listening to the evidence it was scarcely possible to doubt the reality of the positron." Privately, however, Bohr still expressed uncertainty. The day after the wire service report, he wrote to Heisenberg from Pasadena:

> I have also learned a lot of physics, even though regarding the positive electrons I am still inclined to say like Lord Rayleigh, "some of those who know me best, think that I ought to be more convinced, than I am, perhaps they are right."

Not even a visit to the very laboratory from which the discovery was first announced had completely convinced Bohr of its validity.[44]

Bohr's opinion was soon to change, however. On his return to Copenhagen after a stopover in England, he wrote to Heisenberg:

> I have had much opportunity myself to think over the principal questions in America and am enthusiastic about the possibilities for work which the discovery of the positive electrons have opened up, and about which any doubt had of course to silence as soon as it was clear that they could be produced by gamma rays under conditions which can be completely controlled experimentally.

As had happened before – for example when he in the early 1920s gave up his proposal to renounce energy conservation to explain atomic processes – Bohr had become convinced by the steadily accumulating experimental evidence.[45]

Thus, by the time of the annual informal physics conference at the institute in mid-September 1933, which had been postponed from the previous spring, Bohr had become convinced of the positron's existence. Thirteen days after sending the letter to Heisenberg embracing the positron, he wrote to Ellis at the Cavendish Laboratory that he was sorry not to have been able to speak with him in England

> . . . about the wonderful recent progress in nuclear physics and especially about the new possibilities for the theoretical development opened up by the discovery of the positive electron.

Bohr invited Ellis to the forthcoming conference where the "problems of the positive electrons will be the main topics of discussion."[46]

Even though aroused by this new conviction, however, Bohr did not come any closer to a study of the atomic nucleus in its own right. Indeed, at the prestigious Solvay Congress in Brussels in October 1933, which was devoted specifically to nuclear physics, his main contribution consisted of a spirited discussion of the implications of the positron for the creation of a relativistic quantum physics. As in his lecture at the nuclear physics conference in Rome two years earlier, Bohr relegated a presentation of nuclear problems to the last two of thirteen pages.[47]

The new developments also made Bohr enthusiastic about Dirac's electron theory, which he described as inescapable in a lecture at the Physics Society in Copenhagen in late November 1933. The following month he wrote to Klein that "[i]t now really seems as if the hole theory has been brought to a preliminary harmonic conclusion." Yet he still perceived the development of a new relativistic quantum physics as the main task of theoretical physics. Bohr was as far as ever from seeing nuclear physics as an autonomous field worthy of a full-scale theoretical and experimental research program at the institute. In the next section we will see that Bohr's priorities in physics and for the institute did not change for at least another year.[48]

Physics concerns after 1932

In the summer of 1931, Bohr had been provoked by an article written by Lev Davidovich Landau and Rudolf Ernst Peierls, young colleagues and frequent visitors to the institute. Although the article had been worked out at Wolfgang Pauli's institute in Zurich, its conceptual basis originated from discussions in Copenhagen. Landau and Peierls announced in their article that Bohr would soon make public his assessment of their views in *Nature*. It would in fact take more than two years before Bohr's first criticism was published. Typically, Bohr worked out his views laboriously in close collaboration with his helper Léon Rosenfeld, who appeared as coauthor of the resulting paper.[49]

In their article, Landau and Peierls deduced an uncertainty relationship that was valid for relativistic quantum physics, and argued sweepingly on the basis of a theoretical consideration of the possibilities for measurement "that all the physical quantities appearing in wave mechanics are generally no longer definable in the relativistic domain." According to Landau and Peierls, this conclusion explained the difficulties involved in creating a relativistic quantum physics. In particular, they used it to explain the problems of beta decay and nuclear electrons.[50]

Bohr took particular exception to Landau and Peierls's conclusion that in quantum theory electric and magnetic "field strengths are not at all measurable quantities." Although, as he later admitted to Pauli, he did not fully appreciate the limitations

Informal conference at the institute, September 1933.

First row: Niels Bohr, P. A. M. Dirac, Werner Heisenberg, Paul Ehrenfest, Max Delbrück, Lise Meitner. *Second row:* Horowitz, Carl Friedrich von Weizsäcker, Edward Teller, Hans Jensen, Walter Heitler, Otto Robert Frisch, Milton S. Plesset, unidentified, Evan James Williams, Rudolf Peierls. *Third row:* Eugene Rabinowitch, three unidentified persons, Mrs. Nordheim, Lothar Nordheim, Ivar Waller, H. B. G. Casimir, Christian Møller, Felix Bloch, Hans Kopfermann. *Fourth row:* Harald Bohr, Walter Gordon, Fritz Kalckar, Salomon Rosenblum, Savard, Charles Manneback, unidentified, George Placzek, Victor Weisskopf. *Fifth row:* Guido Beck, Johan Holtsmark, Niels Arley, Ramesh Chandra Majumdar, Oskar Klein, Homi Jehangir Bhabha, Léon Rosenfeld. *Last row:* J. C. Jacobsen, Sven Høffer-Jensen, Sørensen, Egil Hylleraas, Bjørn Trumpy, Alex Langseth.

Bohr and "helper" Rosenfeld enjoy the sun, 1931.

of quantum theory at the time, Bohr felt that Landau and Peierls's statement ran counter to the correspondence principle, which expressed the equivalence between classical physics and quantum physics for large quantum numbers, and which for years had served as Bohr's main guide in his persistent search for a new physics. As Bohr saw it, this principle might be inapplicable to the nuclear electrons, but not to a general description of electromagnetic fields.[51]

By October 1932, Bohr and Rosenfeld, as Bohr wrote to Heisenberg, had reached the

> . . . result that the possibilities for measurement, as I was of course convinced in advance that it had to be, [were] in complete correspondence with the quantum mechanical formalism.

Bohr also promised that an article would be ready in a matter of days. Unexpected problems with points of detail in the manuscript, however, delayed its completion, and there is ample evidence that Bohr remained occupied with it into 1934. Thus, Bohr's main creative effort, at least through the year following the miraculous year of nuclear physics, was concerned with the general understanding of the limitation of measurement in relativistic quantum physics.[52]

Bohr's main contribution to the Solvay Congress in October 1933 was formulated in part as a critique of Landau and Peierls's general claims in their 1931 article. As such, it went beyond his paper with Rosenfeld, which limited itself to a discussion of the measurability of electromagnetic field quantities. After having submitted his manuscript for the proceedings of the Solvay Congress, Bohr wrote to Pauli in March 1934 that it must be regarded only as a preliminary discussion, and that he had submitted it in its present form only "in order not to delay further the printing of the Solvay report." Already in mid-January he had reported to several colleagues that he was working on a "small article" pertaining to these problems, and on 15 February he sent a "heartfelt outpouring" [*Hjerteudgydelse*] in a letter to Pauli explaining his present stand on relativistic quantum physics. He wanted Pauli's opinion of his views, he wrote, because he was working on a new article with Rosenfeld. Bohr's concern is underscored further by the fact that he sent copies of the letter to his younger colleagues Felix Bloch and Werner Heisenberg for comment. As late as March 1934, then, rather than planning a concerted redirection of research at the institute toward nuclear physics, Bohr was becoming increasingly involved in the kind of theoretical discussion he had been carrying on since the late 1920s.[53]

The continuation of Bohr's concerns into 1934 is particularly striking in view of a theory of beta decay proposed by Fermi in late 1933, which with hindsight has made Bohr's suggestion of nonconservation of energy, and the consequent need for a totally new theoretical basis, redundant. Instead of resorting to a violation of the time-honored principle of energy conservation, the theory introduced another new particle to account for the experimental anomalies of the nuclear electron.[54]

As early as the end of 1930, in a letter to the participants at a conference he was unable to attend, Bohr's close friend and colleague Wolfgang Pauli had proposed an entirely new particle in order to save energy conservation in beta decay. In particular, Pauli encouraged the German experimentalists Hans Geiger and Lise Meitner to check whether or not a previously undetected electrically neutral particle was emitted together with the electron from a radioactive nucleus during beta decay. Such a particle, Pauli argued, could carry with it the energy not carried by the beta particle; hence energy would be conserved in the decay process. Having made the proposal long before Chadwick's discovery of the neutron in 1932, Pauli called his new particle the "neutron." For the time being, he viewed it as a permanent nuclear constituent, which could also explain the nuclear spin anomaly. He repeated his proposal during lectures in the United States in the summer of 1931.[55]

At Enrico Fermi's conference on nuclear physics in Rome in October of the same year, Pauli discussed his suggestion with Bohr as well as with Fermi. He found Fermi positive to the idea, whereas Bohr preferred his own approach of questioning energy conservation. Samuel Goudsmit, who had heard one of Pauli's lectures in the United States, at Fermi's request formally presented Pauli's idea at the conference. Goudsmit mentioned the idea only in passing in his lecture, which was devoted to the so-called hyperfine structure of atomic spectra. Pauli himself was still too uncertain to present his idea in a context that would lead to its publication.[56]

By the time of the Solvay Congress in 1933, experimental developments had taken place that made Pauli feel more secure about his proposition. Thus, Pauli's own presentation of his idea, as well as Bohr's rejection of it, were published in the proceedings of the congress. After Chadwick's discovery, Fermi had named Pauli's particle the "neutrino," a name it has been known by ever since. Pauli no longer saw it as a nuclear constituent, but only as a particle being emitted together with the electron in beta decay. In his rebuttal of Pauli's idea, Bohr was still open to the possibility of energy nonconservation. Earlier at the conference he referred favorably to work by Guido Beck, who had worked at the institute; Beck had developed a theory of beta decay without introducing a new particle. Bohr was still guided in nuclear questions by his desire to generalize the complementarity argument.[57]

After the Solvay Congress, Fermi developed his elaborate theory of beta decay, a preliminary version of which was published before the end of the year. The new theory presumed energy conservation and predicted the creation of an electron and a neutrino in beta decay. Early the following year, Felix Bloch, a refugee scientist in Fermi's laboratory who had collaborated closely with Bohr, reported favorably on Fermi's theory to his former mentor. He reported that "[a]s far as the experiments suffice for a quantitative comparison, the β-spectrum seems to come out really very beautifully." Still Bohr was unmoved. He replied:

> We have naturally all also been very interested in Fermi's new work, which undoubtedly will have a very stimulating effect on the work with electrical nuclear problems, although I still do not feel fully convinced . . .

Bohr's negative reaction confirms his continued emphasis on establishing a new basis for relativistic quantum physics without limiting his questions to the nucleus as such. In fact, he probably still saw the concept of the nuclear electron as a paradox requiring a wholly new kind of relativistic quantum theory incorporating nonconservation of energy; not until two and a half years later did he publicly renounce nonconservation of energy in favor of Fermi's theory as explaining beta decay. As judged by his work in physics, by early 1934 Bohr seemed completely unprepared for a concerted redirection of the institute toward theoretical and experimental nuclear physics.[58]

Theory and experiment

Like Bohr's own scientific correspondence and publications, the work at the institute did not change substantially from the late 1920s until early 1934. Not only did publications on physical theory continue to reflect the free-wheeling discussion between Bohr and his collaborators about the generalization of quantum physics; the experiments conducted were mostly devoted to spectroscopical investigations of phenomena on the atomic scale, not the nuclear scale. That is, experiments were conducted in accordance with the institute's original tradition of unifying theoretical and experimental investigation of atomic phenomena. Increasingly, however, the theoretical activity did not provide precise direction for the experimental work, which was guided instead by the institute's existing experimental tradition. In short, the unity between theory and experiment, so strongly urged by Bohr at the institute's inception, was being severed.[59]

Certainly, there are exceptions to this admittedly rigid picture; thus, not all theoretical publications addressed relativistic quantum physics, and not all experiments were spectroscopical investigations into the quantum mechanics of the atom. Since 1922, the resident experimentalist J.C. Jacobsen had conducted experiments on radioactivity, obtaining his doctoral degree for this work in 1928. Also, as I will discuss in more detail in Chapter 3, George Hevesy, during his stay at the institute from 1920 to 1926, published the results of experiments involving radioactivity. Until 1928, however, when George Gamow began publishing the results of his effort to apply quantum mechanics to the nucleus, Jacobsen's and Hevesy's experimental studies of nuclear phenomena as opposed to atomic phenomena were fairly unusual at the institute.[60]

In 1930, as many as ten out of twenty-three publications from Copenhagen were devoted to the atomic nucleus, with Gamow continuing his efforts on the theoretical side, and Jacobsen publishing more nuclear experiments. During the next two years, however, work on the nucleus decreased, reaching a low point in 1932. The following year, the activity increased once more; in particular, the young German physicist Hans Kopfermann, visiting the institute on a fellowship from the Rockefeller Foundation, published four articles on nuclear momentum. Still, these publications were only a minor part of the institute's total output. Moreover, Kopfermann's experiments continued the institute's experimental tradition of spectroscopy. His capacity for drawing conclusions about the nucleus therefore owed more to the increasing precision of traditional spectroscopy than to the introduction of any new experimental technique.[61]

In 1934, the number of publications from the institute dealing with the atomic nucleus continued to increase. This increase, however, did not signify a new effort; instead, it arose from continued efforts by Jacobsen, Gamow, and Kopfermann. Until 1934, there is no indication, for example, that Bohr saw this or any other work as a way to change the research emphasis at the institute.[62]

Accordingly, the relatively minor theoretical and experimental effort in nuclear physics at the institute until early 1934 does not signify any new concerted action to introduce nuclear physics as a major field of concentration there. On the contrary, they constitute another reflection of the lack of specific research direction provided by the Copenhagen spirit of open-ended discussion. This work only confirms that the unity between theory and experiment proclaimed by Bohr at the institute's establishment had been replaced by a kind of laissez-faire approach to physics research. Like Bohr's own work, the research publications from the institute through early 1934 gave no indication of the complete turn there to theoretical and experimental nuclear physics research that was soon to come.

At least on one occasion during this period, Bohr expressed awareness that his concerns were becoming overly theoretical, and that research at the institute did not live up to his original concept of a unity between theory and experiment. In mid-December 1931, he provided an unusually candid critique of his own role in the scientific work at the institute, indicating how research could be more ideally organized. His comments were made in a short speech to the Royal Danish Academy of Sciences and Letters in appreciation of the Academy's decision on that occasion to offer him the honorary Carlsberg mansion after the recent death of its former resident, Harald Høffding.

Bohr suggested that moving from his home at the institute would improve conditions both for his own work and for research at the institute generally. First, the continual theoretical discussions had left him too little time and peace to work out the resulting ideas in detail. Second, this heavy concentration on theoretical problems had made it ever more difficult for him to participate actively in experimental investigation, which, he stressed, constituted "an important part of the work at the institute." In this connection, he expressed the hope that some of the younger experimentalists would be allowed to move into his previous home. Bohr's remarks, which amounted to self-criticism, seemed to be calling for more planning and more experiment.[63]

Concluding remarks

After Bohr had moved to the Carlsberg mansion, the experimentalist Ebbe Rasmussen moved into Bohr's former home at the institute. This was the only immediately tangible result of Bohr's self-criticism. Indeed, as we have seen, Bohr was unable, at least into early 1934, to break away from the tendency toward intense discussion of what he and some of his prominent colleagues considered to be the most fundamental conceptual problems of theoretical physics. The effort to understand the atomic nucleus was carried on as part of this effort to understand theoretical physics generally. The work was characterized by frank discussion with younger colleagues of the basic problems of relativistic quantum physics. These ongoing discussions – as Bohr often referred to them – were entirely open-ended, and did not involve any future planning of research at the institute. In particular, as Bohr himself admitted, the original vision of parallel theoretical and experimental investigations at the institute had been relegated to the background. The Copenhagen spirit had evolved into a mode of work on a kind of physics problem uniquely suited to it. In effect, it had taken on a life of its own, dominating work in physics at the institute. Characterized by free-wheeling discussion of general conceptual problems rather than by work within a well-defined theoretical and experimental research program, by early 1934 there was nothing in the pursuit of physics at the institute to foreshadow the imminent redirection toward theoretical and experimental nuclear physics.[64]

INTEREST IN BIOLOGY, 1929 TO 1936

As background for the biological aspect of the concerted changes in research at the institute from the mid-1930s, the second part of Chapter 2 will follow Bohr's concern with biological questions in the period up to his redirection of research. I will show that at the end of 1929, Bohr did indeed begin discussing these questions in his publications and correspondence. Although he spent substantially less time and effort on biological questions than on pondering the atomic nucleus, it is evident that his interest in biology, like the discussions of physics, was motivated to a great extent by a desire to expand his comple-

mentarity argument. Bohr's concern with biology expressed itself in intense discussions with younger physicist colleagues,
thus providing another example of the Copenhagen spirit at
work. As we will see, Bohr's philosophically oriented interest in
biology, like the discussions of the atomic nucleus, did not lead
up to the scientific reorientation at the institute that was soon to
take place.

Background

In a unique autobiographical comment from 1958, Bohr traced
his biological interest explicitly to an early exposure to discussions in his parents' home around the turn of the century. There,
four of the most prominent Danish intellectuals of the time – his
father, the physiologist Christian Bohr; the philosopher Harald
Høffding; the physicist Christian Christiansen; and the linguist
Vilhelm Thomsen – gathered regularly for general discussion.[65]

Like Christian Bohr, Høffding was actively interested in biological questions. He was a founding member of the Biological
Society of Copenhagen, where in 1898, two years after its establishment, he gave a lecture "On Vitalism." The Society was established to further collaboration among professional biologists,
and Høffding was one of the few members who did not work
professionally in that field.[66]

As soon as they were old enough, Niels and his younger brother Harald followed the discussions in the Bohr home attentively.
Yet Niels did not begin publishing on biological questions until
1929, at the age of forty-three. The immediate occasion for this
renewed interest was not the influence of the intellectual milieu
of his father, who had died in 1911, but his contemporary struggle to clarify his complementarity argument. Nevertheless, there
are clear parallels between the concepts and views discussed at
the turn of the century and Bohr's discussion of biological questions beginning three decades later. A brief discussion of the fate
of his father's work and ideas in light of the general development
of biology will therefore provide a useful context for Bohr's published biological statements beginning in 1929.[67]

It is likely that Høffding, who was Bohr's philosophy teacher
at Copenhagen University, had an impact on Bohr's philosophical thinking. Indeed, one of the similarities between the views

Young Niels Bohr and his mother, c. 1900.

of the father's group and the much later statements of his son, was Niels Bohr's use of the term "psycho-physical parallelism" around 1930. Bohr's understanding of this concept is strikingly similar to Høffding's idea of "psycho-physical identity," which the latter discussed extensively from the 1880s. Høffding, in fact, distinguished between psycho-physical parallelism and his own identity hypothesis, finding the former term unsuitable for his way of thinking. The conflation of the two terms in Bohr's vocabulary may stem from his close relationship with his old schoolmate and prominent Danish psychologist, Edgar Rubin, who, in an encyclopedia article from 1925, introduced the two concepts as identical.[68]

Among several possibilities, Høffding concluded that ". . . mind and body, consciousness and brain, are evolved as different forms of expression of one and the same being." The nature of this being, however, "lies beyond our realm of knowledge. Mind and matter appear to us as an irreducible duality, just as subject and object." Høffding's belief in the determinism of physico-chemical processes led him to consider free will as a

The brothers Harald and Niels Bohr proudly show off their student garb, c. 1900.

mere subjective illusion; being identical to the physiological pro-cesses that were determined by physico-chemical laws, mental processes were also predetermined. As we will see, almost half a century after these pronouncements, Bohr and his followers would view complementarity as a means of retaining the possi-bility of free will without abandoning the psycho-physical iden-tity.[69]

In his generally oriented publications, Bohr rarely cited other authors. It is therefore not surprising that he never explicitly referred to Høffding as a source for any of the philosophical concepts or arguments he used in biology. In his autobiographical article of 1958, however, Bohr did quote a specific passage from his *father's* work. In this general statement from 1910, Christian Bohr stressed the importance in practical physiological work of explaining the functions of different parts of the living organism in relation to their purpose of keeping the whole organism alive. Although he viewed classical deterministic physical and chemical laws as indispensable in the study of physiology, he also thought that the investigation of organic processes could not be pursued in isolation from their role in the upkeep of life. For Niels's father, this circumstance made physiology different in principle from physics and chemistry. As Høffding wrote in an obituary for his friend, Christian Bohr:

> It was his task, in one of the most central areas of organic life, to seek the borderline between life and the forces of inorganic nature – to see if there was a set borderline and if so [to see] where it was.[70]

Toward the end of his life, Christian Bohr reported that he had found confirmation of the uniqueness of life science in his physiological research. Since the 1870s, his teacher, Carl Ludwig, the "founder of modern physiology," had carried on a debate with the school of the physiologist Eduard Friedrich Wilhelm Pflüger in Bonn over whether the gas exchange of oxygen and carbon dioxide in the lungs could be described as a simple diffusion process. After a long series of experiments in Copenhagen, conscientiously performed with experimental equipment improved by his student and assistant August Krogh, Bohr made the firm claim, in accordance with the views of his teacher, that lung tissue did play an active part in the gas exchange in the lungs, and that this exchange was not a simple diffusion process. On the basis of experimental work by several investigators, Christian Bohr subsequently concluded in a detailed review article published in 1909 that "a specific cell activity in the gas exchange in the lungs has been established beyond doubt." Bohr interpreted this finding to mean that the behavior of lung tissue was regulated according to the needs of the organism as a whole. Lung tissue behavior could therefore only be understood in relation to the function of the lungs to keep the organism alive.[71]

August Krogh, however, was not satisfied with Christian

August Krogh in 1904. [Courtesy of Bodil Schmidt-Nielsen]

Bohr's results, and independently pursued a series of thorough experiments. Shortly before Christian Bohr's death, Krogh published a rejection of his mentor's view, concluding in no uncertain terms that the

> . . . absorption of oxygen and the elimination of carbon dioxide in the lungs takes [sic] place by diffusion and by diffusion alone. There is no trustworthy evidence of any regulation of this process on the part of the organism.

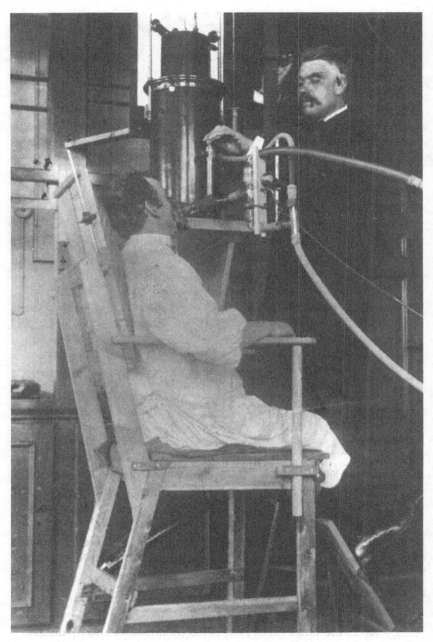

Christian Bohr (right) experimenting in his physiology laboratory.

According to the obituary written by Valdemar Henriques, his successor in the chair of physiology at Copenhagen University, Christian Bohr accepted respiration as a diffusion process before he died. When Niels Bohr began to publish on biological questions in the late 1920s, Krogh's conclusion had long since become generally accepted among physiologists.[72]

The fate of Christian Bohr's views during the intervening period is usefully illustrated by the career of his student and prominent British physiologist, John Scott Haldane. Haldane had worked for several weeks under Christian Bohr's direction in the latter's Copenhagen laboratory in the early 1890s, when he became a close friend and great admirer of his teacher. Although soon accepting the validity of the diffusion theory under normal conditions of respiration, Haldane argued that secretion processes regulated by the organism were crucial in acclimatizing the animal body to low atmospheric pressure and hard work. In an article published in 1927, Haldane described the contribution of August Krogh in the following words:

> He entirely broke away from Bohr on the question of whether the lungs ever secrete oxygen actively. In this I think he made a most serious mistake, though his actual experiments were a model of what experiments ought to be.

Even eighteen years after Krogh's rebuttal of Christian Bohr's views, Haldane still found it essential to point out that Krogh's behavior toward his mentor had been inappropriate. However, Haldane was becoming increasingly isolated in his view of the breathing mechanism, being vigorously opposed by the British physiologist Joseph Barcroft in particular.[73]

Haldane had a greater philosophical investment in confirming the secretion theory than either Carl Ludwig or Christian Bohr. Publicly promoting a philosophy in which the qualitatively different disciplines of mathematics, physical science, biology, psychology, and religion constituted increasingly closer representations of human experience, Haldane considered respiration physiology – his field of specialization – to be a prime example of the difference between a physical approach and a biological approach; in biology, life, defined as the maintenance of the organism, was considered to be a fundamental irreducible fact. In the later part of his life, Haldane presented the secretion theory again and again as an argument for the view that respiration could only be understood in terms of its role in keeping an organism alive.[74]

Just as he became more and more isolated in the biology community as a proponent of the secretion theory, so Haldane experienced increasing estrangement in advocating his general philosophy of life science. Always careful to stress that he was not advocating a special vital force governing the life processes, in his Silliman Lectures of 1915 he characterized his "doctrine" of biology not as vitalism, but as "organicism." In his last book on the philosophy of his science published twenty years later, he in fact denounced as vitalist those who, according to current terminology, were called "organicist" or "organismal" biologists. By 1935, in other words, he considered as vitalist both the currently fashionable concept of life as organization and the attempt to establish a quantitative science of biology on this basis.[75]

For their part, the younger generation of working biologists considered Haldane's concept of life – not as a quantifiable mode of organization, but as an irreducible physiological fact – to be unscientific, and even vitalistic. By this time, Haldane's views belonged to a world of discourse altogether different from the contemporary mainstream of intellectual debate among biologists. As we will see, Bohr would draw similar conclusions to Haldane's from his complementarity argument.

A review article from 1928, "Recent Developments in the Philosophy of Biology," brings the ongoing transformation in general biological viewpoints and Haldane's place in it into particularly lucid perspective. The author, the biochemist Joseph Needham, already distinguished at the age of twenty-eight, took what he considered Haldane's outdated views as his starting point. He criticized particularly Haldane's conception of physical and biological science as qualitatively different, as well as Haldane's intimation that biology constituted the more advanced scientific viewpoint. Needham insisted on the contrary that there was no difference in principle between separate fields of scientific investigation. In line with more modern concepts, he contended that a meaningful distinction existed only between *quantitative* science and *qualitative* philosophy. On the basis of these observations, Needham accused Haldane of holding views contrary to the scientist's obligation toward quantification.[76]

Along with developments in biology itself, the parallel evolution toward a quantitative, yet nondeterministic, physics did not go unnoticed in the general discussion of biology as a science. Haldane, for example, took these developments as a confirmation of his long-held view that physics would ultimately reach the level of biology. Thus he wrote in 1928 that "the disclosure of quantum relations . . . seems to have partially brought physical investigation within sight of distinctly biological investigation."[77]

Needham also took notice of these developments. For example, in his 1928 review article he referred approvingly to a recent paper by the American physiologist Ralph Stayner Lillie proposing specifically how biology could make use of the new physics. In the lead article of the journal *Science* for 12 August 1927, Lillie speculated that the processes of life could be understood in light of the amplification of individual nondeterministic atomic phenomena. Needham did not accept Lillie's specific proposal, but expressed sympathy with his motivation to seek a scientific basis for free will. He referred to Lillie's contribution as "an altogether fresh wind."[78]

First biological statements, 1929 to 1930

Lillie published his paper only a few weeks before Bohr presented his revolutionary complementarity concept in Como in 1927. In turn, Bohr published his first statement on biological questions two years after the Como lecture. Bohr made this statement in one of a series of articles developing the "lesson" of quantum mechanics and complementarity.[79]

Why did Bohr begin publishing statements on biology? In a later article in the same series, he volunteered an explanation. Such discussion, Bohr suggested, would help clarify his interpretation of physics. He surmised that the epistemological implications of his complementarity argument would be more readily understood by referring to more well-known problems in other fields of investigation, such as biology. As Bohr saw it, his first public comments on biological questions did not arise from a radical change of interest, but from a desire to make his complementarity argument better understood by more people. In other words, they were designed to serve as rhetorical ploys or figures of speech, rather than definitive scientific statements.[80]

Bohr wrote his first article introducing his ideas in biology – as well as psychology – for the commemoration in June 1929 of the fiftieth anniversary of the German physicist Max Planck's doctorate. As we have seen, Bohr was also using such prestigious occasions during this period to discuss the problems of the atomic nucleus. Just as he relegated his discussions of nuclear problems to the very end of lectures on the current state of theoretical physics, so his biological statement was added onto a much longer discourse on the complementarity argument.[81]

Although Bohr dealt with the same interdisciplinary problem as Lillie and other contemporaries, he did not seek to understand biology directly on the basis of the new physics. On the contrary, he sought to use his complementarity viewpoint as an overarching argument, exactly as he did to illuminate relativistic quantum physics and the atomic nucleus.

Picking up on the last paragraph of his Como lecture, Bohr thus referred in his Planck commemoration article to the "deep-going analogy [in quantum mechanics] to the general difficulty in the formation of human ideas, inherent in the distinction between subject and object." He then contrasted the feeling of free will in psychic life with the "apparently uninterrupted causal chain of the accompanying physiological processes." He referred to what he considered to be the commonly held opinion that mental processes could be reduced to physiological processes in the brain, and that free will was hence a mere illusion. Although he did not as yet introduce a specific term, or mention Høffding's name, the opinion Bohr referred to reads like a paraphrase of Høffding's psycho-physical identity. Bohr went on to propose by analogy with quantum mechanics that an experiment designed to follow the physiological processes of the brain in detail might affect these processes in an unpredictable way and even "bring about an essential alteration in the process of volition"; the complementarity argument, in other words, might provide an explanation of free will by implying that physiological processes were not deterministic after all.[82]

Hence, although he did not say so explicitly, Bohr's argument implied that Høffding's concept as modified by the "lesson" of quantum mechanics could be upheld without abandoning free

will. In spite of his own statement that he entered the biological –
and psychological – domain in order to explain his complementarity argument to a larger audience, in effect he seemed to employ it in order to come to terms with the kind of problems discussed in his parents' home before the turn of the century.

Bohr's lecture at the Eighteenth Scandinavian Meeting of Natural Scientists, which was held in Copenhagen in August 1929,
was devoted to a general discussion of complementarity. In concluding, he touched once more upon psychological and biological questions. Whereas he had previously implied that free will
might be inferred from an observation problem in physiology
analogous to the observation problem in quantum mechanics,
Bohr now presented both free will and causality as

> . . . idealizations whose natural limitations are open to investi-
> gation and which depend on one another in the sense that the
> feeling of volition and the demand for causality are equally in-
> dispensable elements in the relation between subject and ob-
> ject which forms the core of the problem of knowledge.

With this passage, Bohr seemed to imply that free will in psychic
processes could not be deduced from the application of complementarity in physiology. Instead, free will and causality were
mutually independent and equally useful idealizations.[83]

In this lecture, Bohr also introduced a new argument in his
discussion of biology that he would return to on subsequent occasions. He first noted that experiments had shown that only a
few light quanta were sufficient to stimulate the human eye. Because the description of individual light quanta belonged to the
domain of quantum physics, Bohr argued, it might be necessary
to employ the new physics in order to describe the behavior of
at least some animal organs. Admitting that many problems in
biology could be formulated "with the help of our ordinary visu-
alizable perceptions" – that is, without introducing quantum no-
tions – Bohr continued:

> With regard to the more profound biological problems, howev-
> er, in which we are concerned with the freedom and power of
> adaptation of the organism in its reaction to external stimuli,
> we must expect to find that the recognition of relationships of
> wider scope will require that the same conditions be taken into
> consideration which determine the limitation of the causal
> mode of description in the case of atomic phenomena.

Bohr now seemed to be groping for a more direct connection between quantum theory and biological phenomena.[84]

In a new introduction to a collection of previously published articles, which appeared in 1929 in connection with Copenhagen University's 350th anniversary, Bohr was more explicit with regard to his uncertainty about the precise relationship between the new physics and biology. He reiterated that parallels between "psychological problems" and the recent interpretation of quantum mechanics might prove to be mutually productive for both areas of investigation. "It may well be," Bohr continued,

> . . . that behind these analogies there lies not only a kinship with regard to the epistemological aspects, but that a more profound relationship is hidden behind the fundamental biological problems which are directly connected to both sides.

However, he left it an

> . . . open question whether the information we have acquired of the laws describing atomic phenomena provides us with a sufficient basis for tackling the problem of living organisms, or whether, hidden behind the riddle of life, there lie yet unexplored aspects of epistemology.

As yet unable to state the precise implications of his complementarity argument for the relationship between the sciences, his ideas on the subject were still developing.[85]

Vacationing in Norway in early January 1930, Bohr reported to Werner Heisenberg:

> [U]p here I am really dreaming most of the day about the problem of evolution, on skis, though, only when it goes up-hill, for when it goes down, there is luckily no question about any conscious analysis of the riddles of life.

At the same time he assured Heisenberg that he had not forgotten physics completely; almost apologetically, he stated that his letter's emphasis on other matters was an expression of vacation laziness. Bohr considered his concern with biological questions more a recreational interest than a main concern. Yet toward the end of his lecture to the Royal Society of Edinburgh in May 1930, he used the opportunity to discuss these questions once more, without, however, arriving at any new conclusions. In

Bohr takes time out for a brief ski vacation in Norway, 1930.

other words, for a year after his first publication touching on biological questions, his statements on such matters were brief, tentative, and unspecific, without any firm conclusions about the potential role of the complementarity argument in the life sciences.[86]

The exchange with Jordan, 1931

Bohr's correspondence in 1929 and 1930 contains no discussion of his developing ideas in biology. In May and June 1931, a brief but intense exchange of letters on this topic took place between Bohr and his young admirer, colleague, and frequent visitor to the institute – the German physicist Ernst Pascual Wilhelm Jordan. Pursued in a true Copenhagen spirit, this exchange was instrumental in provoking Bohr to become more precise in his published statements on biology.

Pascual Jordan received his doctorate under Max Born at Göttingen University in 1924 when he was twenty-two years old. The influence of Bohr's writings showed early in Jordan's publications. Bohr first showed personal concern for Jordan in 1926, when he contributed some money for the treatment of Jordan's heavy stammering. The following year he wrote an appreciative letter of recommendation to the IEB supporting Jordan's application for a fellowship, part of which Jordan spent in Copenhagen in the summer and fall of 1927. During this period, Jordan contributed importantly to the formalism of quantum mechanics, and Bohr followed his work closely.[87]

Jordan had expressed interest in the relationship between quantum mechanics and biology before discussing these matters with Bohr. In his inaugural lecture at Göttingen in 1927 on the "Philosophical Foundations of Quantum Theory," which he gave some months before Bohr introduced his complementarity argument, he noted that "the notion of determinism must be formulated differently for biology and for physics." Then, in a letter to Albert Einstein in December 1928, which was part of an exchange on the interpretation of quantum mechanics, Jordan countered Einstein's famous complaint against the Copenhagen interpretation that the Lord does not play dice to settle events in the physical world. "I would not say," Jordan wrote, "that the dear Lord *throws dice* [to decide the movement of] electrons; but I would rather say: He permits the electrons to decide *for themselves.*" He went on:

> This point of view seems to me to be particularly relevant in relation to *biology;* I believe that in organisms truly consistent departures from the anorganic development of physical reactions take place, which *quite primitively* one can interpret to mean that the atoms or electrons belonging to a living individual in a certain way coordinate their otherwise statistically independent decisions.

Jordan's subsequent correspondence with Bohr on biological questions would serve to sharpen both his and Bohr's developing views.[88]

When Jordan visited Copenhagen in March 1931 to participate in the annual informal physics conference, he and Bohr expressed their intentions to publish on biological matters. By the second half of May, Jordan had submitted a manuscript to Bohr for comment. This manuscript was the catalyst for the ensuing correspondence between the two.[89]

Jordan devoted the first four sections of his manuscript to a detailed introduction to Bohr's interpretation of quantum mechanics. Only in the last section, "On the Question of Free Will," did he enter into its implications for psychology and biology. Jordan was determined to show that quantum mechanics undermined the contention that the experience of free will was a mere subjective feeling. Indeed, he claimed that free will was a necessary consequence of quantum mechanics.

Jordan began his last section by discussing whether there was a limit to how precisely a human being could be subjected to scientific investigation before the person was killed. According to Jordan, the common view was that there was indeed such a limit, which, however, was not a limit of principle, but owed itself to the still imperfect means of observation. Contrary to this view, Jordan posited that such a limit was a necessary consequence of quantum theory, which as a matter of principle imposed a limitation on any observation of atoms. Referring to Bohr, Jordan employed light perception to argue that physiological processes took place on a microscopical level. Echoing Lillie's earlier claims without referring to them, as well as paraphrasing his own previous comments to Einstein, he then

Visiting lecturer Pascual Jordan at the institute during the annual physics conference, 1931.

postulated the essential difference between animate and inanimate matter: Whereas in inanimate matter "the statistical acausality of atomic reactions" in effect averaged up to causality at the macroscopical level, in animate matter "the acausality of specific atomic reactions *amplify* to a macroscopically operative acausality." Jordan postulated, in other words, that by some amplification mechanism the acausality of atomic processes became operable macroscopically. To Jordan, this was the basis for free will and the uniqueness of conscious life.

Jordan did not define psycho-physical parallelism as a relationship between physiological processes and psychic experience. Without mentioning earlier presentations of the concept by Høffding and others, he identified it instead directly with the dualism of quantum mechanics, which stated that light (or electrons) could be observed either as waves or particles, depending on the experimental setup. To Jordan, Bohr's complementarity argument as applied to quantum mechanics implied that a sufficiently detailed physiological investigation would destroy this duality, which was responsible for the acausal behavior of the atoms. Because this acausal behavior in turn provided the basis for life, the investigation would kill the patient. To Jordan, Bohr's interpretation of quantum physics had come to imply that the explanation of psychic processes was directly reducible to quantum mechanics.

In the letter accompanying his manuscript, Jordan commented that his recent discussions with Bohr in Copenhagen had reminded him of how inferior his insistence on "clarity and on a certain 'radical' sharpness of formulation" was compared with Bohr's "more careful and more striking form of expression." Characteristically polite, Bohr expressed appreciation in his reply for "the sympathy you demonstrate for my efforts in spite of all imperfections in the form of expression." Yet, although "fully appreciating the clarity of [Jordan's] more forceful style," Bohr went on to address some points that he "would have formulated somewhat differently." When discussing substance, then, he emphasized differences rather than agreement.[90]

Bohr stressed that he had never aimed at a "narrow analogy" between psycho-physical parallelism and the wave–particle duality; he thus disapproved of Jordan's explanation of psychic processes and argumentation for free will directly on the basis of quantum mechanics. Psycho-physical parallelism, Bohr continued, meant to him a *particular* complementarity, irreducible to one-sided explanation in physics. Referring to their previous discussions in Copenhagen, he reiterated his own view that the death of an organism introduced a limitation in principle to the observation of life phenomena. However, he emphasized that life phenomena were peculiar to themselves and could not be reduced to the laws of physics; in particular, they could not be understood in terms of an amplification of atomic phenomena as Jordan proposed. Bohr now suggested that

> . . . just as the stability of atomic phenomena is irrevocably connected with the limitation of the possibilities for observation expressed in the uncertainty principle, so the peculiarity of life phenomena is related to the impossibility in principle [of determining] the physical conditions under which life takes place.

Bohr here seemed to imply that an organism's constant exchange of atoms with the environment made it impossible to study as an isolated atomic system, and that this impossibility could serve as a basis for understanding life phenomena; Bohr's argument was another case of explanation by renunciation. For Bohr, then, biology was qualitatively different from physics and chemistry. He viewed neither the phenomena of life nor free will simply as a consequence of quantum mechanics; rather, they could both be illuminated by the complementarity argument as it distinctively applied to these sciences.[91]

Although he had arrived at his complementarity argument from an interpretation of quantum mechanics, from the outset Bohr had conceived of the argument as having more general validity than quantum mechanics itself; from his earliest statements he was open to the possibility that it might apply to physics, biology, and psychology in different ways. However, the discussions with Jordan in Copenhagen, as well as Jordan's letter, had provoked Bohr to be more specific on the relationship between the sciences. Now Bohr did not just allow for the *possibility* of biology's autonomy, he was positively affirming an empirical criterion for distinguishing physics from biology. As a result, his statements on life science were becoming increasingly similar to those of his father and J.S. Haldane. Like them, he was now clearly considering biology an autonomous field of study, distinct from physics or chemistry because of the irreducible fact of life.[92]

Two more letters were exchanged between Bohr and Jordan on their respective views on biology. These letters, however, do not shed much new light on their differences. Jordan, in response to Bohr's criticism, sought to present his views of the organism in greater detail. In particular, he expanded upon his "amplifier theory," dividing the living organism into three "zones": the actual "centers" of atoms behaving acausally and providing the true basis of life; the "amplifier organs" amplifying the acausal behavior of these centers; and the "tool organs" [*Werkzeugorgane*] whose behavior does not differ significantly from that of inorganic matter.[93]

Moreover, Jordan emphasized the physicist's crucial obligation of enlightening the biologists about possible "applications of the new physical results and points of view." Yet, as Jordan reported to Bohr, the editor of *Die Naturwissenschaften* regarded his paper as philosophical rather than biological. Indeed, together with the positivist philosopher Hans Reichenbach he encountered difficulties getting the journal to publish his paper, the editor claiming that too much had been published on "causality" already. Clearly, Jordan's views were not considered publishable on the basis of their relevance for biology. This incident suggests that Jordan's argumentation, and by implication Bohr's, was not in the mainstream of their contemporaries in biology.[94]

In the final letter concerning Jordan's manuscript, Bohr again expressed general agreement with Jordan's views. In particular, he wrote that in recent years he had often discussed similar ideas to Jordan's amplifier concept with Danish biologists, who responded positively. Bohr emphasized, however, that his main concern was not practical biological research. Rather, such questions provided "the background for my general mode of expression which above all is intended for the epistemological analysis of our general situation" – which was Bohr's polite way of saying that he still disagreed with Jordan's attempt to reduce psychic phenomena, as well as biology, directly to quantum mechanics.[95]

Jordan's paper appeared in *Die Naturwissenschaften* one and a half years after his correspondence with Bohr. Probably following the advice of the editor, Jordan provided an amended title not containing the word "causality." The paper was practically identical to the earlier one sent to Bohr, except for a concluding section, "The Essence of the Organic." Here, Jordan presented an elaboration of the amplifier theory that he had described in his second letter to Bohr. More importantly, he expressed agreement with Bohr's criticism that this theory was insufficient as an explanation of the peculiarity – or as Jordan called it, "stability" – of the living organism. Instead of abandoning his direct reasoning from quantum mechanics to biology, however, he responded to Bohr's objection by ascribing to his first, or "inner," zone "a degree of non-observability of its physical state still higher than what we know from atomic physics." As John Heilbron has pointed out, such an explanation of life by an arbitrary hypothesis amounts to no less than vitalism. Nevertheless, Jordan was once more claiming to employ Bohr's views on biology to bolster his own approach. Indeed, he presented the whole article as a clarification and elaboration of Bohr's views.[96]

This distortion of his views may have hurt Bohr, as may Jordan's early overt Nazi sympathies. In the spring of 1934, for example, Jordan committed what Max Born in a letter to James Franck referred to as a "literary slip" [*literarische Entgleisung*], which seems to have amounted to an argument for the value of pursuing science, even under the Nazi regime. Although I have been unable to trace the direct source of Born's disapproval, Jordan's argument in a booklet published the following year may be indicative of what he wrote. After Born told Jordan that Bohr was also critical of the contribution, Jordan wrote to Bohr to explain himself. Jordan wrote that he had simply argued

> . . . that the sign of this quantity [representing the pursuit of physics] can be shown to be positive also on the basis of a system of axioms which stands in a relationship of complementary exclusion to those axioms from which one otherwise customarily derives this sign.

In his reply, Bohr told Jordan that he understood "very well how relative, if not complementary, [were] all the relations in human life." He expressed his thankfulness

> . . . in these stormy times . . . that we in natural science and epistemology have before us sufficient problems [and that we] in the effort to counter prejudices and passions ever again find new small points which may contribute to the illumination of the whole.

By suggesting that science was a means "to counter prejudices and passions," Bohr seems to have disapproved of Jordan's view of physics as entirely value-free. The anticipation that letters to Germany might be censored may have added to Bohr's usually obscure form of expression. In any case, except for his purely formal return of Jordan's belated greetings on his fiftieth birthday the following year, this is the last known letter from Bohr to Jordan. Jordan did, however, attend the annual informal conference at the institute in June 1936. On this occasion, Bohr seems to have disapproved of Jordan's application of the complementarity argument to parapsychology, for when Jordan submitted an article to Bohr on this topic after the conference, he enclosed a letter predicting Bohr's disagreement and carefully explaining his views.[97]

As we will see, Jordan continued his campaign to present his version of Bohr's ideas, and it may be that this soured the relationship between them. If so, this is one of the rare instances in which forthright discussion between Bohr and a younger colleague led to a disagreement too serious to allow further communication.[98]

Bohr's next published elaboration on his biological views was characteristically brief, but the impact of the exchange with Jordan is readily apparent. At the time of this correspondence, Bohr was convinced by colleagues that his Danish booklet of previously published articles ought to be issued in Germany. In the German version, Bohr added the lecture he had given in Copenhagen in the late summer of 1929. In order to justify the new addition, Bohr prepared an addendum to the booklet's original Danish introduction. Because Bohr told Jordan about the planned publication and its new addendum only in the last letter of their correspondence on Jordan's manuscript, it is likely that Bohr wrote the addendum concurrently with his discussions and correspondence with Jordan. If so, the addendum was Bohr's first published statement on biological matters after he and Jordan had tried out their ideas on each other.[99]

The contents of Bohr's new addition substantiate this order of events. In the two-page addendum, Bohr published for the first time the point – which he had discussed with Jordan – that a thorough physiological investigation would cause the death of an organism. As in his first reply to Jordan, he was careful to present this point as an independent principle not to be identified with the observation problem in quantum mechanics. Elaborating further, Bohr wrote:

> . . . there is set a fundamental limit to the analysis of the phenomena of life in terms of physical concepts, since the interference necessitated by an observation which would be as complete as possible from the point of view of the atomic theory would cause the death of the organism. In other words: *the strict application of those concepts which are adapted to our description of inanimate nature might stand in a relationship of exclusion to the consideration of the laws of the phenomena of life.*

As in the correspondence with Jordan, Bohr now stressed, without qualification, the impossibility of "associating the psychical and physical aspects of existence . . . by one-sided application either of physical or psychological laws." As we have seen, in his earlier publications Bohr had been considerably more uncertain about the relationship between physics, biology, and psychology. The correspondence with Jordan had compelled him to state the distinction in principle between physical and biological science in no uncertain terms. Toward the end of a lecture in August 1931 to the students of the International People's College in Helsingør, Denmark, Bohr repeated this refined viewpoint.[100]

"Light and Life," 1932

Nearly a year went by, however, before Bohr presented his views on biology in a more complete form. Until then, he had found too little time to develop his line of thought at any length in writing, and he may have considered it a welcome opportunity when he was asked to give the opening address at the Second International Light Congress to be held in Copenhagen in August 1932. This congress, which extended over five days and included more than fifty speeches, was devoted to the role of light in "Biology, Biophysics, [and] Therapy."[101]

The first mention of the congress in the Bohr Scientific Correspondence was made by Bohr's longtime colleague and member of the organizing committee, Hans Marius Hansen. On 17 July

1932, Hansen wrote to Bohr in his summer house that "the light people" were still putting pressure on him to ask Bohr to lecture. He noted that Arne Henry Kissmeyer – dermatologist, venereologist, and the secretary general of the congress – was "of the opinion that the Light Congress must be equally close to your heart as the philosophers' congress." This was only one instance of congresses and organizations competing to add Bohr's prestigious name to their activities.[102]

When he answered Hansen's letter on 22 July, Bohr had still not made up his mind whether or not to give a lecture; he did so only about three weeks before presenting it. Probably as a result of the late decision, his talk at the Light Congress is one of the very few of his public lectures of which a manuscript has not been preserved. Moreover, the lecture was subsequently published in several languages and editions containing only minor alterations of formulation. This was quite atypical for Bohr, whose writings were usually drafted several times before publication. It indicates that after his correspondence with Jordan, Bohr's thoughts on biology had matured substantially. Even in this case, however, Bohr could not help expressing his circumspection with regard to the printed word; in a letter to Oskar Klein, he thus deemed the Danish version to be sufficiently improved over the original English to consider it the "original" rendition of his lecture, to be used for subsequent translations.[103]

Although not containing significant changes in his views on biology, Bohr's Light Congress lecture, appropriately titled "Light and Life," was his first presentation devoted mainly to biological questions. As such, it constituted a synthesis of previously scattered remarks, and was also more precise than his earlier presentations.

In the context of the Light Congress, it was particularly apt to employ, as Bohr had done before, the ability of the eye to respond to only a few light quanta as an argument for the biological relevance of atomic physics. At the same time, however, Bohr made it clear that biological explanation could not be reduced to atomic physics, and he took care to treat the two disciplines as mutually autonomous. As before, he argued for autonomy by distinguishing between the different conditions under which inanimate and animate phenomena were studied. In the case of atomic physics, he reiterated his belief that a causal description of atomic phenomena was prohibited by the indivisibility of the quantum of action, which imposed an uncontrollable interaction between the experimental arrangement and the object under study. In biology, on the other hand, any investigation of the phenomena of life depended on keeping the organism alive. On these grounds, Bohr proposed a basic analogy between the fundamental – and in principle inexplicable – quantum of action in physics, and the concept of life in biology. Just as his father and John Haldane had done before him, he concluded that life must be regarded as an "elementary fact." Also like Haldane, Bohr did not seek to base a quantitative description of life on physics or any other science. Instead, the peculiar observational condition in biology – that to study life one must not kill the organism – provided the basis for "the wonderful features which are constantly revealed in physiological investigations and which differ so markedly from what is known of inorganic matter."[104]

Continuing biological interest, 1933 to 1936

Bohr's personal interest in biologically related questions even beyond 1934 was largely unrelated to the scientific redirection of the institute. Although continuing to lament that all-consuming discussions of the "open questions" in atomic physics allowed him too little time for thinking about biology, Bohr was still able to pursue his philosophically oriented biological interest; however, he did not publish again on such matters until 1936.

In part, Bohr's continued concern with biological questions was motivated by negative reactions to Jordan's continuing crusade for what he still considered to be Bohr's views on the relationship between physics, biology, and psychology. Having presented his philosophical views within a positivistic framework in the manuscript he had discussed with Bohr in 1931, Jordan now sought to align himself more closely with the so-called Unity of Science group, centering around Moritz Schlick and others in Vienna and Hans Reichenbach in Berlin. He thus contributed a long article to the group's journal *Erkenntnis* in 1934. In these "Quantum Physical Remarks on Biology and Psychology," Jordan hailed the group's forebear Ernst Mach, and more particu-

larly Reichenbach, for having anticipated what was now regarded as the epistemological implications of the new quantum
mechanics. He repeated his earlier argument for free will at
greater length, adding psychoanalysis as an example of the impossibility in principle of predicting human decisions: Such decisions can only be predicted on the basis of knowledge of the
states of the unconscious; these states, however, are themselves
strongly affected by the act of observation represented by psychoanalysis.[105]

Jordan's views, and by implication Bohr's, were fiercely attacked as unscientific and speculative at the positivists' conference in Prague later in 1934, where Edgar Zilsel, a close ally
of the group, introduced a discussion session on "P. Jordan's
Attempt to Save Vitalism Quantum Mechanically." Finding Jordan's inferences from quantum mechanics to psychology and biology alike entirely artificial, Zilsel concluded that Jordan's argument that an organism would die from a sufficiently thorough
experimental investigation "hence does not prove the vitalist
thesis, but presupposes it." In the discussion after Zilsel's contribution, Otto Neurath, Moritz Schlick, and Philipp Frank – all
prominent members of the Unity of Science group – found Jordan's contentions equally reprehensible. Only Reichenbach,
whom Jordan had praised in his *Erkenntnis* article, was willing
to give him the benefit of the doubt. Jordan was provoked into
formulating a defense, in which he emphasized more than ever
that his contributions constituted a clarification of Bohr's
views.[106]

In the following years, Jordan continued to generalize the application of quantum mechanics and Bohr's complementarity argument. Thus, he continued to promote his "amplifier theory,"
and in 1945 published a book called *Physics and the Secret of
Organic Life*. Two years later, he even produced a pamphlet,
Inhibition and Complementarity, devoted entirely to an argument for the close relationship between these two concepts, introduced by Sigmund Freud and Bohr, respectively. As he had
done before, he went on to explain parapsychological phenomena on the basis of positivism and complementarity.[107]

In the meantime, Bohr was becoming increasingly sensitive to
conclusions allegedly drawn on the basis of his complementarity

argument. In his "Light and Life" lecture, he had included a long defensive paragraph, where he even found it necessary to explain that

> . . . it is impossible, from our standpoint, to attach an unambiguous meaning to the view sometimes expressed that the probability of the occurrence of certain atomic processes in the body might be under the direct influence of the will.

As usual, Bohr did not cite a source. The only clue I have been able to find is in a letter to Otto Meyerhof written more than four years after the "Light and Life" address; here, Bohr criticized the British physicist Arthur Eddington for misusing "quantum physics in a spiritualistic manner." Whomever Bohr referred to in his address, it is clear that he deemed it urgent to discount misrepresentations of his views.[108]

After the demise of the relationship with Jordan, it did not take long before Bohr got another helper in biological questions from among his younger physicist colleagues – the German Max Delbrück. According to his own reminiscences, it was Bohr's "Light and Life" address in 1932, which he attended, that provided the final inducement to turn to biology full time, and to pursue the question for biology, as he understood it, posed by Bohr's complementarity argument. The "intellectual impetus" for the rest of Delbrück's career consisted of the search for an empirical manifestation of complementarity in biology like the experimental evidence in physics that had led Bohr to his argument in the first place.[109]

One of Delbrück's contributions as Bohr's helper in biological questions was to report on a particular instance of Jordan's exposition of Bohr's views. In late 1934, Delbrück informed Bohr that Jordan had given a lecture to the Society for Empirical Philosophy – the Berlin forum of the Unity of Science group led by Hans Reichenbach – in which he claimed to present Bohr's philosophical views. Delbrück reported that many of the assembled biologists had reacted with alarm to Jordan's lecture. According to Delbrück, this was no fault of Bohr's, because Jordan "misrepresented [Bohr's biological] arguments completely, so far as he had mentioned them." In the ensuing debate, the prominent biologist Max Hartmann of the Kaiser Wilhelm Institute for Biology had complained bitterly about the confusion caused by Bohr's and Jordan's publications on biological matters. "In the process," Delbrück reported, he misrepresented also those of your assertions that Jordan had not mentioned. The result was that all biologists scorned all physicists."[110]

Delbrück had even prepared a resumé of Bohr's biological views, which he had handed to Hartmann to clear up the misunderstandings. In this resumé, a copy of which he enclosed with his letter to Bohr, Delbrück sought to rectify the biological implications of Bohr's thoughts as stated by Jordan. He emphasized that the complementarity between the biological and physical sciences ensured that specific life phenomena could *not* be reduced to the laws of atomic theory. Consequently, Bohr and his followers did *not* assert that biologists would have to destroy life in order to study it. On the contrary, because the biologists did not work at the level of individual atomic elementary processes, their descriptions could remain strictly causal. In his reply, Bohr expressed agreement with the contents of Delbrück's resumé.[111]

In contrast to Bohr, Delbrück contributed to genuine biological research. Delbrück's turn to biology had in fact begun even before he attended the "Light and Life" address. Thus, at the end of June 1932 he wrote to Bohr:

> I have accepted Lise Meitner's offer to go to Dahlem as her 'family-theoreticist' in October largely because of the neighbourhood of the very fine Kaiser-Wilhelm-Institut für Biologie, with which I am entertaining very friendly relations.

He went on to describe how he presently pursued his biological interest in Bristol.[112]

In the summer of 1935, Delbrück's change of interest reached a momentary climax. He then published a long article, "On the Nature of Gene Mutation and Gene Structure," in collaboration with the "resident specialist in X-ray mutations" N.W. Timofeéff-Ressovsky and the radiation physiologist K.G. Zimmer. With the help of Delbrück's expertise in physics, the three were able to propose a theory of mutations based on atomic theory. When he sent Bohr a prepublication copy of what later became famous as the *"Dreimännerarbeit,"* however, Delbrück expressed regret about still not having been able to establish empirically a complementarity argument for biology. Bohr, for his part, was sufficiently interested in the article to organize a conference around it. By then, he had begun the concerted scientific redirection of the institute, which justified a conference there devoted to biological research.[113]

In spite of the institute's turn toward more practical matters of biology after the mid-1930s, Bohr's own publications on biological matters continued to be philosophically oriented. Significantly, after the "Light and Life" address, his first published statement touching on biology was a lecture at a conference organized by the Unity of Science group. In a letter expressing his appreciation for Bohr's decision to participate in the conference, Philipp Frank referred to a new pamphlet by Jordan showing "that people are trying to exploit quantum mechanics in the sense of the national-socialistic world view and even their race biology." Frank considered it "extraordinarily important that we agree on a formulation of [quantum mechanics] that excludes such misunderstandings." The positivists' Second International Congress for the Unity of Science, devoted to "The Causality Problem," convened in Copenhagen in 1936. Speaking on "Causality and Complementarity," Bohr used the opportunity to clear up recurrent misunderstandings of his views. He thus deemed it urgent to underscore at the outset that these did not "involve a mysticism incompatible with the true spirit of science." He firmly rejected the "widespread opinion that the recent development in the field of atomic physics could directly help us in deciding such questions as 'mechanism or vitalism' and 'free will or causal necessity' "; Bohr also distanced himself explicitly from the kind of argument for free will promoted by Jordan. Toward the end of the lecture, he repeated a disclaimer from his "Light and Life" address that the renunciation of causal description in atomic physics did not amount to an argument for spiritualism.[114]

In a rejoinder, Philipp Frank, who had been one of the strongest opponents of Jordan's views two years before, judged Bohr's own presentation to be sound. He found, however, that it "differed strongly from that of most of his 'philosophical interpreters.' " Yet he could not see any basis for applying complementarity to psychology or biology. Such a basis, according to the positivist and empiricist Frank, could only be founded on experiment. While this basis was indeed achieved in physics, it was yet to be established in other fields. Just as it had been experimentally established in physics that the space–time localization and causal description of an atom could not be decided at the same time, Frank thus contended,

> . . . so, in the consistent completion of the Bohrian mode of thought, an empirical proof must be given that the precise physical observation of the atoms of the living bodies is incompatible with the known empirical laws for the behavior of the living bodies and with the physical hypotheses regarding their atomic constitution.

Incidentally, this was a precise statement of the intellectual impetus guiding Delbrück for the remainder of his scientific career. According to Warren Weaver of the Rockefeller Foundation, who attended the conference, there was a lively debate between Bohr and Frank after the lectures. Unfortunately, there is no evidence as to the exact points of disagreement.[115]

Bohr's motivation to participate in the positivists' conference seems to have been twofold. First, he felt impelled to rebut interpretations of his views that he regarded as false; second, he sought to spread the gospel of complementarity to a broader audience. In the process, he had moved the discussion of biological problems from the safety of discussion with younger physicist colleagues to the open arena of philosophical debate. Yet Bohr's personal interest in biological questions in 1936 was clearly a continuation of his first stumbling remarks in 1929, and he was still moved by the philosophical implications of the new physics rather than by an urge to begin any well-defined theoretical and experimental research project at the institute.

Response to Bohr's biological views

In spite of Bohr's active participation in the philosophy conference in Copenhagen, and in spite of substantial contemporary attention to borderline problems between physics and biology, the response to his biological statements was generally minimal, particularly among biologists. Although they published the "Light and Life" address and other of Bohr's biological statements, the two widely read general science journals *Nature* in

England and *Die Naturwissenschaften* in Germany contained hardly any reactions to his views on biology. Other scientific and philosophical journals also did not take up Bohr's viewpoints.[116]

Among the few published utterances that indeed were made, most were negative. In a lecture in 1933, for example, the prominent German biologist Otto Meyerhof, who had received the Nobel Prize in physiology or medicine the same year that Bohr received his for physics, criticized Bohr for claiming that it was impossible to make successful detailed biological experiments on living organisms; thus, Delbrück's correction in his resumé the following year seems to have been warranted by more than a single incident. In England, Joseph Needham attacked Bohr's views in his widely read book *Order and Life*, from 1936. Needham considered as completely groundless the analogies between the quantum of action and life – or, as Needham quoted Bohr, between the "insufficiency of the mechanical analysis for the understanding of the stability of atoms" and the "impossibility of a physical or chemical explanation of the function peculiar to life." Like Meyerhof, he provided examples of detailed biological experiments on living organisms as counterevidence. He presented Bohr's views explicitly as an extension of Haldane's concept that life and biological organization were irreducible to physical and chemical terms. For Needham, a view such as Haldane and Bohr's implied a dogmatic and unscientific removal of "biological organization . . . from the realm of experiment."[117]

One possible reason for the scant interest in Bohr's biological statements is his obscure style and his tendency to relegate biological statements to the end of more physics-oriented papers. In addition, however, it must be considered that the subject–object problem and the psycho-physical parallelism discussed by Bohr were not central concerns among contemporary scientists. Nor was Bohr's clarification of his thoughts in the direction of the views held by his father and J.S. Haldane in line with contemporary biological thinking. Indeed, judging from the journals and conferences that Bohr chose as outlets for his ideas, his intended audience seemed to have been physicists first, philosophers second, and only then working biologists. Having visited Copenhagen in the spring of 1933, the prominent biologist Hermann J. Muller reported to his Norwegian colleague Otto Louis

Mohr that he had been glad "to meet the physicist Bohr there, but then I found that his ideas in biology were hopelessly vital-istic."[118]

Concluding remarks

Bohr's genuine interest in biological questions up to the mid-1930s did not augur an imminent change in the theoretical and experimental research at the institute. Like his contemporary interest in the atomic nucleus, his biological concern was theoretical, even philosophical. It was conducted in open-ended discussions with younger physicist colleagues. Bohr's biological concern shared another characteristic with his nuclear interest: It belonged within a substantially broader area of discussion. Just as Bohr's approach to nuclear physics was part of a general thrust toward formulating a new relativistic quantum physics, so his involvement in biology was another aspect of his effort to generalize his complementarity argument. Indeed, his published statements in both nuclear physics and biology were usually brief remarks tacked onto more comprehensive discourses in the broader areas to which they pertained. This fact, in addition to their theoretical and open-ended treatment, made nuclear physics and biology as pursued by Bohr and his disciples unlikely sources for a full-scale concerted theoretical and experimental redirection of research.

Bohr and his collaborators' concern for biological questions constitutes another example of the Copenhagen spirit at work. Just like the contemporary approach toward physics, the biological discussions were general, rather than specialized; they concentrated on the theoretical rather than the practical; they were totally open-ended, without any call for immediate results; and they involved the same unique mode of collaboration between Bohr and his physicist disciples. Indeed, allowing questions to lead into other fields of inquiry was not contrary to but quite in line with the open-ended Copenhagen spirit.

Nevertheless, this stepping onto the biologists' turf led to an intellectual isolation unlike anything the group around Bohr had experienced in questions of physics. As such, the experience of biological discussion can shed light on how the proponents of the Copenhagen spirit faced up to extraordinary situations. As we have seen, Bohr and his collaborators accepted and even thrived on their splendid isolation. This circumstance only underscores the cohesiveness of the group, which did not allow itself to be deterred by conceptual difficulties or outside criticism; it perceived itself to be above soliciting followers, seeking slow, careful, and elaborate discussion of problems to the core. John Heilbron has interpreted this self-assuredness in entering domains outside the field of specialization as the "imperialist" aspect of the Copenhagen spirit.[119]

The origins of Bohr's discussion of biological questions in the late 1920s are intricate indeed, and it would take us too far from the scope of this book to treat them in their full complexity. Nevertheless, the account presented here raises important questions that need to be followed up. For example, I have found it useful to present Bohr's interest in biology in the early 1930s against the background of the intellectual and scientific milieu of his physiologist father. In the development of Bohr's own thought, what was the relationship between his early exposure to philosophical and biological questions in his parents' home and his much later statements on biology based on the complementarity argument? Should the origins of Bohr's biology be linked to his intellectual milieu and influences more generally during his formative years, just as some have sought to explain the origins of his complementarity argument? Can the origin of the complementarity argument itself be traced to Bohr's early exposure to biologically related questions, so that historically Bohr's ideas on complementarity originated from his ideas on biology rather than the other way around? What was Bohr's real motivation for taking up biological questions in the late 1920s? For example, was he motivated psychologically by a desire to redeem the views of his father some twenty years earlier? The evidence suggests that Bohr expressed no biological interest from his youth until the late 1920s. Can some interest on Bohr's part nevertheless be traced in this period, and if so, to what extent can such involvements shed new light on the origins of his activities from 1929? These difficult questions must be left unresolved in this book. Hopefully, someone will be able address them successfully in a different context.[120]

CONCLUSION

In this chapter I have considered Bohr's interest from the late 1920s to the mid-1930s in the physics of the nucleus and biological problems – areas of investigation that would subsequently provide the scientific locus for the institute's reorientation. I have shown that by early 1934, the work at the institute in neither field had heralded a concerted change of theoretical and experimental research.

Both endeavors, in fact, were characterized by open-ended discussion of theoretical matters rather than any effort toward a directed change of scientific research program. Both nuclear physics and biology were closely related through Bohr's complementarity argument, which constituted a source of philosophically inclined discussion rather than a quest for hard scientific facts. This relationship can be seen clearly, for example, by comparing Bohr's first letter to Jordan on biology of early June 1931 to his concurrent statements on nuclear phenomena. In this letter, Bohr suggested that the stability of atomic phenomena and the peculiarity [*Eigenart*] of manifestations of life could be explained in analogous terms. Whereas the former was inseparably connected to the limitation of measurement possibilities expressed by the uncertainty relations of quantum mechanics, the latter was linked to the impossibility in principle of determining the physical conditions under which life takes place. This comparison closely resembles Bohr's suggestion at the conference on nuclear physics in Rome only a few months later. Here Bohr proposed, also by analogy with the complementarity argument, that the stability of the atomic nucleus can be understood only on the basis of a renunciation of the principle of energy conservation. Thus, Bohr implied, a scientific understanding of the atom, the atomic nucleus, and life could only be obtained at the expense of increasingly far-reaching renunciations, respectively, of causality, energy conservation, and detailed physical investigation.[121]

From this perspective, Bohr's ideas on the atomic nucleus and biology constituted two aspects of his extension and elaboration of the "lesson" of quantum mechanics to other areas of knowledge. Although it would be far too simplistic to claim that Bohr's interest in nuclear physics was guided exclusively by his complementarity argument, the two were closely intertwined. This relationship is a clear expression that the emphasis at the institute in the late 1920s and early 1930s was theoretical discussion rather than a coordinated program of theoretical and experimental research. Earlier, however, Bohr's approach to research at the institute had been substantially different. At the time of its foundation, he had preached and practiced a unity between theory and experiment based on spectroscopical investigations on atomic problems.

Why did Bohr change his approach to research at the institute from one based on a unity between theory and experiment to one based on theoretical discussion? The kind of questions taken up at the institute and the way in which they were pursued during the period covered in this chapter were obviously an expression of Bohr's personal inclination. In particular, the complementarity argument was created by Bohr. Moreover, the discussion between Bohr and his disciples epitomized Bohr's instinctive way of working. He depended strongly upon his sounding boards – Heisenberg, Dirac, Rosenfeld, and others in physics, and Jordan and Delbrück in biology – for working out his ideas.

More general developments in physics proper may also have contributed to loosening the bonds between theory and experiment at the institute in the late 1920s and early 1930s. In an interview, the Danish physicist Christian Møller, who worked at the institute during this period, has commented: ". . . spectroscopy, which is what was mostly done in the early days of the Institute, was perhaps already a little outdated when quantum mechanics was finished." Thus, from the perspective of contemporary physics, the original basis for a unity between theory and experiment at the institute may not have been as relevant as in the early days, when Bohr had championed such a unity based on spectroscopical investigations on atomic problems.[122]

Scientifically, the formulation of quantum mechanics in 1925 and Bohr's complementarity argument in 1927 represented a climax for the existing research program at the institute as well as other centers for theoretical physics. During this period, Bohr also worked hard to obtain economic and material resources for the institute, thus showing a skill and enthusiasm for matters of policy rarely noticed by his younger physicist colleagues. After 1927, the road to progress in physics was less clear and a policy of a unity between theory and experiment harder to formulate. At the same time, the number of visitors and publications at the institute declined dramatically, while the economic support leveled off after a period of steady growth since the institute's establishment. Whether this development resulted from a diminished motivation by Bohr to seek funds or from a decreased availability of economic support for basic science, it did allow Bohr to concentrate more on science and less on policy questions. As a result, Bohr's personal inclination toward theory and freewheeling discussion became the leading force at the institute to such an extent that Bohr himself in late 1931 expressed regret about the decreasing emphasis on experiment. Yet by 1934 he had not been able, or willing, to stop the trend.

The approach to research at the institute up to the mid-1930s was facilitated by the policy of the foundations that had supported it since the 1920s. This policy, as we have seen, consisted of providing support to the best institutions without regard for the specific nature of the work being done, as well as of freely exposing the best students to these institutions for limited periods. The policy was stated most explicitly by the International Education Board, but was also subscribed to by the Danish foundations supporting the institute. While not *causing* the approach to research at the institute, then, the reigning policy toward basic science support strongly encouraged it.

This chapter has confirmed the existence of a Copenhagen spirit at the institute largely as described by physicists reminiscing about their visits to Copenhagen between the world wars. However, the Copenhagen spirit at work did not serve as a force for change. On the contrary, the discussions of both physics and biology were rather constant during the period considered, and did not lead to any plans for reorienting research. It also impeded the close unity between theory and experiment that had been built up at the institute's establishment.

Yet almost immediately after the events described in this chapter, Bohr decided on a new concerted theoretical and experimental research effort at the institute. What motivated him to turn away from a laissez-faire approach to a well-defined scientific research program for the institute? The next chapter will

deal with the origins of Bohr's decision to turn the institute toward nuclear physics. We will see that just as the opportunity
to establish a new institute for theoretical physics had moved
Bohr to create a well-defined theoretical and experimental research program around 1920, so external circumstances – this
time connected with the very different problem of the refugee
scientists from Nazi Germany – were instrumental in provoking
Bohr's decision.

3

The refugee problem, 1933 to 1935

Bohr initiated a concerted redirection of scientific research at the institute only after the Rockefeller philanthropies' approach to funding international basic science had undergone far-reaching changes. These changes were contrary to previous arrangements that had favored "best science" for its own sake by providing support to expand and equip prestigious basic science institutions and bringing the best young scientists to these places on a temporary basis. In effect, the changes constituted a challenge to the Copenhagen spirit. Nevertheless, Bohr quickly showed his capacity for policy making by turning the changes to his own benefit.

The change in funding policy that first affected the institute was provoked by the purge of scientists in Nazi Germany on racial and political grounds. Bohr's immediate response to the refugee problem was to provide a temporary haven at the institute for the young and least established of the physicists purged. This was entirely in line with the Copenhagen spirit of collaboration with younger colleagues. By itself, then, the refugee problem was not apt to change the approach to research at the institute.

In May 1933, however, the Rockefeller Foundation established a Special Research Aid Fund for European Scholars favoring help to the most established professors. In spite of his preference for a different approach, Bohr did not hesitate to make use of the new policy. He thus secured positions at the institute for his old friends, colleagues, and experimentalists – James Franck and George Hevesy. In the process, he showed a pragmatism in the face of funding opportunities going beyond the administrative aptitude he had already shown during the early years of the institute. Bohr was acutely aware that as senior scientists and former professors, Franck and Hevesy required more independence, facilities, and assistance than the younger visitors. This realization was crucial for his decision to turn the institute toward experimental and theoretical nuclear physics.

BACKGROUND

On 7 April 1933, the German government promulgated the *Gesetz zur Wiederherstellung des Berufsbeamtentums* (Law for the Restoration of the Civil Service), better known as the *Beamtengesetz*. This action, in particular, alerted the international community to the ominous fate of the German scholars. The new law provided legal sanction for the dismissal of university appointees on the basis of politics and race. The consequences were immediate and far-reaching.[1]

Subsequent developments at the renowned Göttingen University constitute a graphic example of the fate of the physicists. On 11 April, Bohr's close Jewish friend and colleague James Franck decided to retire as professor and director of the University's Second Physical Institute in protest against the current policy toward academics. Franck's step was particularly forceful because as a First World War veteran he could not be dismissed. The decision soon proved to be foresighted, because in an action on 25 April – which was almost certainly unrelated to Franck's departure – six more professors at Göttingen were placed on leave by the Prussian Ministry of Education. These included the theoretical physicist Max Born, as well as the mathematicians Richard Courant and Emmy Noether, whose work was closely involved with questions of physics.

Two of the three physics institutes were hit particularly hard. Of the seven people working in Franck's institute, five – including three Jews – were either dismissed or left on their own accord. At Born's Institute for Theoretical Physics, where all but one of the five scientists were Jews, nobody stayed on. As a result of the purge, physics at Göttingen was practically eliminated. Similar developments, although rarely as extreme, occurred in several other places.[2]

Such systematic persecution was bound to arouse attention and concern in other countries. The front page of the *Manches-*

ter Guardian Weekly for Friday 19 May 1933, for example, made the point simply by showing "A Long List of Dismissals" – 196 in all. Organizations were established in several countries to support scholars forced to set out on new careers outside Germany. Both the Academic Assistance Council in England – with Ernest Rutherford as president – and the Emergency Committee in Aid of Displaced German Scholars in the United States were set up in May 1933.[3]

Among the academic disciplines, physics was one of the most seriously affected; as the historian of science Paul Forman has described it, Nazi Germany lost "at least 25 percent of its 1932–33 personnel [in academic physics], including some of the finest scientists in Germany." Among academics, the international community of physicists was particularly sensitive to such a deterioration of their constituents' situation in one particular country. More than most disciplines, physics had formed a strong international network of journals, conferences, and visiting professors. Not least important, the fellowship system – formalized on a grand scale by the IEB – ensured a wide-ranging international exchange of scientific work, ideas, and personnel. The physics elite of the 1920s and 1930s has been aptly described by the historian of science Charles Weiner as "the travelling seminar," constituting a tightly knit group with strong international, personal, and professional connections, which in a difficult situation could prove stronger than even national or geographical bonds.[4]

THE FIRST YEAR: PREFERRED APPROACH

There is no indication that the problem of the refugee physicists from Germany, beginning in early 1933, affected Bohr's basic approach to either the pursuit or policy of science. This lack of impact, however, does not mean that Bohr was slow to act on the plight of the refugees. On the contrary, his activities on their behalf were substantial. Moreover, his prominence reinforced these efforts. The lack of impact was rather due to Bohr's concentration on the younger refugees. In effect, his help constituted an extension of his efforts to obtain fellowships for the younger guard of physicists. To the extent that the refugees helped by Bohr obtained refuge at the institute, they only added to the established mode of collaboration between Bohr and his younger disciples.

Correspondence with Ebbe Rasmussen, winter and spring 1933

It is symptomatic of the cohesive structure of the international physics community, as well as of Bohr's own concern, that he received first-hand reports of the position of German physicists as soon as Hitler came to power. In particular, Bohr's scientific assistant Ebbe Rasmussen went to Berlin in January 1933 on a one-year fellowship from the Rockefeller Foundation to study experimental spectroscopy under Friedrich Paschen, president of the Physikalisch-Technische Reichsanstalt (PTR). During his stay, Rasmussen wrote detailed reports to his mentor on the deteriorating situation in Germany and the fate of the physicists there.[5]

In a letter dated 23 February 1933, Bohr learned that Paschen had been dismissed two days before, effective 1 May. Even if Paschen had reached the normal retirement age of 68, Rasmussen wrote to Bohr, he had been promised upon becoming president in 1923 "that he would be allowed to continue until he turned 70. But such promises the new government does not heed." Although hoping that he would stay on in Berlin, and that Paschen's application to continue as president until August would be accepted, Rasmussen was obviously upset about the situation. He anticipated, correctly, that the Nazi physicist Johannes Stark was the strongest candidate as Paschen's successor. If Stark was appointed, Rasmussen feared, Paschen's assistants would either lose their positions, or, if lucky, would be asked "to measure the diameter of steel spheres or to calibrate doctors' thermometers"; in short, working conditions in Berlin would become impossible. Rasmussen wrote that in case Stark became president, he would appeal to Wilbur Earle Tisdale at the Rockefeller Foundation's Paris office to be transferred to Göttingen for the rest of his stay in Germany.[6]

On 6 April, the day before the promulgation of the *Beamtengesetz,* Rasmussen sent a new report to Bohr. He began his letter by stating that there was nothing new to report regarding the

Ebbe Rasmussen in Copenhagen, 1934. [Courtesy of John A. Wheeler]

PTR presidency. He then gave a short account of his own scientific work. The introduction was insubstantial because of the censorship; Rasmussen did not want to begin his letter with the news that he considered most immediate: "It concerns an incident in which a young Jewish physicist has been 'chucked out' without any other reason than his descent – an incident that certainly will interest you."[7]

Rasmussen reported that the gifted 32-year-old physicist Günther Wolfsohn had worked under Rudolf Ladenburg at the Kaiser Wilhelm Institute for Physical Chemistry, one of several institutes under the umbrella of the Kaiser Wilhelm Society, an organization funded with endowments from member individuals and firms. After Ladenburg had left for the United States in 1931, however, his department had been abolished, so Wolfsohn lost his institutional affiliation. Wolfsohn had then been able to work under Paschen with a stipend from the *Notgemeinschaft der deutschen Wissenschaft* (Emergency Committee for German Science), which has been described by Paul Forman as "the dominant source of funds [for science] of the Weimar period." Recently, Paschen had received a letter from the Nazi party leadership denying Wolfsohn permission to work at the PTR. "Therefore," Rasmussen explained, "it is not the government that is to blame." Nevertheless, Wolfsohn had been forced to leave. He was, according to Rasmussen,

> . . . entirely without opportunities to earn a living. He is aware that he must leave Germany as quickly as possible in order not to die from starvation; he has been able to save just enough so that he and his wife can live a couple of months.[8]

In addition to the detailed account of Wolfsohn's situation, Rasmussen reported the dismissals of Professor Otto Reichenheim, who had worked at the PTR for twenty years, and Professor Erwin Fritz Finlay Freundlich of the Einstein Laboratory in Potsdam, which Freundlich had established with industrial funds in the early 1920s to test Einstein's general theory of relativity. Clearly, the case of Wolfsohn was closest to Rasmussen's heart; he described the problems of the prominent physicists only summarily, noting that Reichenheim probably had ample personal means.[9]

Concluding his letter, Rasmussen proposed what might be done. He explained to Bohr that he had written in such detail about the situation of the German physicists because he thought "that something must be done for such people, possibly on an international basis." He suggested three concrete possibilities for help: first, it would be a "beautiful task" for the Nobel Committee; second, the Rockefeller Foundation might relinquish its strict requirement that fellowship recipients must have a permanent position to return to; and third, the Rask–Ørsted Foundation might involve itself in rescuing the hard-hit German physicists. Rasmussen's implicit suggestion, of course, was that Bohr use his influence to help the refugee scientists. He promised to write again as soon as he had learned about new "incidents."[9]

On 31 July 1933, Rasmussen submitted a third report to Bohr. By this time, Stark, whom Rasmussen described as "a very amiable man," had taken over the PTR presidency. "After a couple of months' respite," Stark had allowed Paschen to use the PTR's Rowland grating, an important piece of equipment for spectroscopical research. As a consequence, Rasmussen would be able to work with Paschen until the end of his one-year fellowship. Rasmussen also reported matter-of-factly that "[Otto] Stern, [at] Hamburg and [Peter] Pringsheim, [at] Berlin [had] received their dismissal notices." The serious concern of Rasmussen's last letter was now replaced by a more resigned tone.[10]

From the very beginning of the Nazi period, then, a close colleague informed Bohr first-hand about the situation of the refugee physicists from Germany. Being considerably younger than Bohr, it was only natural that Rasmussen expressed the strongest concern for the younger generation of physicists. As we will see, however, Bohr shared this concern.

Early action and further exposure, summer 1933

By the time of Rasmussen's third report, Bohr had begun to involve himself actively in helping the refugee physicists from Germany. Now he did not need reports from Rasmussen or anyone else to alert him to the gravity of the situation.

In the summer of 1933, Bohr attended the World Exhibition "A Century of Progress" in Chicago, which celebrated the centennial of that city; he served as the official Danish representative at the annual meeting of the American Association for the Advancement of Science held in conjunction with the exhibition. This was Bohr's second visit to the United States after his trip in the fall of 1923.[11]

During the visit, Bohr made personal contact with the Rockefeller philanthropies. Before leaving Copenhagen in late April, he had asked for a meeting in New York on 1 May with Warren Weaver, the Rockefeller Foundation's recently appointed Director of the Natural Sciences. With Weaver absent on a tour of Europe, Bohr met instead with the foundation's president Max Mason. In his diary notes from the meeting, Mason wrote that Bohr proposed the abolition of the Rockefeller Foundation's condition that fellows have a permanent position to return to.

This proposal, which was one of the options Rasmussen had already suggested to Bohr, was intended to make young refugee physicists from Germany with no prospect of returning to their home country eligible for fellowships. Thus, Bohr's proposal expressed his particular concern for the younger refugees.[12]

As Bohr reported to his colleague Heisenberg in Leipzig, as well as to Wilbur Earle Tisdale and Harry Milton Miller at the Rockefeller Foundation's Paris office, Mason had promised that the foundation "would be as benevolent as possible." Yet the Rockefeller Foundation continued to follow a rather rigid policy in fellowship matters. Thus, institute directors who requested fellowships for young physicists who had temporary appointments, and not permanent positions as formally demanded, were far from certain of support. As we will see, this approach did work in the case of Edward Teller, originally from Göttingen, who was allowed to visit Copenhagen from London on a fellowship. But it did not work in the attempt to obtain money for Otto Robert Frisch to visit London from Copenhagen.[13]

While in the United States, Bohr continued to receive reports from younger colleagues about the plight of the refugee physicists. One of the reports came from Hans Kopfermann, an assistant to the renowned Jewish chemist Fritz Haber. Just the month before, Haber had decided to resign his directorship of the Kaiser Wilhelm Institute for Physical Chemistry, even though, like James Franck, he was exempt from dismissal because of his service in the First World War. Kopfermann had been in Copenhagen since September 1932 on a Rockefeller Foundation fellowship. He had just returned to Denmark from a ten-day visit to Berlin, Göttingen, and Rostock. He wrote to Bohr that he "now at last" had to report on his visit, so that his impressions should not become "obsolete and antiquated." Clearly, Kopfermann was aware of Bohr's concern, and felt obligated to inform him about his first-hand observations. Kopfermann's five-page typewritten letter testifies not only to Bohr's interest; it also constitutes a revealing judgement by a German physicist about the contemporary situation for academics in his homeland.[14]

The basic message of Kopfermann's letter was that a conflict was taking place between the radical German masses, most ominously represented by the students, and the more moderate Führer and government. Until recently, as Kopfermann saw it, the former had been on the offensive, and had forced the government to introduce radical measures. Kopfermann thought that if the government emerged as the winner, the situation would improve. This assessment was not unusual for the time; indeed, it is compatible with Rasmussen's blaming the Nazi party – and not the German government – for the mistreatment of Wolfsohn.[15]

In their assessments of which forces were most likely to win out, however, Rasmussen's and Kopfermann's views differed fundamentally. Without any claim upon certainty, Kopfermann was inclined to interpret certain moderating tendencies as signs that better times lay in store. He emphasized, in particular, the recent School Law, which defined children as Jewish only if both parents were of such origin.

Kopfermann's optimistic attitude reflected his preferred tactics for improving the situation for the German scholars. He described Max von Laue, a highly distinguished professor at the University of Berlin, as another optimist. Laue's tactics, Kopfermann noted, consisted of postponing action on the dismissal of physicists as long as possible, on the assumption that they would regain their positions as soon as the radical wind had blown over. Kopfermann reported, by contrast, that his own professor, Fritz Haber, was distinctly not of the opinion that the anti-Jewish wave was about to recede; his pessimistic attitude had influenced Haber's own recent decision to resign. Without saying so explicitly, Kopfermann obviously preferred the attitude of Laue and the prescriptions for action based on it.[16]

It was Kopfermann's definite impression that the great majority of scientists were opposed to the government's anti-Jewish policies. He went as far as to judge the widespread tendency among the young non-Jewish scientists to enter the "movement" as being an attempt at change from within. For, he reasoned, the new policy toward scientists did not affect only Jews; fundamentally, it did not have to do with a "Jewish problem," but with a problem of higher education. According to Kopfermann, the German university was presently becoming less of a research institution and more of a teaching institution based on "ethically idealistic [and] nationalistic" values. Kopfermann saw the tendency of young German scientists to join Nazi organizations as an attempt to change this development. He probably thought of Pascual Jordan as a case in point, because he had included Rostock, where Jordan worked, on his itinerary. Indeed, Kopfermann imagined the German intelligentsia to be joined against the regime in countering the development at the universities.[17]

As to the fate of particular scientists, Kopfermann admitted that the situation was uncertain. Those affected by the *Beamtengesetz* had recently filled out a questionnaire that specifically asked for race and ancestry. Dismissals on this basis could be expected in a month's time. Among people already affected, Kopfermann noted, were Wolfsohn – who had obtained a position with Leonard Salomon Ornstein in Utrecht – and the physicists in Göttingen.

The other extant report that Bohr received while visiting the United States came from Werner Heisenberg in Leipzig and constituted a small paragraph in a rather short letter. Heisenberg exclaimed that he "often had the feeling that he ought to have a bad conscience toward [Bohr] for everything that is now happening in [Germany]." At the same time, however, he was optimistic enough about the situation to work with Laue and Max Planck – professor emeritus at the University of Berlin, director of the Kaiser Wilhelm Society, and doyen of German physics – in order to retain the best physicists for Germany. Concerning the Göttingen people, he had given up all hope for Franck. But he considered that Born might possibly be secured for Germany at some point in the future. Heisenberg also wrote that his assistant Felix Bloch, who was presently giving guest lectures in Paris and was seeking to prolong his stay outside Germany further by applying for a Rockefeller Foundation fellowship, would still have been able to work for a long time in Leipzig. Unable or unwilling to accept as a reality the deterioration of German physics, Heisenberg shared Kopfermann's optimism for a brighter future.[18]

Although differing in their evaluations, these letters show that Bohr's correspondents were aware of their mentor's strong concern for the young refugees. They also served to keep Bohr up to date on the refugee problem. Thus, the reports helped prepare him to step up his activities for the young refugee physicists later in the year.

The Danish refugee committee, fall 1933

Bohr escalated his activities for the refugee scientists soon after his return to Copenhagen. Upon its establishment in October, he joined the executive board of the Danish Committee for the Support of Refugee Intellectual Workers, chaired by Aage Friis, the prominent historian of Danish–German relations and president of Copenhagen University. It testifies to the closeness of the Bohr brothers that Harald Bohr – a member of the committee though not of its executive board – often took part in the board's meetings, and participated actively in its work.[19]

Notwithstanding the domination of academics on its executive board, the aim of the committee was to help a broader group of refugees. Journalists, authors, and actors were regarded as "intellectual workers" on equal terms with scientists and scholars. The work of the various agencies to support refugees in Denmark was coordinated in 1935 or 1936. From then on, aid from the Intellectual Workers Committee constituted only a small fraction of the help supplied by all the agencies combined. In all, the Intellectual Workers Committee helped 400 to 500 people, of whom less than ten percent were academics; of the academics, fifteen to twenty may have entered Denmark through Bohr's direct intervention.[20]

There is little detailed information left about either Bohr's or the Danish Intellectual Workers Committee's activities for the refugees; like Bohr's own files, the committee's papers were destroyed when Denmark was occupied in 1940, so that they would not be confiscated by the Germans. Nevertheless, detailed information on two specific cases is retained in the Bohr Scientific Correspondence.[21]

The first of these concerns Guido Beck, whose assistantship with Heisenberg in Leipzig had been terminated after four years in 1932 as a result of German regulations toward foreign university assistants. Beck was able to continue work in physics at the German University in Prague on a stipend from the *Notgemeinschaft*. There, he served as assistant to Philipp Frank, a member of the Unity of Science group debating Bohr's complementarity argument and Einstein's successor in the chair of theoretical physics. German funding of a Jew at a foreign university could not be expected to last, however, and at the informal physics conference at the institute in the fall of 1933, Beck complained to Bohr that his grant would be terminated as of 1 October. Beck's prospects for receiving aid from foreign relief committees were particularly uncertain because he had lost his Leipzig position before Hitler came to power; as a result, his name did not appear on the list of scholars in Germany dismissed under the Nazi regime. Nevertheless, Bohr was able to obtain funds from the Intellectual Workers Committee to prolong Beck's stay in Copenhagen until Easter 1934. As a more permanent arrangement, however, Bohr suggested in a letter to Frank that a position be obtained for Beck in Prague. Bohr found such an arrangement especially appropriate because Beck was born in Czechoslovakia.[22]

Beck's complaint about his situation at the annual informal physics conference at the institute in mid-September 1933, which was devoted to discussions of the positron, was not an isolated event. Indeed, the refugee situation was a central concern at the conference. On this occasion, for example, Heisenberg asked Reinhold Fürth, another professor in Prague, whether he could help secure Beck's future. As a result of this request, a correspondence followed between Bohr in Copenhagen and Frank and Fürth in Prague. The latter promptly informed Bohr that although Beck was born in Czechoslovakia, he was an Austrian citizen. This meant that Beck could give up hope of obtaining a position in Czechoslovakia. For his part, Frank noted that as a Jew belonging to a nation that was linguistically German, Beck fell between the Scylla of Prague's German speaking Nazis and the Charybdis of the anti-German Czechs. Further complicating matters was the fact that no relief committee had yet been established in Czechoslovakia; accordingly, the three professors agreed that funding must be obtained from foreign sources.[23]

Encouraged by Bohr, Frank and Fürth submitted an application on Beck's behalf to the Dutch Support Committee for German Scholars located in Amsterdam, with a copy to Bohr in Copenhagen. Bohr, Heisenberg, and Felix Ehrenhaft in Vienna then wrote letters of recommendation for Beck, while Bohr personally promoted Beck's and others' cases with the Amsterdam committee and at the Solvay Congress in Brussels in October

1933. Nevertheless, the application was refused. Then, on 25 November, Bohr reported that he had successfully used his copy of Frank and Fürth's application to interest the Danish and Swedish relief committees in Beck's case. Soon after the Danes had agreed to provide a specific sum, the Swedes – through the promotion of Klein – provided the same amount. As a result of Bohr's successful approach in Scandinavia, the Amsterdam committee finally also agreed to contribute. Beck made up his mind to go to Prague even before the Swedish committee had made its final decision.[24]

Walter Gordon, a physics professor for three years at the University of Hamburg, was the other refugee helped by Bohr's efforts on behalf of the Danish Intellectual Workers Committee. After his dismissal from Hamburg, Gordon worked with Wolfgang Pauli in Zurich. Gordon was well-known in the physics community for developing, independently of Oskar Klein, the so-called Klein–Gordon equation – a precursor of Dirac's relativistic equation for the electron. Klein, who wanted Gordon to work with him at his institute in Stockholm, turned to the Rockefeller Foundation with a request that a fellowship be granted for that purpose. Somewhat reluctantly, the Rockefeller Foundation agreed to pay Gordon for one year only, stressing that the support could not be renewed. Klein had already received a promise for additional support from the Swedish refugee committee on the condition that a two-year Rockefeller Foundation fellowship be granted. After the response from the Rockefeller Foundation, Klein feared that the Swedish support might not be forthcoming. Consequently, he turned to Bohr, asking for support from the Danish committee. In less than a month, Bohr had arranged support from the Danish committee for two years on the condition that additional support be obtained from other sources. Only days later, Klein reported back that the Swedish committee most likely would provide the sum promised before the Rockefeller Foundation backed out, and possibly even more.[25]

The final example of support from a relief committee to a young physicist after funding had been declined from traditional sources is the case of Otto Robert Frisch. In this case, support was given by the Academic Assistance Council, the British counterpart of the Danish Intellectual Workers Committee. Although Bohr was not directly responsible for this action, he was involved in the case. Moreover, Frisch would play a major role at the institute soon after.

Frisch was born in Vienna, where he obtained a PhD in physics in 1922. In 1927 he moved to Berlin, and from 1930 he worked as assistant to Otto Stern at the University of Hamburg. As all but one of the five physicists at Stern's institute were Jewish, the director, who was exempt from dismissal because of his long tenure, set out to find new jobs for his staff. Before Frisch's dismissal, Stern had been able to obtain a one-year Rockefeller Foundation fellowship for him to work with Enrico Fermi in Rome. This offer was rescinded, however, when it became clear that Frisch would have no position to return to. Stern seems to have interested Patrick Blackett, by then head of the physics department at Birkbeck College at the University of London, in taking on Frisch for a year.[26]

Visiting Blackett's laboratory on the way back from his tour of the United States, Bohr had evidently expressed his intention of arranging a position for Frisch in Copenhagen, for on 13 August 1933, Blackett asked whether this position could be used as an argument to obtain a Rockefeller Foundation fellowship for Frisch's stay in London. However, Bohr's argument to Harry Miller at the foundation's Paris office that Frisch had received a one-year assistantship in Copenhagen was not heeded. Miller replied that it was highly unlikely that Frisch, who had not yet reached the level of *Privatdozent* – traditionally an unsalaried university lecturer paid directly by students' fees – and was only temporarily employed as an assistant, would be awarded a fellowship. In the latter part of September, Blackett applied successfully for a one-year stipend for Frisch from the Academic Assistance Council, and Frisch arrived in October. After a year in London, Frisch went to Copenhagen, where he spent five years on fellowships from the Rask–Ørsted Foundation.[27]

The cases of Beck and Gordon show how Bohr could take action on his own initiative for the refugee scientists from Germany. Whereas Gordon had reached forty years of age, Beck was ten years younger. As a relatively young and unestablished physicist, Beck fit well into the Copenhagen spirit at the institute; he had also spent two and a half months at the institute in the spring of the year before. Of the two, only Beck stayed at the institute, and only for a brief period. In fact, the Danish Intellectual Workers Committee even provided support for Beck to work outside Denmark. This, of course, was no choice of Bohr's, but reflected instead the limited resources available to establish permanent positions in Denmark. Nevertheless, in providing only temporary refuge at the institute, support from the Intellectual Workers Committee resembled the traditional fellowships for younger students. Accordingly, it served to foster the Copenhagen spirit of collaboration between the established Bohr and his younger temporary visitors.[28]

The case of Frisch provides an example of tension between the work of the relief committees and the traditional fellowship approach. Thus, despite Bohr's pleading, the Rockefeller Foundation refused to provide fellowships to young physicists who had no position to return to. Locally, there was similar tension between the Rask–Ørsted Foundation, the major Danish provider of traditional fellowships, and the Intellectual Workers Committee. For example, in late 1933, when the committee applied to the Rask–Ørsted Foundation for a grant, it received the curt response that "support of this form falls outside the purpose of the foundation."[29]

Fellowship support for the refugees, fall 1933 and later

In spite of the tension between organizations set up to help the refugees, on the one hand, and traditional economic supporters of basic science, on the other, Bohr was able to obtain help for the refugees from both types of sources. We have already noted that he secured help for Frisch through a traditional Rask–Ørsted fellowship. In this section, we will see that during the first year of the refugee problem, Bohr made extensive use of the traditional fellowship system in order to help the refugees; in fact, these fellowships seem to have helped more refugees at the institute than any other means of support. In practice, then, traditional fellowships and support from the relief committees constituted complementary sources of help for the younger refugee physicists.

Eugene Rabinowitch, James Franck's personal assistant at Göttingen and the first on Franck's staff to be dismissed as a result of the *Beamtengesetz,* was the first refugee physicist to be helped with a traditional fellowship through Bohr's intervention. In February 1933, Bohr had applied to the Rask–Ørsted Foundation for a fellowship; it was provided three months later. After staying in Copenhagen for the fall semester, Rabinowitch spent four years in London, then emigrated to the United States.[30]

The Austrian Victor Weisskopf obtained his PhD under Max Born at Göttingen in 1931. Unable at first to find a salaried position, he was supported by his parents while he worked with Heisenberg in Leipzig. He then went to Erwin Schrödinger in Berlin as a substitute for Fritz London, who was visiting the United States for half a year. Subsequently, Schrödinger obtained a one-year Rockefeller Foundation fellowship for him. Weisskopf chose to go first to Copenhagen, then to Dirac in Cambridge. He stayed with Bohr from 5 September 1932 until 25 April 1933, making another forty days' visit in connection with the fall conference. Being a theoretician, Weisskopf collaborated closely with Bohr. Before Weisskopf completed his fellowship in Cambridge, Wolfgang Pauli in Zurich asked him to become his assistant. When the assistantship expired in 1935, Weisskopf returned to Copenhagen, where Bohr had obtained a stipend for him. In 1937, after a trip around the world that included the United States, Bohr reported that he had recommended Weisskopf for a position at the University of Rochester, New York, where Weisskopf went subsequently.[31]

Edward Teller, a Hungarian Jew, obtained his PhD under Werner Heisenberg at the University of Leipzig in 1930. He then became assistant to Arnold Eucken at the Institute for Physical Chemistry at Göttingen University. Although not formally dismissed, Teller was advised by Eucken that he had no future as a scientist in Germany, and subsequently he obtained an appointment with the chemist George Frederick Donnan at the University of London. More successful than Bohr had been in the case of Frisch, Donnan was able to convince the Rockefeller Foundation that the appointment, which indeed was temporary, qualified Teller for a fellowship. Teller chose to spend the fellowship at Bohr's institute in Copenhagen from the beginning of 1934 until the end of September the same year. Soon after his return to London, Teller, then only twenty-six, received an offer from George Gamow, whom he had come to know in Copenhagen, to join him as a professor at the George Washington University in Washington, D.C. In August 1935, Teller went to the United States, where he has been living ever since.[32]

The Swiss physicist Felix Bloch moved from Zurich to Leipzig in 1927 to become Werner Heisenberg's first student even before Heisenberg's arrival from Copenhagen to take up his permanent position. Bloch obtained his doctorate under Heisenberg in 1928. After a year as Pauli's assistant in Zurich and another year with Hendrik Anthony Kramers in Utrecht, Bloch returned to Leipzig as Heisenberg's assistant. He stayed in Leipzig for the academic year 1930–31. On Heisenberg's recommendation, Bohr then invited him to the institute as a Rask–Ørsted fellow. While in Copenhagen, Bloch developed a close collaboration with Bohr, which continued in the form of a prolific correspondence after Bloch's return to Leipzig in the summer of 1932 to obtain the right to teach there as a *Privatdozent*.[33]

Although Heisenberg offered a longer stay, Bloch left for his hometown, Zurich, not long after the Nazi takeover in Germany. The day before the enactment of the *Beamtengesetz* – the very same day that Ebbe Rasmussen had expressed similar views to Bohr in a letter from Berlin – Bloch wrote to Bohr about his and Pauli's concern about the plight of Jews in Germany. Bloch had recently submitted an application for a Rockefeller Foundation fellowship, and in this connection he asked Bohr whether he could change his stated place of return from Leipzig to Copenhagen. Although Heisenberg had promised him that he could stay in Leipzig for as long as he wanted, Bloch anticipated that the situation for Jews at German universities would soon become intolerable. Bloch expressed his confidence to Bohr that Heisenberg would understand his preference for Copenhagen under the present circumstances; yet, as we have seen, almost two months later Heisenberg implied in his letter to Bohr in the United States that he would have preferred that Bloch had not left Leipzig and Germany.[34]

Bloch did obtain Bohr's permission to state in his fellowship application that he was going to return to Copenhagen. Indeed, his application was approved, and Bloch's is thus one of the few cases in which the Rockefeller Foundation made an exception to its requirement for a permanent position. Typically, however, the foundation was quick to point out that the exception would not necessarily be repeated.[35]

Before taking up his Rockefeller Foundation fellowship with Enrico Fermi in Rome, Bloch lectured in Paris, then visited Bohr in Copenhagen. While at the institute, he received a telegram from Stanford University offering him a position on its faculty. Bohr, who had recently visited Stanford, recommended the place highly; it is likely that he had already put in a good word for Bloch. On 10 February 1934, Bloch gave Bohr definite notice from Rome that he had accepted the position. The decision, Bloch wrote, was due less to the lure of America than to the ever worsening situation in Europe. Bloch was only able to use five months of his one-year Rockefeller Foundation fellowship, leaving Rome in early March to begin lecturing in California on 1 April. Although at this point he avoided a permanent appointment at Stanford, Bloch worked there until his death in 1983.[36]

Hilde Levi had completed her doctoral dissertation on spectroscopy under Hans Beutler at the Kaiser Wilhelm Institute for Physical Chemistry when Hitler came to power on 30 January 1933. She decided immediately that as a woman of "non-Aryan" background she had no future in German science. Nevertheless, she stayed on to complete the oral examination for her doctorate in January 1934. She then went to Switzerland for a vacation, and from there she contacted the president of the Danish branch of the International Federation of University Women, a Geneva-based organization established in 1919 with several purposes, including the provision of "assistance for university women in need"; Levi hoped that the organization would help her to be admitted to Bohr's institute in Copenhagen. She wanted to go to Denmark because she had acquaintances there and because she assumed that there were fewer refugees looking for work in a small country. Being a physicist she knew Bohr and his achievements; yet she was not cognizant of the unique position of the institute within the physics community and of the special research atmosphere there.[37]

The recommendation of the University Women's organization had its intended effect, and according to the institute's Guest Book, Levi arrived on 24 April 1934, sixteen days after the arrival of James Franck. Although at this point Franck did not know

her personally, he did know her doctoral work. Levi believes
that Franck's need for an assistant was the reason for her acceptance at the institute. She was at first supported by her father.
The following year, however, it became impossible to transfer
money from Germany. She was subsequently paid with fellowships from the Rockefeller Foundation and the Rask–Ørsted
Foundation. Levi was the only one of the early refugee physicists described in this chapter who would make her career in
Denmark.[38]

Concluding remarks

Bohr's first response to the refugee problem was to help the
younger physicists. Having kept abreast of the situation from
the very beginning through letters from younger collaborators,
he sought to make the Rockefeller philanthropies' traditional fellowship system – which had made his collaboration with the
younger eminent physicists possible in the first place – more
flexible. Unsuccessful in convincing the Rockefeller Foundation
to change its fellowship policy, Bohr then became active in the
Danish Committee for the Support of Refugee Intellectual Workers, effectively helping younger physicists who were unable to
obtain support from traditional sources. Like visitors on traditional fellowships, physicists helped by the Intellectual Workers
Committee were generally young and stayed at the institute for
only a short period.

Despite the inflexibility of the Rockefeller Foundation and
other fellowship agencies, Bohr was also able to obtain traditional fellowships for a number of refugees. Indeed, during the first
few years of the Nazi period, the majority of refugees coming
through the institute seem to have been fellows. As is evident
from the case histories discussed, these physicists typified the
kind of visitors that had come during previous years to participate in the Copenhagen spirit at the institute; practically all had
been trained in physics within the circle of physicists centered
around Bohr. Thus, Rabinowitch worked under Franck, who
had collaborated with Bohr at the establishment of the institute.
Weisskopf chose to work with Dirac and Pauli, who were as
close to Bohr as any among that generation of theoretical physicists. Finally, Bloch was a student and assistant of Heisenberg,

who had perhaps been Bohr's closest collaborator during the institute's first decade. Bloch also spent a year with each of Pauli and Kramers, the latter of whom had stayed in Copenhagen as Bohr's first assistant until 1926.

If anything, the refugee problem *added* to the closeness between Bohr and his disciples, encouraging further the mode of work represented by the Copenhagen spirit – an observation borne out by the scientific work carried out at the institute at least through early 1934. By itself, then, the refugee problem – like the concurrent discussions of the atomic nucleus and, to a lesser extent, biological questions – was not the incentive for redirecting the work at the institute.

The strategy of the Intellectual Workers Committee to discourage permanent positions for the refugees was typical of the European approach. By contrast, the United States, with its greater economic and educational resources, from the outset made the best of the sudden availability of German brainpower. Thus, when in mid-1933 the Rockefeller Foundation sought to help the refugees by establishing a Special Research Aid Fund for European Scholars, the foundation decided on an approach quite different from Bohr's suggestion to modify the existing fellowship system. Although primarily designed for the United States, the Rockefeller Foundation's Special Research Aid Fund would also have an effect abroad. In particular, as will be seen in subsequent sections, it played a crucial role in furthering Bohr's redirection of the institute to a concerted effort in theoretical and experimental nuclear physics.

THE ROCKEFELLER FOUNDATION'S SPECIAL RESEARCH AID FUND FOR EUROPEAN SCHOLARS

The Rockefeller Foundation decided to establish its Special Research Aid Fund for refugee scholars from Germany only days after Bohr's meeting with foundation President Mason in New York. Thus,

> . . . on May 12th, 1933, the Executive Committee appropriated $140,000 to be used "for grants to institutions desiring to provide positions for eminent scholars whose careers have been interrupted by the present disturbed conditions."

This was only the beginning of a long-term effort by the Rockefeller Foundation to aid scholars forced to leave Germany; by the fall of 1936, there were 1,639 of them. In 1933 alone, the Rockefeller Foundation provided support for seventy-one dismissed German scholars through its Special Research Aid Fund. By 1939, it had contributed almost $750,000 to aid 193 people, 120 of whom were resettled in the United States. Yet the program was always seen as a temporary measure that could be discontinued at any time.[39]

Like other American agencies helping the refugees, the Rockefeller Foundation stated as a principle that it was up to the institutions of potential appointment to initiate the request for a refugee to join their staff. Hence, an emigré scientist could not apply on his or her own behalf to obtain a position. This system was designed in part to produce as little friction as possible between the indigenous Americans and their recently arrived refugee colleagues. At the same time, however, it amounted to a selection of scientists based on elitist rather than humanitarian considerations.[40]

The Rockefeller Foundation, in fact, went as far as to state that the "emphasis throughout has been on the preservation of scholarship rather than on personal relief for scholars." Likewise, the American Emergency Committee in Aid of Displaced German Scholars – the word "German" was replaced by "Foreign" when the political purge of academics also spread to other countries – with which the Rockefeller Foundation collaborated closely, excluded in its first annual report aid "to younger German scholars of outstanding promise." In 1933, the Rockefeller Foundation provided half the salary for practically all of the thirty-six scholars from Germany placed by the Emergency Committee. Yet this support was far from automatic; as the Emergency Committee acknowledged in its first annual report, the foundation "reserved freedom of action in regard to each application from the universities."[41]

Thus, when Bohr was finally able to meet Warren Weaver in New York on 10 July just before returning to Europe, the Rockefeller Foundation had instigated a well-defined and autonomous policy toward the refugees. This policy was significantly different from Bohr's suggestion to Mason in May to make the fellowship system for younger, less established scientists more flexible.[42]

When he met with Bohr, Weaver had recently returned from forty days in Europe, which were devoted primarily to a scrutiny of the situation in Germany. It was his second tour of Europe together with the more experienced Lauder William Jones, who succeeded Trowbridge as head of natural science at the IEB Paris office even before that office was transferred to the Rockefeller Foundation. During this tour, Weaver met with Bohr's close friend George Hevesy, professor of physical chemistry at Freiburg University, who used the occasion to report his decision to leave Germany.[43]

Hevesy seems to have arrived at this decision sometime during the first ten days of May. During a visit of Lauder Jones on 5 April, Hevesy had reported "great uneasiness among those who were Jews or foreigners appointed to posts in German universities"; no dismissals had occurred in Freiburg, however, and Hevesy, himself Jewish, expressed no desire to leave. On 13 June, the very last day of Jones and Weaver's second joint tour of Europe, Hevesy reported that "after a recent visit to Paris, he found an order to dismiss, the following morning at 8 o'clock, certain of his assistants." Although he disregarded the order, Hevesy had been able to continue his work. He had also been asked to stay on by the "Minister of Baden." Nevertheless, he had definitely decided to leave. He had not, however, communicated his decision to any authorities, colleagues, or students in Freiburg, who, Hevesy reported to Weaver, treated him cordially. It is likely that Weaver informed Bohr of Hevesy's plans in New York.[44]

Although he would have preferred that the Rockefeller Foundation had chosen a different approach toward the refugees, Bohr was quick to take advantage of the new policy. Even before returning to Copenhagen, he met with Lauder Jones in Cambridge, England, to discuss the extent to which the institute could apply for funds to house refugees from Germany. During these discussions, Bohr also asked whether the Rockefeller Foundation would be willing to provide new experimental equipment for the institute. He pointed out that space for this equipment was already available in the newly constructed Institute of Mathematics adjacent to his own institute.[45]

After Bohr's return to Copenhagen, both Hevesy and the established experimentalist James Franck – Bohr's longtime friend, who in April had decided to retire from Göttingen University – made short visits to the institute. In mid-October, on the second day of Franck's visit, Bohr and J.N. Brønsted – whose new Institute for Physical Chemistry adjoining Bohr's had been completed only three years previously – submitted a joint application to Lauder Jones at the Paris office for the support of Franck and Hevesy through the Special Research Aid Fund for refugee scholars. In their application, Bohr and Brønsted expressed their desire to have the two scientists as research professors at their respective institutes for three years. They argued that Franck and Hevesy required professors' salaries, a minor amount of which, they reported, was already promised from the Rask–Ørsted Foundation and the Intellectual Workers Committee.[46]

In their application, Bohr and Brønsted stressed that although Franck and Hevesy were to be affiliated formally with different institutes,

> . . . both scientists should take part in the general scientific activity in these two neighbouring institutes which are so closely connected as regards the field of scientific investigation.

This connection had been a central consideration when the IEB decided to support Brønsted's institute in 1927. The argument of collaboration was still effective, for the passage just quoted was underscored with red pencil in the Rockefeller Foundation's office after the application had been received. Evidently, there was *some* continuity between the Rockefeller philanthropies' earlier funding practice for basic science and that of the Special Research Aid Fund.[47]

In a covering letter, Bohr referred to the application as a "continuation of the encouraging discussions I was privileged to have with you in Cambridge this summer." He also asked for support for a liquid air and hydrogen plant and a photometer. Apart from fellowship applications, these requests constituted Bohr's first approach for support from the Rockefeller philanthropies since 1925. In spite of their statement that Danish support had already been secured, Bohr and Brønsted submitted a formal application to the Rask–Ørsted Foundation only a few days later, noting that "a significant amount" could also be ex-

Franck with Lise Meitner, Gustav Hertz, and Peter Pringsheim at the
Kaiser Wilhelm Institute for Chemistry in the early 1930s.

pected from the Rockefeller Foundation. Obviously, funding
from the Rockefeller Foundation was considered a first priority,
after which support from other sources would follow more or
less as a matter of course.[48]

At the end of the covering letter, Bohr expressed his wish to
visit Jones in Paris in connection with the Solvay Congress in
Brussels, which was to take place in less than two weeks. Bohr
kept his word, visiting the Rockefeller Foundation's Paris office
after the Solvay Congress to ask for help from the Special Research Aid Fund to bring a third well-established German physicist to Copenhagen. The person in question was Lise Meitner,
who was seven years older than Bohr. Although Jewish, Meitner
was still codirecting the Kaiser Wilhelm Institute for Chemistry
together with Otto Hahn. Her privilege to teach at the University of Berlin, however, had recently been revoked. In spite of her
problems, Meitner was not planning to resign her position with

Hahn. Indeed, in conformity with his usual policy in such matters, Max Planck, the director of the Kaiser Wilhelm Society – which was the umbrella organization for all the Kaiser Wilhelm Institutes – had advised her against such action. Yet Meitner had expressed her willingness to accept an appointment at Bohr's institute for one year, provided funding could be made available. In his formal application, submitted on 18 November, Bohr emphasized the severity of Meitner's situation, asking that payment begin in January 1934. Bohr's request was granted just five days later. By early January 1934, however, Planck had advised Meitner that a one-year appointment at Bohr's institute might endanger her position with the Kaiser Wilhelm Society; hence, her stay in Copenhagen came to nothing. Nevertheless, the incident shows Bohr's continued willingness and effort to make use of the Rockefeller Foundation's Special Research Aid Fund to obtain established scientists for his institute.[49]

Bohr and Brønsted's application to the Rockefeller Foundation for support for Franck and Hevesy, as well as Bohr's request for equipment, were approved by mid-January 1934. The salaries were provided under the Special Research Aid Fund for refugee scholars, and the equipment, costing $7,500, was one of the small number of minor Research Aid Grants that the Rockefeller Foundation still provided for general research purposes. The Rask–Ørsted Foundation granted supplementary support for Franck and Hevesy the day after the Rockefeller Foundation's Paris office had recommended approval of Bohr and Brønsted's application to the main office in New York.[50]

An internal Rockefeller Foundation document stated that the equipment would "be especially serviceable because of the call which [Bohr] has made to Professor James Franck . . . and other scientists associated with him in his investigations." Only two years earlier, Franck had succeeded in obtaining a grant from the Rockefeller Foundation for the purchase of a liquid air machine at Göttingen, to be used for research in physical chemistry. Bohr, however, did not go into more detail about the prospective use of the new equipment. In a letter to Lauder Jones, he simply stated that it would be placed in rooms provided for the physicists in the new mathematics institute. Yet it is clear that Bohr at this point did not seek to expand the institute out of a desire to begin a new experimental research program; instead, the expansion resulted rather fortuitously from the availability of space in the newly constructed building for the mathematicians.[51]

THE EARLIER CAREERS OF FRANCK AND HEVESY

Bohr had long held close personal and scientific relationships with both Franck and Hevesy. An appreciation of these relationships, as well as of Franck's and Hevesy's radically different approaches to scientific research, is necessary in order to understand the two experimentalists' crucial role in the reorientation of the institute to concerted research in theoretical and experimental nuclear physics.

Franck

When Bohr and Franck first met in Berlin in 1920, both were established scientists. Franck, an experimental physicist and Bohr's senior by three years, had received his advanced education in the laboratory of Emil Warburg at the University of Berlin. Just before the First World War, Franck and his younger colleague Gustav Hertz began cooperating on collisions between electrons and atoms. Although they started to publish their results several months after the first installment of Bohr's famous "trilogy" in 1913 (which introduced the quantum theory of the atom), they initially expressed no awareness of Bohr's work. Yet, after the war, Franck and Hertz's experiments came to be seen as the major verification of Bohr's theory. At their meeting in Berlin, Bohr and Franck had a lengthy discussion. By that time, Franck had been appointed head of the physics section at the Kaiser Wilhelm Institute for Physical Chemistry under Fritz Haber. His subsequent investigations of the collisions of electrons with atoms and molecules were guided by the predictions of Bohr's quantum theory.[52]

In the first extant piece of correspondence between the two physicists, which dates from the fall of 1920, Bohr expressed interest in Franck's work, and invited him to come to Copenhagen to initiate experiments at his new Institute for Theoretical

Bohr's first meeting with Franck took place in Berlin in 1920, where Bohr was also invited to present his quantum theory to a group "free of bosses" (the occasion was called *das bonzenfreie Kolloquium*). From left to right: Otto Stern, Wilhelm Lentz, Franck, Rudolf Ladenburg, Paul Knipping, Bohr, E. Wagner, Otto von Bayer, Otto Hahn, George Hevesy, Lise Meitner, Wilhelm Westphal, Hans Geiger, Gustav Hertz, Peter Pringsheim.

Physics. Franck's stay lasted about two months, and he was in Copenhagen when the institute was formally opened in March 1921. The presence of the renowned German experimental physicist made the front page of the Copenhagen newspapers. While there, Franck established close scientific relationships with the rest of the group around Bohr, in particular the visitors Oskar Klein from Sweden and Svein Rosseland from Norway. These relationships encouraged Franck to continue his experimental work on Bohr's theories.[53]

After his stay in Copenhagen, Franck went to Göttingen, where he had accepted a chair of experimental physics. The theoretician Max Born, whom Franck had already come to know during his student days at Heidelberg University in 1901 or 1902, had recently accepted the chair of theoretical physics on condition that Franck would also receive a call. Franck accepted the offer, with Rudolf Ladenburg taking over his position at the Kaiser Wilhelm Institute. Thus, Born and Franck had an excellent opportunity to develop further their close personal and scientific relationship. Later in life, Born recalled only one of Franck's traits with disapproval: "Franck," Born wrote in his autobiography,

> . . . regarded our [i.e., the Göttingen theoreticians'] endeavours with suspicion and never accepted our conclusions without having confirmation from the prophet in Copenhagen himself,

a circumstance that Born claimed "even retarded our work to some degree." Franck's own recollection confirms Born's impression: "I never felt," Franck confided in an oral history interview in 1962, " – I can only say such hero worship – as [I did] to[ward] Bohr." Nevertheless, Franck was surprised when at about the time of the interview he finally learnt about Born's earlier frustrations.[54]

Franck was soon considered to be one of the best experimental physicists in Germany. Thus, when the prominent experimentalist Heinrich Rubens died in 1922, Franck received an offer to succeed him in the prestigious chair of physics at the University of Berlin.[55]

Franck's decision to stay in Göttingen proved wise. In the 1920s, Göttingen – along mainly with Copenhagen and Munich – was central in the experimental and theoretical development of atomic physics. During this period, Bohr and Franck continued their close relationship. In the summer of 1922, Bohr visited Göttingen to give a series of lectures that has since become famous as the Bohr *Festspiele*. On this occasion Bohr met Werner Heisenberg, who was then Arnold Sommerfeld's student in Munich, for the first time. Three and a half years later, Franck played an important role in realizing Bohr's wish to acquire Heisenberg, then with Born at Göttingen, as a scientific assistant at the institute.[56]

From the 1920s, more than sixty letters between Bohr and Franck have survived. Most of them contain elaborate exchanges on scientific matters, showing Bohr's interest in and dependence on experiment, as well as Franck's awareness of and aptitude for theoretical questions. As he described it himself, Franck was "always a little bit in the theory and not entirely 100% experimental."[57]

Later in life, Franck recalled that he enjoyed the collaborative character of experimental physics compared with the lonely desk work necessary to succeed as a theoretician. Thus he was happy to teach in the laboratory rather than at lectures and seminars. Franck enjoyed the relationships with his assistants and students, and was in turn valued as a concerned adviser, in human as well as scientific questions. Discussions often took place outside working hours during walks or bicycle rides. Franck was known for never adopting an authoritarian professorial role.[58]

In 1926, Franck shared the Nobel Prize in physics with Gustav Hertz for their experiments corroborating Bohr's theory. During his tenure at Göttingen, Franck continued his experimental research on atomic collisions, gradually approaching more complex atomic and molecular processes. Later in life he recalled that his work went

> . . . automatically . . . from the simpler to the more complicated things. The methods changed, but in principle it was always energy exchange between particles at collisions or internally.

In particular, by the time of his departure from Göttingen, Franck had shown no inclination in his experimental work to attack the atomic nucleus.[59]

After having resigned his position at Göttingen, Franck hoped for some time to be able to continue his career as a physicist in Germany. The leading physicists there had long been planning to offer Franck for a second time the influential physics chair at the University of Berlin, which was soon to become vacant when Hermann Walther Nernst reached retirement age. This position, however, would have made Franck a civil servant, and since the *Beamtengesetz* applied to the civil service, he would not have accepted the position, even if it had been offered to him.[60]

There was another possibility. Indeed, Fritz Haber, who was retiring from his directorship of the Kaiser Wilhelm Institute for Physical Chemistry, saw Franck as his only acceptable successor. As a result of internal problems within the Kaiser Wilhelm Society, however, this possibility also came to nothing. Eventually, Franck gave up hope of obtaining a position in Germany, and accepted an invitation in July 1933 from the Johns Hopkins University to spend three months in Baltimore giving the Speyer Lectures.[61]

In early August 1933, Bohr's mathematician brother Harald told Franck in Göttingen that the two brothers were seeking an arrangement for him in Copenhagen. A few days later, Franck was visited by Lauder Jones, who seemed "very ready" to support such an arrangement. Franck was enthusiastic. "Let me tell you," he wrote to Bohr, "that when I consider my personal fate, nothing would bring me greater happiness than to work in your institute in Copenhagen." While expressing some concern about separating from his family, he wrote that he judged it advantageous in the long run to separate from his grownup children. His real concern was whether he would be able to contribute fruitfully to *scientific* work: "If you really want me to come, then you are making the mistake, not I. I can only warn you." As we will see, Franck would see his dismal prediction fulfilled.[62]

For the moment, however, Bohr ignored Franck's "warning." Instead, a meeting between the two in Germany seems to have taken place as requested by Franck, and in a letter toward the end of October, Bohr only echoed Franck's enthusiasm:

> As you know, I have cherished the hope for years that we would sometime have an opportunity to collaborate, and the prospect that this hope may be realized in the near future I feel is very fortunate.

Despite his preference for a different approach to the refugee problem than the one taken by the Rockefeller Foundation, then, Bohr was not accepting the new funding conditions merely as a necessary evil. On the contrary, he was genuinely delighted about the prospect, made possible by the foundation's new policy, of collaborating with his longtime friend and colleague.[63]

The American desire to obtain the most prominent refugees

made Franck's presence eagerly sought at several universities in the United States. Thus, Karl Taylor Compton, president of the Massachusetts Institute of Technology (MIT), suggested that Franck spend half a year at MIT and Harvard University after lecturing at Johns Hopkins. Stanford and Princeton also expressed interest. None of these offers, however, promised a permanent position. They were all direct responses to the situation in Germany, and would involve temporary support from the American Emergency Committee in collaboration with funding agencies such as the Rockefeller Foundation. Johns Hopkins came closest to a permanent offer, indicating the possibility, though not the certainty, that Franck might replace his old friend Robert Williams Wood when the renowned American experimental physicist retired in only a couple of years.[64]

Even after Bohr and Brønsted had submitted their application for support to Franck and Hevesy, Bohr, the Rockefeller Foundation, and Johns Hopkins all remained somewhat uncertain of Franck's preferences. In a meeting with Franck in November 1933, Harald Bohr reiterated that no arrangement in Copenhagen could be made beyond the first three years; yet Franck expressed his wish to work there for as long as possible. Shortly after, however, he told Harry Milton Miller of the Rockefeller Foundation that he did not know whether he wanted to spend the rest of his life in Copenhagen. The Rockefeller Foundation then received final confirmation that Franck intended to go to Copenhagen, whereupon Warren Weaver, director of the foundation's Division of Natural Sciences, reported definitely to Johns Hopkins that the Rockefeller Foundation would not contribute to Franck's stay in the United States. On 19 November, Franck left Göttingen for his assignment at Johns Hopkins, whose president Joseph Sweetman Ames repeated his confidence that Franck would return to Baltimore after three years in Copenhagen; "he evidently prefers us," Ames wrote to Weaver. The uncertainty over Franck's prospects and motivations reflects both the difficulty of communication under the circumstances, and the strong wish of both Bohr and the Johns Hopkins University to add Franck to their staffs. Most of all, however, it reflects Franck's own exasperation about his uprooted situation.[65]

Thus, when Franck moved to Copenhagen in early April 1934 as planned, the Rockefeller Foundation officers accompanying him and his wife on the train noted that he was "extremely depressed and not at all certain what he is to do at the end of the year." They noted, furthermore, that there was "no certainty and not much possibility of his remaining in Copenhagen permanently." Franck told them that he was not sure whether he wished to succeed Wood at Johns Hopkins; nor did he know what indeed his chances were of obtaining this position in the first place.[66]

What in fact were Franck's prospects at the institute? First of all, he was an experimental physicist perfectly suited to Bohr's concept of a unity between theory and experiment directed by theory. Indeed, for most of his career, Franck had devoted his experimental research to testing theoretical predictions, notably Bohr's. At the same time, however, in a letter to Bohr he had expressed concern about not being able to contribute fruitfully to research at the institute. As we will see later on, this doubt sprang from Franck's concern with Bohr's intellectual vigor and superiority. Furthermore, at the institute, Franck was almost certain to lose yet another aspect of his independence. In Göttingen he had thrived as a professor with his own institute, assistants, and students. Such a position would hardly be available for him in Copenhagen. Finally, the fate of his immediate family was an important consideration. Thus, even if his stay could be extended on the same terms, Franck could hardly be trusted to stay in Copenhagen.

During the period immediately preceding his settlement in Copenhagen, Franck found it difficult to concentrate fully on scientific research. Yet it is evident that his interest continued to develop toward larger physical systems rather than smaller ones; his last papers before his departure from Germany dealt with molecular rather than atomic processes. The two papers deriving from his visit to the United States also concerned these questions. Like Bohr, whose theories had provided motivation for much of Franck's experimental work, Franck did not even *contemplate* a full-scale study of the atomic nucleus for its own sake. Whatever can be said of Franck's prospects in Copenhagen, he was certainly an unlikely candidate to instigate research in nuclear physics there.[67]

Hevesy

Having met at a formative stage of their personal and scientific development, Bohr and Hevesy retained a close friendship and deep mutual respect for each other's scientific activities. Yet Hevesy's personality and approach to scientific research were radically different from Bohr's, as well as Franck's. Whereas Franck always directed his experiments toward resolving the most urgent problems of theoretical physicists, Hevesy was inclined to transpose experimental techniques from physics to other fields of investigation. His different attitude made him considerably less dependent than Franck on Bohr's guidance.

Manchester, Vienna, Hungary, 1911 to 1919

George Hevesy – also known as George de Hevesy or Georg von Hevesy – was born into a Jewish family in Hungary that had been ennobled in the latter part of the nineteenth century. He shared this background with several prominent intellectuals of his generation, including the scientists Theodore von Kármán and John von Neumann, both of whom ended up in the United States.[68]

Hevesy was just two months older than Bohr, and the two first met as junior scientists in Ernest Rutherford's laboratory in Manchester. Arriving more than a year before Bohr, Hevesy began studying there in early 1911. Whereas at that time Bohr was still developing his doctoral thesis in Copenhagen, Hevesy had already worked in Berlin, in Freiburg, where he obtained his doctorate in 1908, in Zurich, and in Karlsruhe. Hevesy's publication experience was also greater than Bohr's: Between the completion of his doctoral dissertation and his arrival in Manchester, he had published nine scientific articles, and before meeting Bohr he had published two more. Bohr, for his part, had two scientific articles to his credit when he moved to Manchester in April 1912, each of which came out of his work on the surface tension of water that had earned him the Gold Medal of the Royal Danish Academy of Sciences and Letters as far back as 1905.[69]

Even before Bohr's arrival, Hevesy had done work that would typify his subsequent approach to scientific research. During a three months' stay in Europe, he spoke at the nineteenth meeting of the Bunsen Society for Applied Physical Chemistry,

Hevesy with Margrethe Bohr and husband Niels in Manchester, 1912/13.

which was held in Heidelberg in mid-May 1912, on "Radioactive Methods in Electrochemistry." He argued that measuring the radioactivity of deposits from an active solution on a piece of metal constituted the best means for determining differences in electric potential between the solution and the metal. He called the deposits "radioactive indicators." Instead of asking about the basic scientific nature, origins, or implications of radioactive phenomena, as was common in Manchester, Hevesy sought to use these phenomena to obtain better techniques to resolve traditional electrochemical problems.[70]

While together in Manchester, Bohr and Hevesy participated in vigorous discussions on the place of radioactive substances in the periodic system of elements. Yet neither scientist figured among those who contributed – around New Year 1913 – to solving this problem in published articles. With their respective concentrations on the outer part of the atom and the application of physical techniques in other fields, Bohr and Hevesy did not represent the mainstream of scientists at Manchester.[71]

During this period, Bohr found at least one of Hevesy's publications "in perfect agreement" with his own thoughts on atomic constitution. Within months, however, the contrast between Bohr's and Hevesy's approaches to science would show clearly in their work. Back in Copenhagen, Bohr developed his revolutionary quantum model of the hydrogen atom based on spectroscopical evidence. This was the first in a series of theoretical achievements for which Bohr received the Nobel Prize in physics in 1922. It also constituted the beginning of the intimate relationship between experimental spectroscopy and theoretical atomic physics that eventually provided the rationale, described in Chapter 1, for the establishment of the institute in 1921.[72]

During the same period, Hevesy showed his aptitude for expanding techniques from physics to other fields. While in Manchester he made an unsuccessful attempt to separate the radioactive substance radium D (RaD) chemically from ordinary lead. Rather than looking for the implications of his negative result for physics, he shrewdly turned it into a useful experimental technique to probe questions in his own field, physical chemistry. By early 1913, Hevesy had decided to cooperate in Vienna during the spring semester with Friedrich Adolf (Fritz) Paneth toward this end. Proposing a rather specialized application of his technique, Hevesy volunteered optimistically in a letter to Paneth at the Radium Institute that "behind this apparently trivial problem hide ones of greater scope." Paneth was two years younger than Hevesy, born and raised in Vienna. The son of a noted Viennese physiologist, he was educated as an organic chemist before turning to radiochemistry.[73]

In a small but influential book published in 1911, Frederick Soddy, a previous collaborator of Rutherford and a leading researcher in radioactivity, had already suggested that RaD and lead were chemically identical and hence inseparable. Yet Hevesy made his proposal for applying the chemical identity of ordinary lead and RaD before the isotope concept had been fully clarified; it was thus not yet fully realized that the atoms of a chemical element, while chemically identical, comprise different entities – soon to be called isotopes – with different atomic weights, some of which may be radioactive. Hevesy's proposal is particularly remarkable in view of the fact that he was still under some pressure from Stefan Meyer, the director of the Vienna institute, to make a second attempt at separating RaD chemically from lead. Only after Hevesy had begun his research with Paneth, then, did the scientific basis for his proposed technique become clear. This testifies to Hevesy's foresight in the usefulness of his technique, as well as his talent in applying new insights rather than working them out conceptually.[74]

By the end of February, Hevesy and Paneth had used Hevesy's technique successfully, employing RaD as an indicator of lead in a precise determination of the extent to which some salts of extremely low solubility dissolved in water. The collaboration with Paneth was the first of a number of applications of the radioactive indicator technique for which Hevesy was awarded the Nobel Prize in chemistry in 1943. It is a testament to the success of Rutherford's laboratory that Bohr and Hevesy were able to use their exposure to radioactivity work there toward widely different ends, which subsequently brought them Nobel Prizes in entirely different areas of research.[75]

At the annual meeting of the British Association for the Advancement of Science (BAAS) in Birmingham in September 1913, the effusive Hevesy gave a lecture, whereas Bohr did not. Although formally part of a "Discussion of Radio-active Elements and the Periodic Law," Hevesy's presentation typically dealt with his own subject of "Radio-Elements as Indicators in Chemistry and Physics."[76]

Hevesy's self-assuredness at the Birmingham conference contrasted with Bohr's timidity. Joseph John Thomson, the celebrated discoverer of the electron and previously Bohr's own professor in Cambridge before Bohr went to Rutherford in Manchester, misinterpreted Bohr's question after his lecture, and rebuked him for disregarding a fundamental physical argument. Hevesy reported in a letter to Rutherford that he felt obliged "to stick up for Bohr," explaining to the audience what Bohr had really meant. In the same letter, Hevesy expressed profound admiration for Bohr's ability for scientific theorizing. The two men complemented each other in a way that would continue to serve as an important basis for their relationship.[77]

Subsequently, Hevesy sought to promote his technique before a larger audience by giving special attention to his own contribution in an account of the Birmingham meeting's physics and chemistry sessions written for a German scientific journal. Hevesy would repeat this strategy on several later occasions; it would become his standard way of announcing the beginning of a considerable research effort. Although the outbreak of the First World War slowed down his scientific activities significantly, before the end of 1920 he had succeeded in applying his indicator technique to several problems. All of these contributions were carried out in Budapest, where a chair in physical chemistry was set up for him in March 1919.[78]

The chaotic political situation in Hungary after the war induced Hevesy to break with his homeland. In early March 1919, less than three weeks before Béla Kun proclaimed Hungary a Soviet republic, Hevesy signaled his concern to Bohr, who responded with an invitation. In the fall of 1919, after the Soviet republic had broken down and Admiral Miklós Horthy was preparing to consolidate his anti-Jewish regime of White terror, Hevesy set out on a tour of European research centers, beginning with Bohr's institute in Copenhagen. Upon his return in October, Hevesy reported to Bohr that his two Jewish assistants had been dismissed, and that he found the situation so impossible that he had decided definitely to resign and leave the country. By the time Hevesy arrived in June 1920, Bohr had arranged a Rask–Ørsted fellowship for him. There was no prospect of a permanent position in Copenhagen, but Bohr was able to renew Hevesy's fellowship annually during his six-year stay.[79]

Copenhagen and Freiburg, 1920 to 1934

Before his arrival in Copenhagen, Hevesy had no specific plans for research there. As only four inactive elements with radioactive isotopes were known, the applicability of his indicator technique was severely limited. Arriving several months before the opening of Bohr's new institute, he began collaborating with Bohr's colleague and subsequent neighbor of the institute, the physical chemist J.N. Brønsted. By developing a new technique, they were the first to separate different isotopes of a chemical element – first mercury, then chlorine – "by evaporating [it] at low pressure and condensing the evaporated atoms on a cool surface." This work, which was done entirely outside the context of the institute, illustrates Hevesy's independent approach of developing new improved techniques for traditional problems in a variety of fields, rather than probing a specific discipline, as Bohr and the physicists working with him tended to do.[80]

Hevesy's independent approach is shown more fully in his main effort in Copenhagen – namely experimental work in x-ray spectroscopy. During the first years of Hevesy's stay, Bohr realized that such experiments constituted the best means for testing his developing theory of the relationship between atomic structure and the periodic system of elements. In November 1922, Hevesy joined the Dutch experimentalist Dirk Coster to learn the technique of x-ray spectroscopy.[81]

Working in Manne Siegbahn's laboratory at the University of Lund in Sweden, just across the sound from Copenhagen, Coster had established the first independent experimental support for Bohr's theory, whereupon he had been invited to Copenhagen in the fall of 1922. Collaborating on a long paper specifying the significance of x-ray spectroscopy for establishing the atomic basis for the periodic system of elements, Coster was the first scientist ever to share authorship with Bohr. Typically, Bohr was motivated to introduce experimental x-ray spectroscopy at the institute by its potential relevance in resolving his most urgent theoretical questions.[82]

At the end of November 1922, x-ray equipment was obtained, in part from Siegbahn's laboratory. Coster and Hevesy set to work immediately. Soon they detected the characteristic x-ray spectrum of element number 72 in zirconium minerals. In so doing, they established that this element was not a rare earth, as maintained by opponents of Bohr's theory. The discovery was made exactly in time for Bohr to break the news in his Nobel Prize lecture on 11 December. In addition to conducting x-ray spectroscopical investigations of the nature and abundance of the new element, Hevesy was also able to produce pure samples. The work on hafnium, as the new element was called, constituted by far the largest research effort conducted by Hevesy at the institute in the 1920s.[83]

Characteristically, Hevesy's first involvement in x-ray spectroscopy was not uniquely motivated by a well-defined theoretical concern. In statements published later in life, Hevesy traced his interest not only to his wish to confirm Bohr's theory, but also to his simultaneous desire to penetrate the structure of the atomic nucleus, as well as to his awareness of x-ray spectroscopy's potential in mineral analysis. Subsequently, x-ray spectroscopy was to constitute a substantial part of Hevesy's work. Like his application of the radioactive indicator technique, this work was not directed toward the solution of problems in physics; instead, Hevesy employed x-ray spectroscopy successfully to refine geophysical methods for identifying mineral samples. By contrast, Bohr turned to x-ray spectroscopy only when he was convinced that it would prove fruitful for his own theory, and never sought to pursue it for any other reason. The two scientists' application of this technique supplies an excellent example of their fundamentally different approaches to scientific research.[84]

Even Hevesy's third line of investigation in Copenhagen involved the extension of an existing technique to another field of investigation. Seeking to expand the application of his radioactive indicator technique further, he made contact early during his stay with the Institute of Plant Physiology at the Veterinary and Agricultural College in Copenhagen. Here he used a lead indicator to investigate the distribution and exchange of lead in the *Vicia Faba* (horse bean) plant. The results were published in 1923.[85]

About one year later Hevesy, in collaboration with two Danes – the dermatologist and venereologist Svend Lomholt and the chemist Jens Anton Christiansen – published two papers on the distribution and exchange of bismuth and lead in the animal organism, specifically in rabbits and guinea pigs. The first paper was intended as a contribution to cancer treatment with bismuth; the second shed new light on chronic lead intoxication. Hevesy was thus able to demonstrate the applicability of his radioactive indicator technique even for medical questions.[86]

Because of the scarcity of radioactive indicators suitable for biological studies, even Hevesy found it difficult to develop this application of his technique further. His only other paper on medicine or biology from the Copenhagen period merely reviewed his previous results. Like his collaboration with Brønsted, Hevesy's biological work was pursued outside the institute, and there is no contemporary indication that Bohr showed interest in it. Indeed, all of Hevesy's work in Copenhagen in the 1920s was entirely separate from the discussions pursued in the Copenhagen spirit, and there is no sign in Hevesy's correspondence during the period that he participated in these discussions.[87]

In 1926, Hevesy accepted a call to become a professor of physical chemistry at Freiburg University in Germany, thus obtaining his own laboratory, assistants, and students. He used the opportunity to step up work on the technique of x-ray spectroscopy. Typically, he promoted the technique – as he had promoted the indicator technique in the first decade of the century – first in a published article, then at the annual BAAS meeting in Johannesburg, South Africa, in August 1929, and finally in a lecture series at Cornell University the year after.[88]

In addition to his efforts in x-ray spectroscopy, the prolific Hevesy found time for other projects while in Freiburg. Most significantly, he was able to confirm the radioactivity of the only known isotope of the rare earth samarium. This result came out of a systematic investigation of the rare earths; while in Copenhagen, Hevesy had obtained specimens of all the rare earths from his friend Auer von Welsbach – industrialist, chemist, and Austrian nobleman.[89]

In spite of the limited number of stable elements with known radioactive isotopes, Hevesy continued to expand the application of his radioactive indicator technique. This work even included a small-scale continuation of the application of radioactive indicators in biology. Thus, in 1930, Hevesy and his collaborator O.H. Wagner investigated the distribution of thorium in the animal organism in order to see whether this element was taken up in different amounts in sound and cancerous tissue.[90]

In 1931, Harold Clayton Urey at Columbia University in New York City discovered heavy hydrogen, a nonradioactive isotope of hydrogen distinguished from regular hydrogen by its greater mass. Because any two isotopes of the same element are chemically identical, the newly discovered isotope could be used as an indicator of hydrogen, albeit with a different method than Hevesy's radioactive indicator technique. In view of the scarcity of known radioactive isotopes, it is therefore not surprising that Hevesy soon sought to use deuterium, as the new isotope came to be called, as an indicator in biological investigations. According to Hevesy's autobiographical account, Urey, an old friend who like Hevesy had spent time at Bohr's institute in the early

1920s, "promptly supplied us with some liters of waters [sic] containing 0.6 per cent of deuterium oxide." By early 1934, Hevesy, in collaboration with his student Erich Hofer, was able to measure quantitatively the exchange of water in goldfish. In a letter to Rutherford dated 1 April 1934, Hevesy noted this work last among his many scientific projects in Freiburg, describing it as a "very modest topic compared with the great problems of atomic physics." In his correspondence with Bohr, Hevesy did not mention the project at all, and gave no indication that he wanted to pursue biologically oriented research in Copenhagen. Clearly, Hevesy did not think of his work in biology as his first priority. Nor did he think that it ranked high among Rutherford's and Bohr's interests.[91]

In fact, just as he had proposed no specific research items before coming to Copenhagen in 1920, so Hevesy uttered no preferences after support for another stay had been obtained from the Rockefeller Foundation in 1934. Indeed, even though in their grant application Bohr and Brønsted had anticipated that Hevesy would work at the Institute for Physical Chemistry, Hevesy wrote to Bohr only three weeks later that he expected to collaborate at the institute with J.C. Jacobsen – the resident Danish experimentalist – primarily in order to save apparatus and space. In the same letter, Hevesy asked whether his arrival could be postponed from early spring to summer, as he wanted to complete a variety of "not so important" work in Freiburg. In subsequent correspondence with Bohr, Hevesy pushed his date of arrival further forward. In these letters, too, Hevesy refrained from stating specific wishes for one or more research projects in Copenhagen. In fact, as late as the end of July 1934, Hevesy contemplated accepting an offer from Chandrasekhara Venkata Raman to direct the Chemical Institute in distant Bangalore, India, from October to May the following year. After having solicited advice from Franck, however, Hevesy decided to come to Copenhagen after all, and was thus spared from telling Bohr about his alternative plan. On 11 August, Hevesy confirmed that he would arrive in September, and reported to Bohr that he had finally informed his Freiburg colleagues and authorities of his decision to resign. He was glad to have received "a very friendly treatment from all sides" and to take leave "in a friendly way."[92]

What were Hevesy's prospects in Copenhagen compared with Franck's? Unlike Franck, Hevesy had not devoted his career as an experimentalist to testing the conjectures of theoretical physicists. On the contrary, once having developed or learned an experimental technique in physics, he was inclined to extend its use to other fields of investigation. Whereas this tendency made him less close to Bohr scientifically, it also made him more independent of Bohr's theorizing. Thus, Hevesy never expressed any alarm that he might be restrained by Bohr's intellectual superiority. Indeed, he had already demonstrated his capacity for working in the same environment as Bohr for an extended period. Moreover, although he had directed his own institute in Freiburg since 1926, Hevesy was significantly less accustomed to the life of a full professor than Franck was. Besides, there is no indication in his correspondence that he worried about losing the freedoms and responsibilities that a full professorship entailed. Although he had pondered Raman's offer to work temporarily in India, he had made less of an effort than Franck to investigate other possibilities for work. Having left his family behind in Hungary long ago, and even marrying a Danish woman, Pia Riis, during his last stay in Copenhagen, Hevesy would hardly jeopardize family relations by moving to Denmark. For these reasons, Hevesy seemed more likely than Franck to stay in Copenhagen.[93]

Whereas Hevesy's approach to experimental research was fundamentally different from Franck's, it was similar in one respect: Although experienced in experimental radioactivity research, Hevesy had not contributed significantly to the knowledge of the atomic nucleus. In spite of their differences, Hevesy seemed as unlikely as Franck to instigate a concerted effort in experimental nuclear physics at the institute.

ORIGIN OF EXPERIMENTAL NUCLEAR PHYSICS

Between the miraculous year (1932) of nuclear physics and Franck's and Hevesy's arrivals in Copenhagen, there had been new significant developments in the experimental study of the atomic nucleus. As we will see in this section, the presence of the prominent refugee scientists induced Bohr to act. Instead of continuing to employ these developments as fodder for purely theoretical discussion, Bohr now decided to take up experimental research on the atomic nucleus at the institute.

Background

In a letter to Rutherford in early February 1934, Bohr emphasized more than previously the experimental work at the institute. He noted, in particular, J.C. Jacobsen and Evan James Williams's use of cloud-chamber photographs of positrons to establish the interaction between gamma rays and atomic nuclei. However, he also wrote:

> At the moment all the theoretical people here are occupied with the puzzles of the electron theory, but it does not mean that we do not all follow the marvellous developments of nuclear problems with the keenest interest and were most delighted to learn about the many new results of which you so kindly told me in your last letter.

In the letter to Felix Bloch in Rome two weeks later – which contained Bohr's negative evaluation of Fermi's new theory of beta decay and with which he enclosed a copy of his "heartfelt outpouring" to Pauli – Bohr referred once more to the rapid developments in experimental nuclear physics, noting that "it must often seem very trivial to speculate on the paradoxes of electron theory." For the first time, Bohr was explicitly contrasting experimental work on the nucleus with the theoretical concerns at the institute. Although he had referred to the latter work somewhat defensively, however, Bohr had yet to express any intention to change the research direction at the institute. As we saw in Chapter 2, Bohr's main scientific activity continued to be the discussion of theoretical topics.[94]

Among the new experimental developments, Bohr referred in his letter to Bloch particularly to "the last discovery in Paris." In late 1933, Irène Curie and her husband Frédéric Joliot had found that the three elements aluminum, boron, and magnesium, when bombarded by alpha rays from radioactive polonium, were transformed into hitherto unknown radioactive isotopes of phosphorus, nitrogen, and silicon, respectively. This result was an unexpected outcome of the Joliot-Curies' program to produce positrons by irradiating several elements with alpha particles from radioactive polonium; after irradiation, the samples continued to emit particles, and the Joliot-Curies claimed to have found "a new kind of radioactivity," which was soon to be known as "artificial" or "induced." The two scientists also suggested that more radioactive isotopes than the three already established might be produced artificially.[95]

In a letter written in March – which incidentally introduced Aage Friis, chairman of the Danish Committee for the Support of Refugee Intellectual Workers – Bohr told Marie Curie, Irène's mother and a living legend of French physics:

> It is the cause of very great pleasure, indeed, that just now, where [sic] the human aspects of science are so regrettable, the scientific work itself progresses so marvellously. Especially I need not say with what pleasure and interest we have here as anywhere else learned about the recent wonderful discoveries of your daughter and son-in-law, which mark a new and most promising epoch of atomic theory.

Just as in his last letter to Jordan, written less than four months later, Bohr contrasted contemporary scientific and political developments.[96]

Bohr was not alone in his enthusiasm for the new discovery. The prospect of creating radioactive isotopes at will carried vast implications for both basic and applied science. The new possibilities for employing the radioactive indicator technique, which we will see was not lost on Hevesy, is only one particularly relevant example. It is not surprising, then, that the Joliot-Curies had to wait less than two years, an unusually short period, to receive the Nobel Prize in chemistry for "their synthesis of new radioactive elements."[97]

In March 1934, Enrico Fermi set out with his collaborators in Rome on an ambitious experimental research program to follow up the discovery of the Joliot-Curies. In his Bakerian Lecture of 1920, Rutherford had suggested that because of its net zero electricity, his hypothetical neutral "atom" would penetrate the atomic nucleus much more easily than a charged particle. Combining this idea with the recent discovery of the Joliot-Curies, Fermi was led to expect that artificial radioactivity could be produced more readily by bombarding elements with neutrons than, as the Joliot-Curies had done, with alpha particles. Moreover, because lower energies were required to enter the nucleus, it seemed more likely that a systematic study could be undertaken without using expensive and elaborate accelerator equipment like that used to accelerate protons at the Cavendish Laboratory. Fermi and his collaborators obtained neutrons from a radon–beryllium source, the radon being supplied by the Laboratory of Public Health next door. They then set out systematically to bombard all elements in the periodic system they could lay their hands on. The first positive result was obtained with element number 9, fluorine, and this, together with the production of radioactivity by bombarding aluminum, element number 27, was announced in the first paper – dated 25 March 1934 – of a long series that would appear in the Italian journal *Ricerca Scientifica*.[98]

The roles of Franck and Hevesy

When he settled in Copenhagen only two weeks after the publication of the first results from Rome, Franck had limited access to experimental equipment and scientific assistance. Moreover, he was still pondering whether he should move to the United States. In his effort to find a research project that would consume as few resources as possible, Franck and his recently acquired assistant Hilde Levi set out almost immediately after Franck's arrival to study the fluorescence of green leaves. This was a natural extension of the previous research of both scientists. For Franck it meant continuing his studies of energy exchange in complex molecular systems, while Levi would be able to build on her doctoral work on experimental molecular spectroscopy at the University of Berlin.[99]

Within a couple of weeks of Franck's arrival in April 1934, however, Bohr wrote to Werner Heisenberg that Franck was directing experimental investigations along the same lines as the work on radioactive phenomena in Paris and Rome. Like Fermi in Rome, Bohr obtained his radioactive sources from a local institution, the Radium Station – the main Danish center for research and treatment of cancer, established in 1913. Moreover, it was Franck who only a few days later wrote to Frisch in London asking on Bohr's behalf whether Frisch would accept the invitation to work in Copenhagen. In his reply, Frisch accepted the invitation and expressed his satisfaction that Franck was planning to work in nuclear physics, his own area of interest. He was excited about his recent accomplishment in Blackett's laboratory to induce radioactivity by alpha irradiation in another two elements (sodium and phosphorus). He was only sorry that his fellowship in Copenhagen would last no longer than a year. In mid-July, Franck reported to Bohr that he had written to Frisch and received a positive response; the long interval between Frisch's letter and Franck's report to Bohr in his summer house was probably a result of Bohr's visit in the meantime to the Soviet Union. In any case, it seems that Bohr was directing Franck to make the first modest steps at the institute toward work in experimental nuclear physics.[100]

At about the same time, Hevesy wrote to Franck from Freiburg about his recent trip to London. At Bohr's request, he had visited the Metropolitan-Vickers Company, starting negotiations to acquire high-voltage equipment for nuclear research at the institute. Metropolitan-Vickers, which had been involved in installing this kind of equipment at Rutherford's Cavendish Laboratory, offered to have the director of its own high-voltage laboratory, Thomas Edward Allibone, visit Copenhagen in May or June to see the facilities, and to discuss the details of a possible installation. In his letter to Franck, Hevesy, who had a high opinion of Allibone's competence, strongly recommended that the offer be accepted. Evidently Bohr, who was visiting the Soviet Union at the time, was seriously contemplating introducing equipment for nuclear research at the institute, and he had involved his senior colleagues Franck and Hevesy in the new effort.[101]

Upon his return from the Soviet Union, Bohr expressed his enthusiasm about research on induced radioactivity in a letter to Torkild Bjerge, a promising Danish experimental physicist studying at the Technical University in Copenhagen who, on Bohr's recommendation, was spending some months in Cambridge working under Rutherford. Bohr wrote:

> . . . and also here in the laboratory, where professor Franck's collaboration has meant so much for us, we have from the very first day we heard about Fermi's discovery pursued the subject as best we could both theoretically and experimentally.[102]

As a particular example of the efforts in induced radioactivity at the institute, Bohr referred in his letter to the work of Johan Ambrosen, another young Danish experimental physicist. Ambrosen's work was published soon after. In his research, Ambrosen set out to determine whether the same radioactive isotope could be produced by irradiating two different elements with neutrons. For such a determination to be possible, the induced radioactive isotope must have a sufficiently long half-life to allow for chemical separation. Fermi had recently been able to produce sufficiently long-living phosphorus isotopes by neutron irradiation of sulfur and chlorine. Ambrosen produced these isotopes, and compared their half-lives and beta decay. He concluded that the isotopes were identical, having a half-life of seventeen to eighteen days. At the end of his article, Ambrosen wrote that it was his "pleasant duty" to thank Bohr for his "constant interest" and Franck for "valuable discussions." Moreover, in the letter in which Franck finally reported to Bohr on Frisch's decision to come to Copenhagen, Franck wrote that in Bohr's absence he had taken the liberty of submitting Ambrosen's paper to *Zeitschrift für Physik*. Franck was now becoming increasingly involved in administering experimental nuclear physics research at the institute.[103]

Soon Bohr also began to express himself on the *theoretical* implications of the experiments of Fermi and his group. Nine days after writing to Bjerge, he told Rutherford that he had no new experimental results to report from Franck's neutron impact investigations. Instead, he noted that the people at the institute disagreed with Fermi's interpretation that some of the impinging neutrons attached themselves to the bombarded nuclei. He found it more probable, he wrote, that "the collision resulted in the expulsion of two neutrons from the nucleus instead of the attachment of one." This was a point of contention even in Rome. Edoardo Amaldi and Emilio Segrè hence joined forces with Bjerge and H.C. Westcott in Cambridge in order to establish whether their bombarded nuclei emitted gamma rays, as predicted by Fermi, or two neutrons, as predicted by Bohr; they obtained ambiguous results. Bohr reached his preferred conclusion on theoretical grounds: If the nucleus, as generally assumed in analogy with the outer part of the atom, was a system of largely noninteracting particles, it was difficult to understand how it could absorb an impinging neutron. Whereas Rutherford did not fully agree with his criticism, Bohr regarded it as so important that he sent a copy of it to at least one colleague, George Gamow, for comment. In addition to having decided to make a small-scale effort in experimental nuclear physics at the institute, Bohr was now also involving himself in the theoretical interpretation of these experiments.[104]

A report by an American physicist touring European research centers confirms the new attention to experimental nuclear research at the institute. In June 1934, T. Russell Wilkins of the University of Rochester in New York reported back to his university:

> Work in Copenhagen, as in nearly every center of physics in Europe, is centered at the moment in the astounding discoveries made within the last few months in Radioactivity.

The "discoveries" he referred to was the production of artificial radioactivity in Paris and Rome. In addition, he noted:

> The Institute at Copenhagen is undoubtedly destined for even greater days, for two of Germany's great physicists are joining Bohr – Franck from Copenhagen [sic] and Hevesy from Freiburg.

Even at this early time, the emerging transition at the institute was obvious to the visiting American.[105]

In early July, the Bohr family received a serious blow. During a sailing trip with his father, Bohr's oldest son Christian, who had just completed high school, fell out of the boat and drowned. Quite untypically for Bohr, his scientific correspondence does not contain any outgoing letters from the time of Christian's death until 8 September, when he informed Heisenberg that the informal annual conference would be canceled. Because of the accident, Bohr did not go to the third major – and by far the largest – international conference on nuclear physics, which took place in London in early October; the two previous ones had been Fermi's conference in Rome in 1931 and the Solvay Congress in 1933. The death of his son had a strong impact on Bohr's personal and scientific life. Yet before the end of the year he participated in the institute's work with his usual energy.[106]

On 15 October 1934, Enrico Fermi wrote a letter to Hevesy. By addressing his letter to Freiburg, Fermi revealed that he was not yet informed of Hevesy's move. He wrote that he had been referred to Hevesy by Fritz Paneth, codiscoverer with Hevesy of the radioactive indicator technique. Obviously, then, there had been little previous contact between Fermi and Hevesy. Fermi's letter contained a request for samples of rare earths for his induced radioactivity investigations. As these were difficult to obtain, Fermi had previously reported on the irradiation of only three of the fourteen known rare earths. In addition, he asked for scandium and hafnium; the existence of the latter Hevesy had established with Dirk Coster in Copenhagen more than ten years before.[107]

Fermi's letter cannot have taken long to reach Copenhagen, for Hevesy's reply is dated 26 October. Hevesy emphasized that a research program similar to Fermi's had been conducted at the institute since April. The reason that only one publication – Ambrosen's article on radioactive phosphorus – had come out of this effort, Hevesy explained, was that the Rome group had consistently published the same results as those obtained by the Copenhagen researchers before the latter had a chance to write them up. Because of "the close relationship between this element and Bohr's institute," Hevesy continued, he had begun neutron bombardment of hafnium immediately after his arrival in Copenhagen. He provided Fermi with the preliminary results of this investigation, without, however, supplying any hafnium. Regarding Fermi's requests for other elements, Hevesy claimed that his samples of rare earths not already studied by Fermi were too small to be employed in an investigation of induced radioactivity.[108]

Hevesy, then, was unwilling to provide Fermi with any of the samples asked for, suggesting instead that the chemist B. Smith Hopkins in faraway Illinois might help in the matter. Hevesy's answer can hardly be called helpful, especially since he had ample supplies of all the elements asked for. By the end of 1936, Hevesy, in collaboration with Hilde Levi, whom Bohr in the meantime had persuaded to become Hevesy's assistant, had published a paper on the induced radioactivity of all the elements Fermi had requested.[109]

Milton S. Plesset, Bohr, Fritz Kalckar, Edward Teller, and Otto Robert Frisch (reading from the left) discuss the day's events at the institute, c. 1934. [Courtesy of John A. Wheeler]

Three days after Hevesy had answered Fermi's letter, Franck wrote about his work and general situation to his close and long-time friend Max Born, providing a similar impression of the slow progress of experimental nuclear physics research at the institute. In his letter, Franck confirmed that he was pursuing such work. Before so doing, however, he mentioned his forthcoming papers with two German collaborators, which took off from his previous interest in fluorescence in fluids. He wrote that he felt compelled to turn to nuclear physics full time only because he was in the process of losing these collaborators. Obviously frustrated, Franck exclaimed that this was the only subject the young Danish experimentalists could handle. He wrote that he was happy to learn, and that he was getting "unbelievably much" from Bohr. Yet he complained that he was progressing only slowly in his new area of investigation. He admitted that he had poor facilities and slow co-workers. Indeed, he was seriously considering the advice of Born, his previous colleague at Göttingen Richard Courant, Wilbur Tisdale of the Rockefeller Foundation, "and in a certain sense Harald Bohr" to settle in the

United States. Contrary to his earlier report to Bohr, Franck now told Born that his most important consideration was the opportunity to live in the same country as his daughters. It was unrealistic to hope that his sons-in-laws would obtain positions in Denmark. Moreover, one of them had almost secured a job in the United States. Franck thus considered the possibility of remaining in the United States after participating in the summer school at Cornell University in 1935. On the very same day that Franck wrote his letter to Born, Frisch wrote to Emilio Segrè, Fermi's collaborator in Rome. While otherwise enthusiastic about his stay at the institute, Frisch described the Copenhagen work on artificial radioactivity as unsystematic and primitive.[110]

Despite these reports of insufficient progress in the pursuit of experimental nuclear physics at the institute, Bohr wrote to his colleague Ralph Fowler in Cambridge on 12 December 1934, typically contrasting the concurrent scientific and political developments:

> The anxiety and trouble to which the whole German problem gives rise are indeed so great, that they almost overshadow the delight that everyone of us takes in the marvellous development of atomic physics. Still we do also here try to take a modest share in the work, and at the moment the whole laboratory is busily occupied with research on nuclear physics under the effective guidance of Franck and Hevesy.

In contrast to his earlier concentration on theory, Bohr now considered the activity in experimental nuclear physics, led by Franck and Hevesy, as essential for the development of the institute. His insensitivity to the problems expressed by Franck and Frisch only substantiates further his genuine dedication to the new venture.[111]

Another letter from Franck to Born in early January 1935 confirms the increasing attention at the institute to theoretical and experimental questions about the atomic nucleus. By now, Fermi and Bohr's disagreement on the scattering products from nuclei bombarded with neutrons had been resolved in Fermi's favor. In October 1934, the Rome group had made a far-reaching discovery entirely by chance. By placing paraffin between the neutron source and the target, they had observed that the induced radioactivity increased tremendously with *decreasing*

Bohr, Franck, and Hevesy at the institute, 1935. [Courtesy of AIP Niels Bohr Library, Margrethe Bohr Collection]

neutron velocity, the neutrons being absorbed at a substantial rate. This result was completely unexpected and made untenable the analogy between the outer and inner part of the atom for which Bohr had argued in his letter to Rutherford. Franck noted this anomaly in his letter to Born, thus suggesting that it was a

Niels and Harald Bohr after bidding a sad farewell to Franck, September 1935.

point of serious concern in Copenhagen. As we will see in Chapter 5, however, it would take another year until Bohr presented a solution to the problem that would revolutionize nuclear physics.[112]

Concerning his own experimental work, Franck was less than enthusiastic: "My nuclear physics exhausts itself at present in work which is just about to be completed, when someone else publishes it in *Nature*." He added drily that such experiences could best be considered as his dues for entering a new field. Franck could now inform Born definitely that he would move to Baltimore in the fall.[113]

In spite of the frustrations expressed by Franck and Frisch, which in the case of Franck even contributed to his decision to leave Copenhagen, Bohr was clearly leaning on the services of these two scientists, as well as of Hevesy, to begin a redirection of research at the institute toward nuclear physics. Before continuing our narrative, it will be useful to pause in order to consider the relationship between the potential program of experimental research led by Franck and Hevesy and the established mode of work at the institute that we have identified as the Copenhagen spirit.

Franck, Hevesy, and the Copenhagen spirit

When interviewed about his career toward the end of his life, Franck noted that "Bohr was so superior that some people had difficulties with him." He continued:

> [This] I understood only after Hitler came to power, and I was for one and a half years in [Bohr's] laboratory. Bohr did not allow me to think through whatever I did to the end. I made some experiments. And when I told Bohr about it, then he said immediately that might be wrong, what [sic] might be right. And it was so quick that after a time I felt that I am [sic] unable to think at all.

Franck concluded his assessment of Bohr's influence with the following words:

> Bohr's genius was so superior. And one cannot help that one would get so strong inferiority complexes in the presence of such a genius that one becomes sterile. You see? Therefore, I would like to say no one fitted really in that circle the whole time. But to be there for a time and to learn and to see this man and to understand the greatness of him and the goodness. He is really something. Yes, a man for whom one can feel only a hero worship.[114]

Franck thus found confirmation for his doubts about his ability to pursue fruitful work in Copenhagen, which he had expressed to Bohr even before his arrival. After devoting a substantial part of his career to testing experimentally the predictions of theoretical physicists, Franck felt too constricted working on a day-to-day basis under the master theorist. Whereas most young students considered the opportunity of cooperating closely with Bohr as the greatest honor and experience, to the established professor James Franck, who was used to having his own stu-

James Franck, 1935.

dents and laboratory, it meant losing his independence. In this light, Franck's personal recollection constitutes a striking confirmation of the Copenhagen spirit of devoted collaboration between young, brilliant physicists and their master, Bohr. Indeed, Franck's experience indicates that the Copenhagen spirit *depended* on Bohr's collaborators being younger and less established.

In contrast to Franck, Hevesy, who before his arrival in October 1934 had already spent six years at the institute, never expressed worries about Bohr's superiority. This difference between Franck's and Hevesy's reactions was closely bound up with the two scientists' divergent styles of research. Although he admired Bohr's theoretical insight tremendously, Hevesy aimed in his practical work not at confirming theories of Bohr and other theoretical physicists, but sought rather to apply physical experimental technique to other fields of investigation. In the same vein, Bohr's and Hevesy's approaches to writing were entirely opposite; whereas Bohr labored incessantly over his manuscripts, Hevesy published his results as quickly as he possibly could.

There is an anecdote exemplifying Hevesy's approach to writing. When he was in Stockholm after the Second World War, he is reported to have written a letter to the secretary at the institute requesting that some tickets be ordered for his travels. In the course of writing the letter, however, he decided that he did not need the tickets after all, concluding his letter by canceling the tickets. He then sent off the letter containing both his request and its cancelation. Whether or not this anecdote about Hevesy is true, it is unthinkable that Bohr would have mailed such a letter without rewriting it at least once. Because of Bohr's and Hevesy's very different approaches to their work and writing, there was less occasion for a direct scientific confrontation between these two than between Bohr and Franck.[115]

The difference between Franck's and Hevesy's approaches to research is reflected in the extent of their prior contact with each other. Indeed, aside from their brief encounter in Copenhagen, they were never close scientifically. Thus, in the interview quoted earlier, Franck described his relationship with Hevesy as follows: "I had a kind of friendship with him, but it was never that we were really too much in the same field." For his part, Hevesy visited Göttingen, a mecca for theoretical physics and Franck's base from 1921 to 1933, for the first time only in the summer of 1932.[116]

Hevesy showed his scientific independence during both of his extended stays with Bohr. Victor Weisskopf, who worked with Bohr in Copenhagen from 1935 to 1937, contrasts Hevesy's "autocratic" approach to his students and collaborators with Bohr's "productive anarchy," and confirms that Hevesy rarely participated in the discussions embodying the Copenhagen spirit. Unlike Franck's approach, Hevesy's approach to science was largely independent of Bohr's. This independence was crucial for the long-term coexistence of Bohr and Hevesy in the same research environment.[117]

In contrast to Franck, Hevesy would stay in Copenhagen for several more years. In subsequent chapters, we will see both how Hevesy secured his scientific independence from Bohr, and

George Hevesy, c. 1935.

how the Copenhagen spirit survived in spite of Bohr's concerted redirection.

CONCLUSION

The Rockefeller Foundation's Special Research Aid Fund for European Scholars was instrumental in persuading Bohr to turn the research at the institute toward experimental nuclear physics. Unlike the traditional funding policy for basic science, this program concentrated on supporting the most established scientists. In so doing, it inspired Bohr to acquire his good friends James Franck and George Hevesy for the institute. The stature of the two former professors contrasted sharply with that of the traditionally young and unestablished visitors to the institute who were supported by conventional funding programs; Bohr therefore thought it imperative to define an independent research project for them. This was an unusual decision for Bohr, who for several years had not contemplated introducing any well-defined theoretical and experimental research program, but instead was used to discussing general problems in a free-wheeling manner with the younger guard of eminent physicists visiting the institute on a regular basis. In view of contemporary developments, it was only natural to settle on experimental nuclear physics as the pertinent research program for his two colleagues.

Prior developments indicate that Bohr would not have instigated this research effort had not the new policy brought Franck and Hevesy to Copenhagen. In Chapter 1, we saw that Bohr did not campaign for any new substantial research support until presented with the opportunity to obtain funding for Franck and Hevesy. Furthermore, Chapter 2 showed that the physics at the institute into the year 1934 was pursued in an entirely laissez-faire manner, without any thought of launching a concerted theoretical and experimental research effort. Bohr's early response to the refugee problem confirms this observation. Like previous visitors, the refugees helped by Bohr through the Intellectual Workers Committee and traditional fellowship support were young physicists, visiting on a temporary basis, who were eager to participate in the renowned debate with Bohr.

Even when deciding on a particular research program for his friends, it was not theoretical discussion that moved Bohr. Indeed, when Bloch in February 1934 wrote enthusiastically from Rome about Fermi's new theory of beta decay – which provided a framework for the neutrino particle that Pauli had introduced to rescue energy conservation – Bohr was still doubtful. In his reply, he was considerably more excited about the recent *experimental* discovery in Paris of artificial radioactivity. In his subsequent enthusiasm for nuclear research, it was this discovery and its further elaboration in Rome, rather than traditional theoretical discussion at the institute, that compelled him to act. Only then, and on the basis of the experimental developments, did Bohr begin to address theoretical questions in nuclear physics.

The Copenhagen spirit, then, constituted a regressive rather than progressive force in the first movement toward a concerted nuclear physics program at the institute. Indeed, it contributed to Franck's decision to leave Copenhagen even before the program had gotten off the ground, while Hevesy survived in the Copenhagen environment in part because of his independence from the Copenhagen spirit. Yet, once Bohr was bent on introducing experimental nuclear research at the institute, it was he, and not his two experimentalist friends, who was most enthusiastic. Indeed, whereas Franck and Frisch expressed frustration about the lack of progress of the new program, and Hevesy was yet to settle in Copenhagen, Bohr was writing to colleagues about Franck and Hevesy's efforts in experimental nuclear physics with increasing excitement. Thus, Bohr seemed entirely unaware of the potential for conflict between the mode of work represented by the Copenhagen spirit and a concerted experimental effort led by his professional peers. This aloofness in relation to the social realities at the institute may have been crucial in guiding the effort in experimental nuclear physics through its first difficult stages.

When he was promoting the idea of a new institute for theoretical physics around 1920, Bohr used as his rationale the idea of a unity between theory and experiment. During the next few years he continued to create opportunities to expand the institute. In the process, he successfully developed the unity between theory and experiment he had originally sought. Subsequently, however, after his original research program had been consolidated, Bohr reduced his activities as policymaker and fund-raiser for research at the institute. At the same time, concerted work in physics based on an equal combination of theory and experiment was replaced by freewheeling discussions of theory; the Copenhagen spirit had come to reign. This mode of scientific work was strongly encouraged by the predominant policy toward basic science funding.

The Rockefeller Foundation's response to the refugee problem confronted Bohr with a new approach to such funding. The opportunity to seek support for Franck and Hevesy persuaded

Bohr to consider reintroducing an experimental program at the institute, this time in nuclear physics. The possibility of a new unity between theory and experiment based on nuclear physics began to emerge. Without tracing this new unity to the presence of Franck and Hevesy at the institute, the Danish physicist Christian Møller recognized these new developments. In the same interview in which he observed that the experimental technique of spectroscopy was "perhaps already a little outdated when quantum mechanics was finished," he thus noted that about this time a new unity between theory and experiment based on nuclear physics was beginning to take precedence over Bohr's philosophical interest.[118]

Although the original introduction of experimental nuclear physics at the institute owed itself to external circumstances, it was now up to Bohr to consummate it. In the following chapter, we will see that just as he did in the very first years of the institute, Bohr was able to make the best of contemporary funding opportunities for basic science to accomplish his mission.

4

Experimental biology, late 1920s to 1935

The Rockefeller Foundation's Special Research Aid Fund for the refugees was only temporary. Furthermore, although breaking with the traditional fellowship arrangement for visiting scientists, it was not in conflict with the "best science" aspect of traditional funding policy, which helped Bohr in developing the Copenhagen spirit. Like previous basic science funding, the Special Research Aid Fund looked to quality and prestige rather than to specific scientific projects and their application. Indeed, in seeking more prestigious scientists than did the traditional fellowship policy, the Special Research Aid Fund can even be argued to represent a further move toward a best science approach. Its challenge to the Copenhagen spirit, therefore, was only limited.

In the 1930s, the Rockefeller philanthropies also changed their overarching policy for basic science funding. Taking over the IEB's responsibilities for supporting natural science, the reconstituted Rockefeller Foundation thus launched its "experimental biology" funding program. The ambitious new program sought to improve what was considered backward biology by introducing the most advanced mathematical, physical, and chemical methods and techniques into the life sciences. As such, it was totally independent of the temporary and makeshift Special Research Aid Fund for the refugee scientists. With the introduction of the new program, the promotion of best science as an end in itself was replaced with the vision of a specific kind of scientific project meriting investment.

This chapter describes the origins of the Rockefeller Foundation's experimental biology program and Bohr's response to it. Considering his commitment to the Copenhagen spirit, it is not surprising that he initially reacted with hesitation toward the new scheme. As shown in Chapter 2, Bohr's prior interest in biological questions, although genuine, was entirely philosophical and carried on through freewheeling discussion rather than concerted research. As such, it was not consistent with the kind of directed, experimentally based biological research project that the experimental biology program seemed to demand.

However, after the acquisition of Franck and Hevesy for the institute and the resulting effort to introduce experimental nuclear physics research there, Bohr's response to the Rockefeller Foundation's new funding policy changed completely. After the foundation had sought for some time, persistently but unsuccessfully, to recruit Bohr and his Institute for Theoretical Physics for its new program, Bohr's attitude suddenly changed. Thus, he and Hevesy proposed an experimental project in biology requiring the same large and expensive apparatus as did a full-scale redirection to nuclear physics research. Having thus begun formulating an experimental project in line with the foundation's wishes, Bohr then entered into extensive negotiations with the foundation. In the process, he was able to continue the scientific redirection of the institute, not only introducing biological research, but, more enduringly, making possible a full-scale transition to theoretical and experimental nuclear physics.

REORGANIZATION OF THE ROCKEFELLER PHILANTHROPIES

The change in the Rockefeller philanthropies' approach to basic science funding accomplished in the mid-1930s had its origins in the previous decade. Toward the late 1920s the funding policy of the IEB, and with it Wickliffe Rose's ideal of supporting basic science, had become debatable issues within the Rockefeller philanthropies. There was increasing uneasiness, for example, that the organization of the Rockefeller philanthropies was becoming too unwieldy. Not only did it consist of a wide variety of largely autonomous agencies that were less than willing to cooperate with one another, but the areas of interest of the individual agencies were beginning to overlap, thus leading to unfruitful and uneconomical competition. The shared interest of the IEB and the Rockefeller Foundation's Division of Medical

Education in developing Copenhagen physiology, described in Chapter 1, was but one of several instances of this general development.[1]

In early 1927, a committee was established to look into the possibility of reorganizing the Rockefeller philanthropies. The committee was led by Raymond Blaine Fosdick – John D. Rockefeller, Jr.'s lawyer and a chief trustee of the Rockefeller Foundation. A major debate arose between those in the committee promoting "horizontal" divisions between levels of education and those preferring a policy for the "advancement of knowledge," with one agency for the sundry "vertically" divided areas of scholarly investigation. In line with his elitist approach to education, Rose strongly championed the second alternative. After months of vacillation, Fosdick decided in October 1927 upon a scheme emphasizing the advancement of knowledge. At the same time, he resolved that the large majority of the Rockefeller philanthropies would be unified under the auspices of the Rockefeller Foundation. The IEB, in particular, would lose its independence and be absorbed. The old Rockefeller Foundation had concentrated on specific projects, especially in the medical field, rather than on the advancement of knowledge for its own sake. It was hence not to be expected that the reorganized Rockefeller Foundation would concentrate exclusively on the advancement of general knowledge. Having reached retirement age, Rose was undoubtedly disillusioned by the prospect of the IEB's demise. Even though the organization he had created existed formally until 1938, its activities for the rest of its existence were minuscule. An era in basic science funding, with Wickliffe Rose personally presiding over the funding of major institutions and individuals, as well as providing a model for other funding agencies, was over.[2]

A full-fledged alternative, however, was slow to emerge. George Edgar Vincent stayed on as president of the foundation. The new Division of Medical Sciences continued the work of the former Division of Medical Education – which had provided funding for August Krogh and Copenhagen physiology in the 1920s – with Richard M. Pearce remaining director. The Division of Natural Sciences, which was envisioned to take over the responsibilities of the IEB, was established under the leadership

Max Mason, 1932. [Courtesy of Rockefeller Archive Center]

of Max Mason only a year after the decision to reorganize. Subsequently, divisions for the social sciences and for the humanities were also set up.[3]

Max Mason was born in 1877 and received his doctorate in mathematical physics under David Hilbert at Göttingen University in 1903. After four years at Yale University, he was appointed a professor at the University of Wisconsin, where he remained until 1925, when he became President of the University of Chicago. He was at Chicago when he accepted the directorship of the Rockefeller Foundation's Division of Natural Sciences. At the time he took up this position, on 1 October 1928, it was clear that he would succeed Vincent as president.[4]

At its establishment, the Division of Natural Sciences took over the IEB's Paris office, including its staff, thus providing further continuity in Rockefeller funding of natural science. The Paris office, for example, continued to accept fellowship applications on the same basis as before. This, and the fact that Mason's tenure as director was only temporary, meant that the Division of Natural Sciences, in particular, suffered from a combination of insufficient leadership and organizational inertia that hampered the early establishment of a strong new funding policy.[5]

At a conference of the officers and trustees of the Rockefeller Foundation held at Princeton University in October 1930, Mason pleaded that funding be concentrated on more well-defined areas. However, within only a year Herman Augustus Spoehr – Mason's successor as natural sciences director from the fall of 1930 – decided to return to his previous position as plant physiologist at the Carnegie Institution's Laboratory of Plant Research in Stanford. It is hardly surprising, then, that a concerted change of direction for natural science funding did not take place even during Spoehr's tenure.[6]

A glance at the Rockefeller Foundation's grants for natural science during this period shows that the lack of innovation in policy was reflected in practical work as well. Although total expenditures for natural science were maintained at the same level as before the reorganization of the Rockefeller philanthropies, the relative emphasis on different kinds of funding changed. Thus, the reorganized Rockefeller Foundation's Division of Natural Sciences spent significantly less to comply with individual research institutions' substantial requests for new "construction and equipment" than the IEB had done previously. Of the larger grants, the Rockefeller Foundation spent a relatively bigger share on "endowments" for less specific purposes. The grants for specific activities that nevertheless *were* provided were significantly smaller on the average than previously and were spread over a broader range of scientific fields. At the same time, the reorganized Rockefeller Foundation continued, albeit on a reduced scale, the IEB's previous practice of granting and administering fellowships for the most promising European scientists. At least through 1931, then, the new Division of Natural Sciences had been unable to establish a working alternative to Rose's best science policy, which continued to be applied, but on a less enthusiastic and aggressive scale.[7]

For our purposes, the support in 1930 of George Hevesy's Institute for Physical Chemistry at Freiburg University constitutes a particularly relevant example of the kind of grants supplied by the Rockefeller Foundation's Division of Natural Sciences during its early years. After a preliminary approach in March 1930, Hevesy submitted a detailed application to the Rockefeller Foundation for new apparatus for his institute. The variety of work for which he sought support was reflected in the headings of the application's four chapters: "The problem of nuclear stability approached from considerations of the abundance of the elements"; "Transport of matter in crystalline bodies (Diffusion and electrolytic conduction)"; "Radioactivity of potassium and rubidium"; "Biological problems." Toward the end of the application, Hevesy wrote that

> . . . a number of other researches are in progress and contemplated dealing with the chemistry of hafnium and of the rare earth elements, with the electro-chemistry of molten salts and the crystal structure of oxides . . .

The variety of projects testifies not only to Hevesy's approach to research; it also shows that the Division of Natural Sciences continued to encourage general purpose applications from well-reputed laboratories just as the IEB had done in the 1920s.[8]

The handling of Hevesy's application also exemplifies the unsettled state of Rockefeller support for natural science during this period. At the end of September 1930, Hevesy arrived in New York on the way to Ithaca to give the George Fisher Baker Lectures in chemistry at Cornell University. Before his departure from Freiburg, Lauder Jones at the Rockefeller Foundation's Paris office had advised Hevesy to look up Herman Spoehr, who just at this time was taking over as natural sciences director, in New York. Jones thought such an action wise, even though he had "not yet been able to present the case thoroughly to Doctor Spoehr." Evidently, the inconstant leadership hampered communication between the New York and Paris offices of the Rockefeller Foundation, and made it difficult to carry out an effective policy.[9]

A meeting with Spoehr was arranged on 29 September, and the proposal was accepted two and a half months later, while Hevesy was still in Ithaca. As had been common for the general purpose grants provided previously by the IEB, the support was matched by support from German sources. Obviously, Mason's plea for greater concentration, expressed only two months before Hevesy's application was accepted, could not by itself change the main course of Rockefeller funding of natural science. From 1929 to 1932, Rockefeller support for natural science suffered from lack of direction.[10]

EMERGENCE OF A NEW POLICY

In the fall of 1931, President Mason was able to persuade Warren Weaver to become natural sciences director of the reorganized Rockefeller Foundation. Born in 1894, Weaver had received his undergraduate training in physics at Wisconsin, where, according to his own recollection, Mason was his most respected and important teacher. Having worked under Robert Andrews Millikan at the California Institute of Technology (Caltech) from 1917 to 1919, Weaver returned to join the Wisconsin physics faculty, where he spent substantial time and energy developing a textbook, *The Electromagnetic Field,* with his former teacher Max Mason. Their book, which was first published in 1929, proved to be influential; it was subsequently used for teaching at the undergraduate level for some twenty years. Clearly, Mason and Weaver had been close and successful collaborators even before either of them went to the Rockefeller Foundation.[11]

Both Mason and Weaver rejected the new quantum theory, which during those years was establishing its position as the very basis of modern physics, and which was taken up enthusiastically by the vanguard of American physicists. Yet their decision to give up active research for administration seems to have been unrelated to their attitude toward the contemporary development of physics. According to his own reminiscences, Weaver proposed during his first interview in New York that biology, not physics, ought to serve as the future field of concentration

Warren Weaver, natural sciences director of the Rockefeller Foundation in the 1930s in a photograph taken in 1953. [Courtesy of Rockefeller Archive Center]

for Rockefeller support of natural science. He was appointed by the Rockefeller Foundation soon after.[12]

In April 1932, less than three months after taking up work in New York, Weaver set out, together with the more experienced Lauder Jones at the foundation's Paris office, on a grand tour to visit "essentially all of the university centers of western and south-western Europe." Twenty-five years older than Weaver, Jones had obtained his doctorate in chemistry from the University of Chicago in 1897, and had been a chemistry professor at Princeton University since 1920. As Trowbridge's successor, he had served as Associate Director of Natural Sciences at the Paris office since 1929. Weaver and Jones made the same kind of extensive tour as Wickliffe Rose had almost eight and a half years earlier when he developed the program of the newly established IEB. Because there had been little time to design a detailed program for the division's work, the purpose of the trip was to obtain a general overview of the situation rather than to evaluate specific requests for support.[13]

As described in Chapter 1, Rose granted his first institutional support to the institute even before setting out on his exploratory trip to Europe. Although not quite as generous, Weaver and Jones did choose the institute as their first stop in Copenhagen. The stop was made even though Bohr had already informed them that he would return from a sailing trip only after their arrival. According to Jones's diary, "vacation had commenced, Professor Bohr was absent and the laboratories closed." The eagerness to visit the institute, despite its director's absence, highlights the Rockefeller philanthropies' continued special interest in Bohr.[14]

A letter from Bohr written almost nine months later shows that a meeting between Bohr and Weaver indeed took place during Weaver's stay in Copenhagen. In this letter, Bohr expressed his wish to continue their discussions during his forthcoming visit to the United States. He enclosed "a few articles dealing with the general physical problems we discussed in Copenhagen." There is no indication that Bohr and Weaver discussed biology at their first encounter, even though the meeting coincided with Bohr's making a decision on whether or not to give a biologically oriented lecture to the International Light Congress. This suggests both that the Rockefeller Foundation's funding program in biology was only in its beginning stages, and that Bohr was attractive as a recipient of support by virtue of his standing in physics alone.[15]

Only after returning to New York did Weaver complete his first draft for a new funding program. As major categories he proposed the mathematics, physics, and chemistry of "vital processes" as well as of the earth and the atmosphere. To this list he added genetic biology and quantitative psychology. As minor areas of concentration he suggested fundamental physics and chemistry of matter, along with probability theory and statistics.[16]

The Rockefeller Foundation's annual report for 1932 did note that Weaver had taken over as director of the Division of Natural Sciences. In view of the slow developments it is not surprising that there was no indication of a new funding policy. Introducing his report on natural science funding, Weaver simply noted:

> During 1932 the work of former years was continued. The main features of the program were aid to a number of institutions and organizations, to individual research projects, and to a system of fellowships, travel grants, and grants in aid of research.

The detailed contents of the annual report confirms Weaver's statement.[17]

In his draft for a new policy and in subsequent refinements, Weaver was naturally constricted by the more general goals of the reorganized Rockefeller Foundation, which were also in the process of being reformulated. Thus, a staff meeting held in mid-March 1933, almost a year after Weaver's first trip to Europe, proposed

> . . . two major categories of operations and functions: (1) Conscious control of race and individual development with rather particular reference to mentality and temperament; and (2) study and application of knowledge of social phenomena and social controls.

The first of these items was meant to represent the concern of the Divisions of Medical and Natural Sciences; the Divisions of Social Sciences and Humanities were to deal with social and cultural problems, respectively. Rhetorically, at least, the Rocke-

feller Foundation was striking out in a new direction quite different from that of the IEB. In particular, Rose had been entirely
negative about supporting the social sciences.[18]

The staff meeting in March was held to prepare for the meeting of the foundation's trustees on 11 April 1933. This meeting
was a milestone in the Rockefeller Foundation's formulation of
a new funding policy. The foundation's officers presented an
elaborate report in which they carefully laid out the goals of the
reorganized foundation in the explicit context of the history of
the Rockefeller philanthropies' past efforts in the relevant fields.
In this report, Weaver gave a detailed presentation of his program in relation to the general funding policy of the reorganized
Rockefeller Foundation.[19]

In order to present the general aims of the foundation, President Mason had written a "proposed future program" for the
report. Here, he characterized the past, present, and future contributions of the foundation as "promoting procedures in the rationalization of life." After reiterating the Rockefeller Foundation's role as serving the progress of mankind, Mason concluded
that it

> . . . would be as unwise for the Foundation to relinquish the
> support of the fundamentals on which in the long run control
> of man's destiny depends, as it would be to remain indifferent
> to the pressing problems of today.

This passage states clearly the major difference between Mason's Rockefeller Foundation and Rose's IEB. Despite his own
personal feelings, Mason had to strike a delicate balance to satis-
fy the expectations of his foundation's trustees.[20]

After his general introduction, Mason became more specific,
presenting three arguments for restricting the range of interests
of the foundation. First, its limited resources would be effective
only if "applied over a narrow front." Second, administrative
efficiency would increase if the staff could "reasonably be expected to reach a high competency" in the field to be funded.
Third, the foundation should use its independence to "choose
just those difficult and critically important fields which other
agencies must or do neglect." These statements amounted to an
elaboration of the plea for concentration and specialization that
Mason had uttered in October 1930, immediately following his

appointment as president. As may have been expected, the statements also amounted to a clear departure from Rose's concept of basic science funding as promoting general higher education. Specialized scientific results rather than general educational values were to serve as the criteria for funding.[21]

Mason suggested that the specific fields of concentration should comprise "the general problem of human behavior, with the aim of control through understanding." In conformity with the statements from the staff meeting the month before, he wrote that the Division of Social Sciences should be concerned with "the rationalization of social control," whereas the Divisions of Medical and Natural Sciences should approach the problems underlying "personal understanding and personal control." Echoing the concern about unwieldiness that had motivated the reorganization of the Rockefeller philanthropies in the first place, he emphasized the "structural unity" of the total program, clearly implying that no costly and unnecessary competition for projects and for individual autonomy among the divisions would be tolerated. Such an intention was also reflected in Mason's suggestion that the allocation of funds to the different sciences was to be proposed by the officers of the different divisions in cooperation. In short, Mason demanded that all activities of any individual division comply strictly with the foundation's general purpose and guidelines.[22]

Weaver molded his report on the natural sciences around Mason's general schemes. Whereas in previous natural science funding, Weaver wrote, the quality of scientists and institutions had counted more than particular fields of interest, from now on these priorities would be reversed. He went on to define and explain his special choice of concentration:

> The welfare of mankind depends in a vital way on man's understanding of himself and his physical environment. Science has made magnificent progress in the analysis and control of inanimate forces, but science has not made equal advances in the more delicate, more difficult, and more important problem of the analysis and control of animate forces. This indicates the desirability of greatly increasing emphasis on biology and psychology, and upon those special developments in mathematics, physics and chemistry which are themselves fundamental to biology and psychology. Similarly, it is desirable to

emphasize those studies of the earth, sea, and air, which furnish information concerning the physical background for the development of man.

Weaver implied that atomic physics and astronomy were of no concern to the Rockefeller Foundation, thus abandoning these minor items of his previous proposal. Moreover, he now presented "earth science" as a field of minor interest compared with biology. Consequently, Weaver's views conformed more closely than before with Mason's general scheme. In the process, his proposed program had become more focused, allowing less of a promotion of science for its own sake.[23]

Whatever pressure Weaver may have felt from the foundation's trustees, however, it is clear from his statements that he at least indirectly assigned the physical sciences a crucial role in his scheme. Not only did he assume that the physical sciences had developed to a more advanced stage than the life sciences; he also believed that the salvation of biology lay in the introduction of concepts and techniques from the more advanced physical sciences. Historians of the Rockefeller philanthropies disagree as to whether Mason and Weaver's approach reflects the two physicists' desire to "colonize" biology, or a genuine interdisciplinary sentiment rooted in their shared experience at the University of Wisconsin.[24]

When arguing for the need to introduce physical concepts and techniques into biology, Weaver relied on the statements of prestigious scientists. In his report, for example, he quoted three out of "some twenty of the outstanding experimentalists of the world in biology, biophysics, physical chemistry, etc." who had been solicited by a "biological institute [i.e., the Rockefeller Institute], wanting advice as to the most fruitful field to develop." In his argument for providing support for the life sciences, Weaver, like Rose before him, viewed physics as the most fundamental science. Also like Rose, he thought that the scientists themselves – especially the most prestigious ones – could be trusted for the best advice on these matters.[25]

In accordance with the substantial activity within the foundation to define a new approach to philanthropy, the Rockefeller Foundation's annual report of 1933 did proclaim a change of policy. This proclamation of general change figured much more prominently than the announcement of the Special Research Fund for European Scholars, which was also made in this report. In his brief foreword, President Mason restated some of the points he had made previously, explaining the change in terms of the "economic, social, and political stress in many parts of the world, giving rise to pressing problems of national and international scope." In his section on natural science funding, Weaver introduced for the first time two "programs of specific concentration"; in his general introduction to the section he presented "vital processes" as the principal program, with earth science" chosen as a "modest complement."[26]

Weaver had designed the substance of his section to bear out this statement. Thus, seven grants were presented within the program of vital processes, the three largest ones going to Thomas H. Morgan at Caltech; to a large group at the University of Chicago led by Frank Rattray Lillie – the older brother of Ralph, who also profited from the grant; and to the Committee for Research in Problems of Sex of the National Research Council (NRC), which oversaw sex research in more than sixty laboratories. Although all these grants came within the purview of vital processes, the latter two were continuations of support begun before the new program had been defined. Moreover, the program in earth science provided only one minor grant. In all, the so-called programs of specific concentration received $244,300, which included small amounts for grants-in-aid to already ongoing projects in vital processes and earth science. By comparison, the Special Research Aid Fund for European Scholars, which paid for Franck and Hevesy's stay at Bohr's institute, received $100,000. New grants under the rather self-contradictory rubric of the "former program" amounted to $372,950, while what was called the "general program," which consisted of traditional fellowships and grants-in-aid for research, took up $200,000 of the Rockefeller Foundation's budget. It should also be borne in mind that with the exceptions already noted, all these grants were new appropriations; in addition, the foundation continued to pay for ongoing projects that were outside the new program and that had been started before the new funding policy had crystallized. As yet, the Rockefeller Foundation had certainly not fully implemented its new funding policy; at best, it was at the beginning of a transition period.[27]

Bohr's visit to the United States in the summer of 1933 took place soon after Mason and Weaver had presented their ideas for a new funding policy to the Rockefeller Foundation's trustees. As indicated in Chapter 3, during this visit Bohr focused his main attention on the problem of the refugee scientists from Germany. In view of the ongoing discussions within the Rockefeller Foundation, however, it was only natural that the question of support for biology was brought up at Bohr's meeting with Warren Weaver on 10 July. According to Weaver's diary, Bohr suggested the possibility of bringing together at the institute some

> . . . able and thoroughly trained young men in mathematics, physics or chemistry, who under B[ohr]'s direction turn their attention to some quantitative phase of important biological problems.

Obviously, this vague statement was not intended as a definitive request for funding. Despite its vagueness, however, it did reflect the Rockefeller Foundation's concern with providing a secure experimental and quantitative basis for biology. One can only guess to what extent Weaver's diary entry reflects a conscious strategy on Bohr's part to keep potential funding channels open, or a purposeful attempt by Weaver to recruit a scientist of Bohr's stature for his foundation's new program in biology.[28]

The latter interpretation is supported by Weaver's continued reliance on prestigious scientists for an evaluation of his program. Such a reliance, which Weaver had expressed in his statement to the trustees of April 1933, was particularly evident when Weaver met with the illustrious Raymond Fosdick in March of the following year. Fosdick was heading a "committee of appraisal" recently appointed by the trustees to consider whether new "sailing directions" for the foundation were required. Weaver's vigorous argumentation for his cause at this meeting testifies to his reliance on the opinion of prestigious scientists, as well as to his continued sense of a need to assert his case within the foundation.[29]

In his memorandum prepared for the meeting, Weaver bolstered his cause by quoting from ten of the twenty letters from outstanding scientists that had been solicited by the Rockefeller Institute a year earlier. The first of these was from Winthrop

John Vanleuven Osterhout of the Rockefeller Institute itself. Osterhout emphasized the importance of getting physicists and chemists interested in biological problems in order to make biology a truly quantitative science. He noted that a particularly successful outcome of such an approach was the method of measuring the conductivity of blood corpuscles developed by Hugo Fricke.[30]

Fricke was a Dane who had received his physics education at Copenhagen University in the first decade of the century, after which, on Bohr's recommendation, he had gone to the United States in 1919 to specialize in experimental spectroscopy. Bohr hoped in this way to obtain a well-trained experimentalist as a collaborator at the institute. Fricke, however, never returned to his homeland. Instead, he became a prominent researcher in the United States into questions on the boundary between physical and life science, contributing, in particular, to radiation chemistry. The experience with Fricke made Bohr aware at an early stage of the interconnections between atomic physics and biology.[31]

It is doubtful that Osterhout or Weaver knew about Fricke's connection with the institute. Even so, Weaver found the case of Fricke to be important enough to use it as an argument for his funding program.

In Weaver's own words, the second letter quoted came from "A[rchibald] V[ivian] Hill, Physiologist of the University of London and Nobel Prize Laureate"; Hill, incidentally, shared his 1922 Nobel Prize in medicine or physiology with Otto Meyerhof. In his letter, Hill expressed enthusiasm for

> . . . the application of exact quantitative methods, physical and chemical, in biology: not in the direction of making models of supposed biological processes, but in the careful investigation of actual living phenomena.

He insisted "that biology needs the *best* brains in physics and chemistry, not the second best."[32]

The third letter came from August Krogh in Copenhagen, whom the Rockefeller Foundation had supported in the 1920s as part of the effort to develop physiology at Copenhagen University. Krogh could not

. . . doubt for a moment that for the progress of biology as a measuring science the closest possible contact with chemistry, physics and certain branches of mathematics is absolutely essential.[33]

Views similar to those of Osterhout, Hill, and Krogh also appeared in the remaining seven letters "from a world of leaders in biology." The Rockefeller Institute had thus succeeded admirably in obtaining expert support for its own – and Weaver's – views. As shown, in particular, by his introduction of A.V. Hill, Weaver found it decisive that the arguments were formulated by some of the most prestigious scientists.[34]

As if to emphasize this point, Weaver added in his memorandum four more quotations from prominent biologists. He wrote to Fosdick that these statements had "more or less accidentally come to my attention during the last months." In fact, Weaver had collected them from various readily available published sources.[35]

The shortest of the quotations, which came from a monograph by the endocrinologist Roy Graham Hoskins and from a *New York Times Magazine* article on the recent election of James Bryant Conant as president of Harvard University, emphasized the value of bringing chemical laboratory techniques into biological science. Two longer quotations were taken from a voluminous popularization of modern biology by Herbert George Wells, Julian Sorell Huxley, and George Philip Wells, and from the biochemist Frederick Gowland Hopkins's 1933 presidential address to the British Association for the Advancement of Science (BAAS). Hopkins, in particular, regretted the strong lead physics and chemistry had taken over the life sciences. More importantly, in both publications Weaver found arguments for his program that were entirely in line with the broader goals of the reorganized Rockefeller Foundation; both stressed the need for an improved biology to control man and ensure progress in an increasingly complex society.[36]

Taken together, then, the four quotations provided a case for the principal program of Weaver's two programs of specific concentration on its own merits as well as from the general perspective of the Rockefeller Foundation. In addition, Weaver could not help pointing to the prominence of the scientists as a further argument for the validity of their views. Thus, he was particularly elated because

> . . . perhaps the most extreme and positive view is expressed by Sir Frederick Gowland Hopkins[, whose] position at Cambridge as well as his simultaneous presidency of the Royal Society and of the British Association for the Advancement of Science surely testify to his scientific competency.[37]

From his appointment as natural sciences director, then, the physicist Weaver had been developing a funding program with particular emphasis on the cultivation of biology with physics, chemistry, and mathematics. While asserting that his program was in line with the general goals of the reorganized Rockefeller Foundation, he nevertheless found it advantageous to quote prestigious scientists in order to argue for his scheme within the foundation. In this sense, Weaver retained one element from Rose's best science approach. Yet Weaver's emerging policy, unlike that of Rose, was clearly directed toward specific scientific goals. Hence, the Rockefeller philanthropies' funding policy for basic science no longer harmonized with the freewheeling Copenhagen spirit at the institute. We will now turn to how the Copenhagen scientists first reacted to Weaver's new program.

THE NEW POLICY MEETS COPENHAGEN SCIENCE

The current state of the Rockefeller Foundation's new funding program was presented to the Copenhagen scientists when Wilbur Earle Tisdale and David Patrick O'Brien from its Paris office visited from 8 to 11 April 1934. Tisdale, who was just eight days older than Bohr, obtained his PhD in physics from the University of Iowa in 1915. He had worked for the National Research Council's fellowship program from 1921, before moving to Paris and becoming administrator for the IEB's fellowships in 1926. O'Brien, who was nine years younger than Tisdale and Bohr, had received his MD from the Johns Hopkins University in 1920 and had worked as a fellow at the Johns Hopkins Hospital before becoming assistant at the Rockefeller Foundation's Division of Medical Education in 1926. In 1929, he became the foundation's Assistant Director of Medical Sciences.[38]

During their visit, which coincided with Franck's arrival in Copenhagen, Tisdale and O'Brien were prepared to discuss specific proposals for projects to be funded by the Rockefeller Foundation's new funding program in biology. Meeting with the two, Bohr expressed his general desire to "get in touch with biologists, and biologists in touch with him." To this request, Tisdale replied that young biologists visiting the institute, as well as physicists from Copenhagen visiting biological research centers, could most probably be supported as part of the fellowship system; this system was being retained on the same basis as before, and was entirely separate from the new funding program in biology. Similarly, he indicated that Bohr's wish to visit biological centers, "and even perhaps have some outstanding biologists visit him for discussion," could be funded from outside the funding program in biology.[39]

As Tisdale noted,

> . . . Bohr envisages a similar activity to that which he carries out in theoretical physics and, without commitment, I told him that we could finally make an arrangement, if a definite plan seemed to be desirable.

Thus, Bohr proposed an activity emphasizing learning by discussion rather than a well-defined research project in biology. Bohr's interest in biology, even as expressed to the Rockefeller Foundation officers, still corresponded to the Copenhagen spirit of open-ended scientific debate. While listening with interest to Bohr's expression of general interest, Tisdale and O'Brien found it too vague to merit full-scale support from the principal program – subsequently to be called "experimental biology" – of the program of specific concentration.[40]

At the meeting with Tisdale and O'Brien, Bohr even expressed distrust in his own biological judgement. He thus thought it desirable to talk to "Morgan and certain other biologists, perhaps including Haldane," before committing himself to a project. Bohr had become acquainted with the noted geneticist Thomas Hunt Morgan during his visit to the United States the year before. In their subsequent account of the meeting, Tisdale and O'Brien expressed agreement with Bohr's unpresumptuous self-assessment as a biologist. Bohr, they wrote, "worked up a very real enthusiasm but it was based on a very meager knowledge of biology." Evidently, the two foundation officers were not impressed with Bohr's biological expertise.[41]

In spite of this rather severe judgement and the vagueness of Bohr's proposals, both Tisdale and Weaver were enthusiastic about Bohr's interest in biology, and eager to see it pursued further. Accordingly, after learning in Stockholm that Morgan was going to present his Nobel Prize Lecture there during the first week of June, Tisdale forwarded this information to Bohr "on the almost impossible chance that you have not learned of it from other sources." Moreover, reporting his meeting with Bohr to Weaver in New York, Tisdale informed his superior that Weaver's discussions with Bohr the year before had "apparently planted a seed which [had] germinated and produced some off-shoots." Indeed, Tisdale wrote, if Bohr's

> . . . enthusiasm increases as much after his talk with Morgan as it did after his talk with D.P. O'B[rien] and me, I can imagine that he will not want to wait to propose entering seriously into the question until he shall have returned from Japan in the summer of 1935.

The visit to Japan referred to by Tisdale was part of a planned trip around the world that actually took place only in 1937. The Rockefeller Foundation officers, then, not only approved of Bohr's biological interest; they actively wanted him to submit an application in line with the foundation's new general approach. As might have been expected from Weaver's way of promoting his program, Bohr's prestige was evidently still crucial. With the new project-oriented policy, however, it was not a sufficient condition for support.[42]

In his response to Tisdale's report of the meeting with Bohr, Weaver made the continued importance of Bohr's distinction explicit. In his report, Tisdale had written that if Bohr was helped, both Bohr and the Rockefeller Foundation must expect criticism from "biologists for thinking that a mere physicist or mathematician can add anything to biological disciplines." He noted that he had heard such criticism "very gently proposed" already, but emphasized that it had "no deterrent effect" on him. As a direct response, Weaver replied:

> It is undoubtedly true that some of the old line conservatives would consider this proposed development [of economic sup-

port for Bohr's biological interest] somewhat askance. On the other hand, B[ohr]'s outstanding position will probably protect us from anything that would reach embarrassment.

In effect, the natural sciences director implied that Bohr's scientific authority was at least as important as the validity from a biological point of view of a proposed program. Weaver's reliance on scientists' prestige was stronger than ever. This reliance, however, seems not to have been successfully communicated to Bohr, who continued to favor general problems over well-defined projects, and balked at posing as a biologist.[43]

Nevertheless, Bohr sought to maintain contact with the Rockefeller Foundation even after the visit of its officers. Only a few days later, Bohr wrote to Tisdale that he

> . . . was most interested to learn about the new great schemes of the Rockefeller Foundation. Surely it shall be a pleasure to me if I can in any way assist in making the theoretical insight required through the studies of atomic theory fruitful in the field of biology, and some of your proposals in this respect are very tempting indeed. When I have thought more of the matter I shall write you again.

Clearly, Bohr wanted to keep his options open. His conscientiousness in formulating a project only confirms the Rockefeller Foundation officers' difficulty in communicating to Bohr their interest in him. This communication problem would only disappear slowly, and would remain evident in the course of continued deliberations between Bohr and the Rockefeller Foundation even after Bohr had submitted a well-defined proposal.[44]

While in Copenhagen, Tisdale and O'Brien also had an extensive interview with August Krogh, director of Copenhagen University's Zoophysiological Laboratory, which had previously been constructed with support from the Rockefeller philanthropies. The Rockefeller Foundation officers noted that Krogh's work was strictly physiological, and as such did not fit into the foundation's new schemes. There is no indication that the Rockefeller Foundation was interested in funding Krogh's work. Moreover, when Krogh's assistant Paul Brandt Rehberg asked whether the foundation would be willing to contribute a grant so that he could relieve Krogh of teaching – which Krogh strongly disliked – the response was immediate and negative.[45]

At the end of the interview with Krogh, Tisdale and O'Brien confronted him directly with the Rockefeller Foundation's real interest, posing a question similar to that which the Rockefeller Institute had asked him before:

> Krogh was asked if he had any suggestions as to how [to] bring the discipline of chemistry and physics more effectively into biology. He thinks it a very important question but prefers to write or talk later about it after further deliberation.

Characteristically, the Rockefeller Foundation officers left it to Krogh to formulate a specific project. Even though his letter to the Rockefeller Institute had expressed sympathy for the application of physics and chemistry to biology, Krogh, like Bohr, was unable immediately to come up with a specific proposal. His response, in fact, suggests that he was totally unprepared for the request. By asking for more time, Krogh showed that he was positive in principle to the Rockefeller Foundation's ideas. Naturally, he too wanted to keep the channels to possible Rockefeller funding open.[46]

Meanwhile, Hevesy was preparing for his move to Copenhagen. Indeed, as Tisdale and O'Brien noted in their diary, Hevesy was "in Copenhagen for a few days in order to find an apartment and to otherwise get oriented for his stay." Only months before, Hevesy had learned that the support granted him in Freiburg in late 1930 probably could not be transferred. This attitude toward Hevesy's support contrasted sharply with the Rockefeller Foundation's statements of only a year before, which spoke approvingly of Hevesy's slow and careful expenditure of his grant – of which as much as $17,000 remained. By mid-October 1933, however, Lauder Jones, on Hevesy's inquiry, could promise no more than "a possibility of . . . a very limited sum for very highly specialized equipment not available in Brønsted's Laboratory." The following day, Hevesy reported to Bohr that Jones had also used Bohr's application for research expenses for the refugee physicist James Franck as an argument against a similar grant to Hevesy. In the same letter, Hevesy reported that prospects for transferring the remaining $17,000 of his Freiburg grant were scant. In effect, then, the foundation used Hevesy's move from Freiburg as a means to invalidate a grant based on a funding policy that had since been discarded.[47]

Tisdale and O'Brien made use of Hevesy's presence in Copenhagen in April 1934 to interview him. With obvious reference to the Rockefeller Foundation's reluctance to transfer his Freiburg grant, Hevesy complained that the lack of equipment in Copenhagen would prevent him from carrying on his x-ray research, which had been his major concern in Freiburg. As an alternative, Hevesy suggested that the recent discovery of induced radioactivity might have biological implications. He expressed hope for more fruitful cooperation in this area in Copenhagen than in Freiburg. Tisdale and O'Brien's report of the interview continues:

> Due to the new developments in physics whereby nearly any metals can be made radioactive for short periods, [Hevesy] hopes to be able by these indicators to trace their distribution in tissue without subjecting these tissues to the dangers of radioactive emanations which exist with the more stable forms of radioactive compounds.

Characteristically, Hevesy sought to benefit from new developments in experimental nuclear physics in order to apply his radioactive indicator technique further to biological questions. However, despite Hevesy's optimistic assessment of the situation, a suitable radioactive isotope of central biological importance was yet to be found, and no publications were to come out of the specific suggestions made by Hevesy in the interview.[48]

On the other hand, although the diary entry does not mention it, Hevesy by this time had begun using heavy hydrogen as a nonradioactive indicator. Hence, although his assessment of a specific line of research may have been overly optimistic, the more general tenor of his argument was promising. Moreover, his call for a specific experimental program was more in line with the Rockefeller Foundation's funding program in biology than anything hitherto proposed by Bohr or Krogh.

Unlike his two Danish fellow scientists, however, Hevesy was not encouraged to follow up his ideas. In fact, Tisdale and O'Brien referred in their diary to Hevesy's statements as mere matters of fact, and did not relate them to the funding program in biology at all. In late April, moreover, Tisdale wrote to Weaver in New York expressing a favorable attitude toward limited research aid "even though [Hevesy's work] is not within [the biology] program.'' He also observed that Hevesy's Freiburg appropriation would lapse despite Hevesy's explicit wish to the contrary. Evidently, the Rockefeller Foundation officers saw general statements of interest coming from established institute directors such as Bohr and Krogh as more important than the more concrete project proposal coming from the recently uprooted Hevesy.[49]

CONSOLIDATION OF THE NEW POLICY

When the Paris officers visited Copenhagen in April 1934, the Rockefeller Foundation's natural science funding policy was still being formulated. For example, the policies toward natural and medical science were still seen as being intimately related. In an officers' report to the trustees in December 1933, the two programs had even been presented as one single unit. Although they approved the content of the report, however, the trustees still found it necessary to appoint its committee of appraisal.[50]

Fosdick's committee was helped in its considerations on natural science by a subcommittee of experts, which in Fosdick's view only proposed irrelevancies. Only one of the subcommittee's suggestions would have lasting effect. On its suggestion, the biological funding program in the natural sciences was renamed from "vital processes" to "experimental biology."[51]

Fosdick found more help in the letter he had solicited from David Edsall – dean of the Harvard Medical School and an active Rockefeller Foundation trustee supporting Weaver's program. Referring specifically to some disheartening aspects of the medical sciences program, Edsall warned against pushing too strongly in specific directions in an exaggerated hope for quick success. Fosdick related Edsall's criticism to Weaver, without, however, seeking to restrict either his or other officers' power to make their own decisions. Instead, he urged the directors to use their own critical judgement to greater effect. As a result, the relationship between the Divisions of Natural and Medical Sciences became looser, and Weaver could concentrate more fully on his own goal of introducing mathematics, physics, and chemistry into biological research.[52]

In line with Weaver's proposed program, Fosdick's committee defined experimental biology as "the application of experimental procedures to the study of the organization and reactions of living matter." Moreover, the report recommended scrapping the complementary program in earth science previously proposed by Weaver. This recommendation, which was carried out subsequently, added to the specificity of Weaver's program. By the end of 1934, then, when the committee presented its final report, the Rockefeller Foundation had clarified its aims for natural science funding substantially.[53]

The Rockefeller Foundation's annual report for 1934 provides an impression of what the changes meant in practice. As he had also stated in his foreword the year before, Mason emphasized the continued narrowing of the foundation's efforts, due in part to the economic situation. In his section, "The Natural Sciences," Weaver reiterated the need to specialize. He even announced that the foundation's European fellowship and research aid program, which traditionally had been generally oriented, in the future would be wholly devoted to what Weaver now termed experimental biology. He presented this program as the only program of specific concentration; earth science had been dropped. The items funded in the program had increased in number and decreased in size; while there were twenty-one such grants, the average amount was reduced to just over $13,500. In accordance with the increasing emphasis on projects in specific research areas, the grants were categorized according to topics: genetics, physicochemical biology, physiology and endocrinology, quantitative biology. The NRC's Committee for Research in Problems of Sex received the largest appropriation ($80,000), and a few institutions continued to receive funding for research instituted before the experimental biology program had been defined. These included the University of Uppsala in Sweden, the only grant recipient outside the American mainland; the University of Chicago; and Caltech, where research was conducted under Linus Pauling.[54]

Nevertheless, such continuity was the exception rather than the rule. One of the largest grants went to research on the biological effects of heavy hydrogen under Harold Urey at Columbia University in New York City. According to the annual report, the program of specific concentration, which included experimental biology and quantitative biology, as well as fellowships and research aid grants for 1934 and 1935, received $528,850. By comparison, what was still termed the "former program" received $93,600. Although amounting to a full $428,760, the "general program" was for the most part supporting scientific publication and not research itself. Both in absolute and relative terms, the experimental biology program had become substantially stronger and well-defined.[55]

The financial crisis experienced by the Rockefeller Foundation in 1933 and 1934 contributed to the development of a more independent and specific funding program in the natural sciences. Weaver's original plan had been to supply institutional grants and fellowships that did not require detailed administration and know-how on the part of the foundation officers. In the new economic situation, however, such an approach was considered extravagant. Now the majority of support would go to specific projects, which meant that the officers *were* required to demonstrate some scientific expertise; these grants were seen as more expedient than the larger institutional grant, which was considered to be too expensive, and the smaller general purpose aid, which tended to become too scattered.[56]

The specialization continued in 1935, when the appropriations for experimental biology projects nearly doubled to $992,950, whereas the Special Research Aid Fund for European Scholars amounted to a mere $14,675. The "general program" no longer appeared as a rubric on the list of appropriations. Although somewhat more than $1 million was granted under the "former program," a full million of this amount was spent to terminate one particular project.[57]

The 1934 report of Fosdick's committee of appraisal noted explicitly that the change of emphasis from an "extension of the pure sciences . . . [to] a concentration within each field on certain specific objectives . . . in a sense reversed the program of Dr. Rose." Robert E. Kohler, the foremost historian of the Rockefeller philanthropies' support of natural science, has perceptively described Weaver as the first large-scale manager of natural science, and the program itself as a pioneering attempt to guide basic science research in preferred directions. Rose's funding policy, which had been based on educational ideals, was definitely being supplanted by a policy seeking to produce particularly important scientific results. For all practical purposes, by 1935 it seems that the transition to a funding program concentrating on experimental biology was completed.[58]

Whereas in 1934 only one of the experimental biology grants had gone to Europe, in 1935 as many as nine of the thirty-six appropriations were made to European institutions. Europe's relative share of the appropriations remained at this level until the Second World War. The very first grant in experimental biology described in the 1935 annual report, which was also the third largest, went to Bohr's institute in Copenhagen. Because the work embodied in the Copenhagen spirit was antithetical to project-oriented research, and was furthermore devoted to physics rather than biology, this may seem surprising. Indeed, with the possible exception of Urey, Bohr and his institute had little or no connection with any of the many other individuals or research institutes supported by the Rockefeller Foundation's experimental biology program. One might think, therefore, that by now Bohr would have considered the Rockefeller Foundation's new funding program in biology a less likely source than ever for support of his own scientific activity, including his biological interest. Nevertheless, he decided to appeal to the experimental biology program for substantial economic support to the institute. We will now consider how this happened.[59]

THE COPENHAGEN EXPERIMENTAL BIOLOGY PROPOSAL

As we have seen, Tisdale and O'Brien's visit to Copenhagen in April 1934 coincided with Franck's arrival. The next recorded meeting between Bohr and the Rockefeller Foundation officers occurred a few months after the project led by Franck and Hevesy had come into being. On 29 October, three days after Hevesy wrote his negative response to Fermi's request for rare earth samples, and on the same day that Franck wrote to his friend Max Born about his limited progress in experimental nuclear research, Tisdale interviewed Bohr in Copenhagen. Since their last meeting in April, a watershed had occurred in Bohr's thinking about the experimental biology program. Tisdale's extensive diary notes read:

N. BOHR and I discussed again the direction his thinking had taken since my talk with him in April on the possibilities of bringing physics or theoretical physics disciplines into biology. At that time Bohr's interests were mainly on the philosophical side, but in recent months he has come to feel that it may be possible to do more effective work at the present time in connection with definite problems. Due to the presence of VON HEVESY in Copenhagen, Bohr has decided to undertake a cooperative problem with KROGH, VON HEVESY, and himself. They now have an arrangement with the Radium Institute [i.e., the Radium Station] for the loan of a small fraction of a gram of radium at intermittent periods, but this has proved so unsatisfactory an arrangement that lacking the possibilities of a better cooperation with the Radium Inst., B[ohr] has decided to install a high-tension equipment for the production of artificially stimulated radioactivity. With the elements so made radio-active, Krogh, von Hevesy and Bohr are to enter upon the cooperative problem of the identification of the processes with which these elements are concerned. Krogh has dropped all of the work which he told us in April he was going to carry on and shows a remarkable enthusiasm for the cooperative possibilities with v.Hevesy and Bohr. . . . V.Hevesy is naturally elated, and Bohr tells me that after the spring he hopes to give more and more of his time to these problems of biological interest. . . .

They propose to begin their work with heavy hydrogen and then sodium chloride, using first artificially prepared membranes, and then probably the most impermeable of body membranes, the bladder, eventually to come to cell walls.[60]

Obviously, the Copenhagen scientists had made a serious effort to coordinate their previously disparate suggestions to the Rockefeller Foundation for work in biology. Bohr, the scientist whom the foundation was most eager to secure for its experimental biology program, had turned completely from a philosophical and discussion-oriented approach to biology to an applied and project-oriented one. He now formulated a proposal that would fit easily into the Rockefeller Foundation's new program. In the process, he prepared the ground for substantial funding for biological research by adding his authority to Hevesy's plans for expanding physical techniques to biological problems; in addition, the proposed project required precisely the kind of expensive high-voltage – or "high-tension," as it was called at the time – accelerator equipment that was necessary for a full-scale research effort in nuclear physics. Indeed, as we saw in Chapter 3, Hevesy, with Bohr's sanction, had consulted with the British Metropolitan-Vickers Company about the acquisition of such equipment half a year before.

It seems that Bohr saw the Rockefeller Foundation's new funding program as a means to realize his ambitions for nuclear physics at the institute. Bohr's radical turnabout in his approach to the experimental biology program from April to October 1934, then, substantiates the conclusion in Chapter 3 that his decision to redirect the institute to nuclear physics research took place soon after the arrival of James Franck. In conformity with his distrust of his own biological judgement, Bohr let Krogh's permeability problem provide the framework for the proposed project. With Bohr's proposal, the sundry experimental biology interests of the Copenhagen scientists had been coordinated – with Bohr providing general leadership, Hevesy the experimental technique, and Krogh the biological insight.

As to how August Krogh became involved in the project, the evidence is ambiguous. In a 1938 publication, Krogh wrote that research in his laboratory "with heavy water was begun on the initiative of Bohr and Hevesy," and that he felt "very grateful because they asked [him] to take part in this line of study." Hevesy, on the other hand, in 1958 remembered the origins of Krogh's participation with the following words:

> When I returned to Copenhagen in the fall of 1934, August Krogh called upon me immediately upon my arrival. He wished to apply labelled water in his permeability studies.

On 10 September 1936, Krogh stated in an address that he had been "exceptionally fortunate in having become associated with Professor Hevesy and through him also with Bohr." This statement confirms the observation in Chapter 1 that Bohr and Krogh never were close, either scientifically or personally. Whoever made the first approach in the collaborative venture, these quotations do indicate that Hevesy and Krogh became close collaborators, whereas Bohr remained at a distance from the practical pursuit of the project.[61]

On the same day that he talked to Bohr, Tisdale also interviewed Krogh. In this interview, Krogh reiterated that perme-

August Krogh provided the biological expertise for the experimental biology project proposed by Bohr. [Courtesy of Bodil Schmidt-Nielsen]

ability problems would define the central focus of research in the proposed project. Hence, Krogh confirmed the impression that it was he who provided the biological expertise for the planned cooperative venture. Tisdale noted in his diary, however, that neither Bohr nor Krogh was as yet ready to present a definite proposal for the Rockefeller Foundation's experimental biology program. Although having changed his approach to the Rockefeller Foundation's demand for experimental biology completely, then, Bohr was still cautious about formulating a specific biological project.[62]

Bohr's and Krogh's lack of precision in formulating a biological problem contrasted with the exact specification, stipulated by both Bohr and Hevesy, of the kind of high-voltage apparatus required. "If and when Bohr gets the new artificial radiation activity equipment," Tisdale wrote in his report of the interview with Krogh, "they will do the same type of thing with radioactive substances." When interviewed in Stockholm during Tisdale and O'Brien's visit there a few days before, Hevesy had let it be known that a representative from the Metropolitan-Vickers Company had estimated that a suitable installation would cost £2,000 ($9,900). Hevesy had informed Tisdale that half of this amount was promised from Danish sources, a statement Bohr repeated in Copenhagen.[63]

As we will see, Bohr did not apply formally to a Danish agency for this support until months later. Nevertheless, he immediately used the alleged promise of Danish supplementary funding as an argument to obtain more support from the Rockefeller Foundation outside its experimental biology program. In so doing, he showed his astuteness as a fund-raiser. As noted in Chapter 3, in the fall of 1933 he had received support for equipment from the Rockefeller Foundation's Paris office in connection with the increased space provided by the newly constructed Institute of Mathematics. At dinner on 30 October 1934, Bohr told Tisdale that because the dollar was falling in relation to the German mark, the amount granted had proven to be insufficient. Tisdale's diary continues:

> He feels a certain moral obligation not to approach the R[ockefeller] F[oundation] to complete the job, but on the other hand he explains that his desire to enter into a cooperative problem between physics and biology such as outlined above with

> Krogh, has necessitated the installation of high-tension equipment for the production of artificially stimulated radio-active substances. He has found that the Carlsberg Foundation, or some other source, considering the prominence of the three men involved, are willing to contribute some [Dkr] 100,000 [$22,100] for their cooperative work, but Bohr feels that if he is to complicate this by asking [the Danish source] for $2,000 to finish the liquid air installation, he may spoil the whole arrangement.

Clearly, Bohr was fully aware that substantial funding from the Rockefeller Foundation could only be obtained by reference to the experimental biology program, and he formulated his request accordingly. The strategy proved to be successful. In mid-December, Tisdale informed Bohr that the Paris office had decided to provide an additional sum to complete the payment for the equipment requested the year before.[64]

In internal memoranda, the Rockefeller Foundation officers showed full awareness, even approval, of the prospective dual use of the new high-voltage equipment asked for in the experimental biology program. Thus Tisdale wrote to Weaver in mid-November:

> In order to have the equipment necessary to produce artificially prepared radio-active substances for their studies, they must have high-potential equipment of the type of either [Ernest] Lawrence, [Robert] Van de Graaff, or [John] Cockcroft. Naturally this equipment would not be limited in its usefulness to the single purpose of preparing radio-active materials for the cooperative problem, but would also permit of studies in nuclear physics from the physics point of view, and it would be used for such biological aspects as were necessary.

The three types of equipment are also referred to in this book as the cyclotron, which is based on resonance acceleration and does not require high voltages; the electrostatic belt-charged generator; and the voltage-multiplier circuit or, more generally, the high-voltage apparatus. These devices were developed, respectively, by the scientists mentioned in Tisdale's letter. Although the experimental biology program required that the equipment be used for biological research, the Rockefeller Foundation did not object to its use also in the basic physics questions under study at the institute. Weaver was probably only happy that funding could thus contribute indirectly to prestigious research in physics proper.[65]

After Bohr's formulation of a proposal for cooperative work, the foundation's judgement of Krogh's candidacy for the experimental biology program had changed entirely. Tisdale was now ready to reconsider Krogh's request to be released from teaching obligations, "which," as Tisdale wrote to Weaver, "we [in April] discouraged to the extent that it is not now under consideration from their side." As an explanation of his new stand, Tisdale stressed that Krogh had "relinquished all thoughts of" the research endeavors he proposed in April, and was "planning to devote himself in his research exclusively to the cooperation with Bohr and v.Hevesy." In his reply to Tisdale's letter, Weaver asked: "Does Krogh swing enough weight with the biologists . . . ?" Obviously, however, this was a rhetorical question on Weaver's part, for he too expressed support of Tisdale's view that Krogh be freed from teaching. The Rockefeller Foundation officers' reversal of its judgement of Krogh, then, indicates that they trusted the schemes and considerations of prominent physical scientists to apply their new techniques in biology more than purely biological or physiological arguments.[66]

In January 1935, Harry Milton Miller, another Rockefeller Foundation officer, visited Copenhagen. Miller, who was 39 years of age, had obtained his PhD degree in zoology from the University of Illinois in 1923. He had taught at Washington University in St. Louis before joining the Rockefeller Foundation's Paris office in 1932. Contrary to his expectation, Miller found that Bohr and Hevesy had slightly modified the formulation of their biological project. He wrote in his diary that the two "stressed the fact that they would work with artificial radioactive substances such as P[hosphorus] and trace their fate in the body." Indeed, this phosphorus isotope, which Johan Ambrosen had already studied earlier in the year, was the first artificially produced radioactive substance to be specified by the Copenhagen group as useful in its biological work. Moreover, Krogh's interest in permeability, which had provided the focus in previous descriptions of the cooperative biology project, was not even mentioned in Miller's diary. Instead, the project was presented in terms of Hevesy's original biological framework, as it had first been formulated in Copenhagen a decade before, of the elements' "fate in the body."[67]

Hevesy now had access to a radioactive isotope of an element that was substantially more important biologically than any that had previously been available for his radioactive indicator technique. He and Bohr may therefore have felt less dependent upon Krogh's specific permeability problem in formulating a biological project. Although still requiring the expertise of real biologists, Bohr may have come closer to realizing that evaluations by prominent physicists such as himself played a particularly crucial role in the foundation's judgement, and that the potential for future application of a physical technique counted for more than actual biological content.

Although paying particular heed to the evaluations of prestigious physicists, the Rockefeller Foundation continued to distinguish between Bohr's original philosophically inclined biological concern, which they deemed to be unfit for support, and the kind of experimental biology it actively sought to promote. The observation in Chapter 2 that Bohr had maintained this concern is supported, in particular, by his correspondence with his younger colleague and admirer Max Delbrück. Motivated by his personal interest in a philosophically oriented discussion of biological problems, Bohr asked Miller whether Delbrück and Victor Weisskopf – who were not then, nor would ever be, involved in the experimental biology project – could be granted fellowships as part of what Bohr now called the "bio-physics program." According to Miller's diary, it was agreed that Bohr "would be able to finance that sort of thing himself if the five-year support [provided by the experimental biology program] is given." Thus, while eager to fund prominent physicists – on whom it also relied heavily for its judgement – the Rockefeller Foundation officers continued to support only those projects that conformed to the experimental biology program.[68]

THE CARLSBERG FOUNDATION'S SUPPORT OF NUCLEAR PHYSICS

As indicated, the Rockefeller Foundation now considered supporting the Copenhagen experimental biology project for a period of five years. Bohr had discussed this timetable with Miller before. He had previously estimated that $40,000 would be required for the whole period. Now, however, he considered this too much. Miller reported in his diary that Bohr would "be very happy with $5,000 [per year], for four or five years, from the R[ockefeller] F[oundation] for work in biophysics." Bohr explained his reduced needs by stating that he would receive more support from other sources than previously expected.[69]

Bohr probably meant the prospective support from the Carlsberg Foundation, which was also mentioned elsewhere in the interview, for a new high-voltage laboratory, which would contain a Cockcroft–Walton generator. As we have seen, Bohr had suggested the possibility of such support to Tisdale in October 1934, when he used the prospect of such Danish support to argue for additional funds for equipment from the Rockefeller Foundation outside the experimental biology program. Moreover, in November 1934 he had written to Tisdale:

> I hope also very soon to write to you about the favourable decision of the Carlsberg Foundation regarding their support of our plans [for] the high tension laboratory, necessary for the extension of our nuclear researches and our cooperation with biologists in this field.

He went on to write that he would provide the details of his needs to the Rockefeller Foundation after the Carlsberg Foundation had made its decision. A week later, he told Tisdale that he hoped to obtain a positive response from the Danish agency after a short visit to Sweden.[70]

However, it was not until 25 January 1935 – the day after Miller's visit – that Bohr prepared a draft for his application to the Carlsberg Foundation, and he submitted the final version two days later. In contrast to his deliberations with the Rockefeller Foundation, he formulated his Carlsberg Foundation application entirely in terms of his interest in physics. Thus, he made a clear distinction between the experimental investigation of electrons orbiting the atomic nucleus and experiments on the atomic nucleus itself. Bohr wrote that emphasis in atomic physics research was now turning increasingly from the former to the latter. It was therefore crucial, he argued, that the institute obtain, in addition to its spectroscopical apparatus for the study of the orbiting electrons, Cockcroft–Walton high-voltage equipment to bombard nuclei. Bohr thus applied for a grant of Dkr 150,000 ($32,500), to be divided equally between high-voltage equipment and a new building in which to house it. He noted that the Rockefeller Foundation during the previous year had provided

> . . . $10,000 for the completion of our facilities for spectro-scopical investigations, which are still continued in an effec-tive way, and that the Rockefeller Foundation has held out the prospect to support our work for the coming years to a signifi-cantly increased extent, provided that the enlarged framework for the experimental investigations laid out here can be real-ized.

Nowhere in the application did Bohr even hint at the Rockefeller Foundation's provision that the project comply with its experi-mental biology program; he merely cited the foundation's condi-tion that funding also be obtained from other sources. Indeed, the application to the Carlsberg Foundation is one of the most unambiguous statements by Bohr that he sought to complete the redirection at the institute – begun after the arrival of Franck and Hevesy in 1934 – to theoretical and experimental nuclear physics. Having previously given Tisdale the impression that the Carlsberg Foundation's funding of the high-voltage facility was intended to supplement the biological cooperation, Bohr now suggested to the Danish agency that both its own and the Rocke-feller Foundation's contributions would be used for research in nuclear physics.[71]

Bohr's seemingly contradictory statements to the American and Danish foundations may be explained in several ways. For one thing, it was only natural for Bohr to suppose that the Carls-berg Foundation, although involved in biological work in other connections, viewed the institute's continued success in the field of physics as most important. After all, this was the kind of work the foundation had been supporting since the establishment of the institute. Unlike the Rockefeller philanthropies, the Carls-berg Foundation had not changed its funding policy for basic science.

Bohr had a substantially closer relationship with the leader-ship of the Carlsberg Foundation than with the officers of the Rockefeller Foundation. His first support from Carlsberg had been for his visit to England in 1912. His formal relationship with the foundation climaxed in 1931 when he moved into the Carlsberg mansion. In 1934, both Johannes Hjelmslev and Valdemar Henriques, who in 1923 had successfully pleaded for an improvement in Bohr's situation in Copenhagen, were still among the foundation's board of five directors. Furthermore, Bohr's close friend, the chemist Niels Bjerrum, had joined the board in 1931, and the philologist Johannes Pedersen, the chairman of the board, was also a friend of Bohr's. The fifth member, Poul Tuxen, a scholar of Eastern philology, had joined the board only the year before. Since 1917, Bohr had been a prominent member of the Royal Danish Academy of Sciences and Letters, which elected the Carlsberg Foundation's board of directors.[72]

Considering Bohr's close relationship with the Carlsberg Foundation, it is unlikely that he would have hidden his real intentions from its officers. Indeed, his relationship being so close, his report to the Rockefeller Foundation of promised Danish support for high-voltage equipment may well have been true; the application to the Carlsberg Foundation in late January 1935 may thus have a mere formality confirming an unwritten understanding made long before. This interpretation is supported by the fact that the Carlsberg Foundation granted the support only fourteen days after Bohr had submitted his application. Indeed, Bohr's closeness to the Carlsberg Foundation makes it more likely that he expressed his central objective when he stated that the new equipment would be used for research in nuclear physics. He may thus from the outset have conceived of his approach to the Rockefeller Foundation's experimental biology program primarily as a strategy to obtain equipment for a new research program in nuclear physics, formulating his requests in accordance with the foundation's formal requirement of a biological project. Indeed, it may well have been that the Rockefeller Foundation officers' own strong desire to recruit this prestigious scientist for the experimental biology program influenced their perception of Bohr as primarily motivated by his biological interest.[73]

The question of Bohr's real motivation may never be resolved, and there is little use speculating further. What the reported contradictions in Bohr's statements do indicate is that once he had decided to change research policy for the institute, he turned into an efficient fund-raiser, who either instinctively or consciously knew how to present his research to the funding agencies in the most advantageous light.

The Carlsberg Foundation made its approval of Bohr's application for high-voltage equipment conditional upon the acquisition of operating expenses for the new high-voltage laboratory from other sources. Bohr explicitly employed this condition as an argument when toward the end of February 1935 he finally submitted an elaborate letter of application to the Rockefeller Foundation's experimental biology program. Again Bohr made the best of the propensities of one foundation in order to obtain the most from another. The next section will discuss Bohr's application to the Rockefeller Foundation.[74]

THE FORMAL APPLICATION FOR EXPERIMENTAL BIOLOGY SUPPORT

In order to present the application at the April 1935 meeting of the Rockefeller Foundation's trustees in New York, Bohr and Hevesy spent their vacation together in the Swiss winter resort of Arosa to work it out. In their application, they described the proposed project as

> . . . the planned cooperation between the Institute for Theoretical Physics and Danish biological institutions to utilise the new possibilities for the investigation of fundamental problems in biology opened by the recent advances in atomic physics.

Hence, Krogh's laboratory was now conceived of as only one of several participating institutions. At the same time, Bohr expressed his strong confidence in Hevesy's work by emphasizing the tradition at the institute of "experimental work in collaboration with biologists." However, as noted in Chapter 3, in the 1920s Hevesy had worked in biology entirely outside the context of the institute. There is, furthermore, no indication from that period that Bohr was enthusiastic about this aspect of Hevesy's work; nor is there any indication that he showed subsequent interest until the advent of experimental biology. Nevertheless, Bohr's new emphasis on Hevesy's place in the tradition of the institute would serve as an important argument within the Rockefeller Foundation for supporting the Copenhagen venture.

Again Bohr had made a successful tactical argument in his dealings with the Rockefeller Foundation.[75]

Bohr and Hevesy did not present a well-defined biological project in their application. Krogh's permeability problem, for example, was not mentioned. Nor was deuterium referred to as a possible biological indicator. Instead, the application stated generally that the Copenhagen scientists would primarily employ induced radioactive isotopes as indicators in biology. In this connection, Bohr and Hevesy noted the importance of

> . . .apparatus for producing artificial radioactive sources stronger than can be obtained through the action of radium emanation, kindly supplied us for our present researches by the medical radium institute of Copenhagen.

Physical apparatus rather than biological content continued to figure most prominently in the formulation of the Copenhagen experimental biology project.[76]

In addition, the support applied for was intended "to secure qualified assistance in the physico-biological work and to cover also other expenses connected with this work." Because of the Carlsberg Foundation's requirement that operating expenses be obtained from other sources, this portion of the support was argued to be particularly important, and Bohr and Hevesy asked that it be granted for five years. All told, the request to the Rockefeller Foundation amounted to:

> $15,000 for a cyclotron, which produced particles at higher energies than the high-voltage equipment obtained from the Carlsberg Foundation;
>
> $6,000 per year for five years, half of which would be used as salaries for two assistants; the other half would "cover running expenses like the purchase of minor apparatus, chemicals, glass and so on in connection with it."[77]

After formulating the application together, and after Bohr had signed it alone, Bohr and Hevesy parted company. Bohr continued his vacation with Werner Heisenberg in the German Alps, and Hevesy traveled to Paris to deliver the application personally to Tisdale. During this meeting, he clarified and augmented some of its points. Despite the application's lack of a clear-cut biological project, it appears from Tisdale's report to Weaver of 27 February 1935 that Hevesy argued successfully that the proposed work fitted into the Rockefeller Foundation's experimental biology program:

> [Hevesy] assures me that this project, involving as it does so much of physics, is completely oriented towards bio-physical problems and is in no wise an attempt on Bohr's part to obtain equipment to permit him to in any wise compete with the Rutherfords, the Lawrences, and others who are working in the field of pure physics.[78]

Tisdale's letter to Weaver reporting on Hevesy's visit continued:

> I sounded out v[on] H[evesy] on the question you raised in one of your recent letters as to whether Krogh filled the bill from the biological side. v[on] H[evesy] expressed the opinion that he could envisage no more suitably qualified person, because he believes the man in question ought to be a physiologist and a chemist, both of which qualifications K[rogh] fulfills. v[on] H[evesy] told me that he had previously cooperated with a physiologist interested in endocrines, and another who lacked chemistry, neither of which were from his point of view satisfactory. I realize this does not answer the question from the biologist's standpoint, but I think it does give an added assurance of the desirability of consummating the scheme.

As in his exchange of letters with Weaver just after the Copenhagen experimental biology project had first been proposed, Tisdale preferred to rely on Hevesy's judgement of Krogh's capabilities in the physical sciences and of the usefulness of the experimental technique employed rather than on an evaluation of Krogh's standing among biologists. Along the same lines, Tisdale reassured Weaver that the annual sum provided for assistants should be used for work on the preparation of radioactive compounds, which was to be done in Bohr's, not Krogh's, laboratory. As if to emphasize further the priority of physical technique, Tisdale underscored Hevesy's statement that Krogh would not require new assistants.[79]

In his discussion with Tisdale, Hevesy mentioned the Finsen Institute – an important center for medical research in Copenhagen – as a specific example of an institution other than Bohr's and Krogh's that was interested in cooperating on the projected biological indicator work. Hevesy reported that Ole Chievitz and Svend Lomholt, who worked at the Finsen Institute, were interested in using induced radioactive isotopes in the treatment of tuberculosis and syphilis, respectively. As we saw in Chapter 3, Hevesy had cooperated with Lomholt in the 1920s. Chievitz was a classmate and close friend of Bohr's from high school who in a personal letter to Bohr several years before had described enthusiastically the potential biological and medical usefulness of modern physics research. Hevesy was beginning to find a receptive interdisciplinary audience for his technique in Copenhagen.[80]

Concerning the experimental apparatus, Hevesy explained at his meeting with Tisdale that the high-voltage equipment supplied by the Carlsberg Foundation and the cyclotron supported by the Rockefeller Foundation were intended for voltages below and above one million volts, respectively. Although claiming that several important radioactive isotopes could be obtained only at the higher voltages, Hevesy implied that the cyclotron was the least urgent request in the proposed package. The reason for this, Hevesy reiterated, was that the Carlsberg Foundation would not release its grant for high-voltage equipment until the Rockefeller Foundation had agreed to provide the annual operating expenses. Therefore, the operating expenses were the most important item. In his letter to Weaver reporting these developments, Tisdale concluded that particularly because the subsequent meeting of the Rockefeller Foundation's trustees would not take place until October 1935, the application ought to be presented at the April meeting.[81]

The content of the application, as well as the fact that Krogh was not at all involved in its preparation, shows that experimental biology in Copenhagen was increasingly the domain of Bohr and Hevesy. In fact, Krogh's name was not even mentioned in Bohr's elaborate application. Several factors were responsible for this development. First, the discoveries of new radioactive isotopes made Hevesy's technique seem more relevant and applicable to many fields of investigation, including biology. Second, Hevesy had now settled in Copenhagen, and was beginning to sense that other scientists and institutions in biology besides Krogh and his laboratory could be persuaded to cooperate.[82]

Moreover, Bohr and Hevesy were increasingly aware of two

main priorities of the Rockefeller Foundation once a project had been provisionally accepted as part of the experimental biology program. One of these priorities was the prestige of the scientists involved. Thus, the Rockefeller Foundation, like the IEB in the previous decade, valued highly the recruitment of Bohr to its program. The second priority was the potential for applying a physical technique, which counted for more than the importance of a specific problem from a biological point of view. Once his project had received Bohr's blessing, Hevesy was the beneficiary of a complete reversal in the Rockefeller Foundation's evaluation of his technique's relevance for experimental biology. As Bohr became aware of these priorities, communication between him and the Rockefeller Foundation improved substantially.

EXPERIMENTAL BIOLOGY SUPPORTED

In spite of the improved rapport between Bohr and the Rockefeller Foundation officers, another problem had to be overcome before funding was granted. Tisdale seemed to approve wholeheartedly of the substantial changes in the Copenhagen suggestions for an experimental biology project that had taken place between the first proposal in October 1934 and the application in February the following year. In the view of other Rockefeller Foundation officers, however, these changes aroused suspicion as to the coherence of the project. This suspicion soon came to the surface in a particular incident. Although it is doubtful that the experimental biology project in Copenhagen was ever in real danger of being scrapped, this incident sheds light on the Rockefeller Foundation's demands on an experimental biology project once it had been proposed.

In early February 1935, when he encouraged Bohr to submit an application as soon as possible, Tisdale had asked Bohr confidentially about his opinion regarding funds to free Krogh from teaching. Two days before leaving for Switzerland to write the application with Hevesy, Bohr replied that he was strongly in favor of such support. When he delivered the application in Paris, Hevesy provided the additional information that Copenhagen University had accepted Krogh's request to be freed from teaching, if funds could be obtained for his assistant Rehberg from private sources. No wonder, then, that the following telegram, submitted by Hevesy in early March, caused some consternation among the Rockefeller Foundation officers:

> MONEY FOR REHBERG NOT WANTED BUT KROGH WISHES THREE-THOUSAND CROWNS [$640] FOR ASSISTANCE FOR WORK WITH US FOR TWO PREFERABLY FIVE YEARS.

Nor did a supplementary letter from Krogh dated 4 March make it clear to Tisdale what had "happened to the Prof. Krogh – Dr. Rehberg arrangement." As Tisdale complained to Hevesy, Krogh's letter did not state unambiguously "whether the proposal to free Prof. Krogh from his teaching duties has been given up, or whether it has already been taken care of by the University."[83]

Immediately upon receiving Hevesy's telegram, Tisdale wrote to Weaver in New York for permission to provide the requested support for Krogh, even if the Danish government had already provided the required funds to relieve him of teaching. Tisdale's letter to Hevesy, written the day after, was hence intended as a reproach rather than representing a real possibility that support might not be forthcoming. This, however, Hevesy could not know.[84]

In his long letter, Tisdale stated that the vacillation in funding requests from Copenhagen

> . . . has suggested to certain of my colleagues here the possibility that the cooperations which we have envisaged under the proposal sent us by you and Prof. Bohr have not been as certainly planned between you and Prof. Krogh as we should have hoped for.

Tisdale reiterated that the cooperative efforts promoted by the Rockefeller Foundation between physicists, chemists, and biologists were expected to "be whole-heartedly entered into by the representatives of these several disciplines." He then presented a necessary criterion for support within the experimental biology program:

> I realize that at the present moment the Physics end of the cooperation is by far the more prominent, but the policy of the Foundation is so firmly decided not to make appropriations for the furtherance of work in pure Physics and Chemistry, that a

cooperation such as envisaged in Copenhagen must give at least as much attention to the Biological side, as it gives to the Physics and Chemistry phases, even though the immediate future in the biological phase may not be as clear as it is in the Physics and Chemistry involved.

Tisdale ended his letter by emphasizing once more that he did not personally doubt the Copenhagen project, but that a clarification was required in order to alleviate the suspicion of some of his colleagues that what was still being referred to as the "Krogh–v.Hevesy–Bohr proposal" did not constitute a sufficiently coherent project.[85]

Naturally, the perceived prospect of losing the funding at the very last minute caused an immediate reaction in Copenhagen. While Bohr corresponded with the Paris office on entirely different matters and hence may not have been informed about the problems caused by Hevesy's telegram, Hevesy wrote to Tisdale immediately, explaining that Krogh had indeed been freed from teaching because of the availability of Danish funds. Hevesy promised to write again after "further conferences with Professor Bohr and Professor Krogh."[86]

Subsequently, as many as three detailed letters of explanation were submitted to the Rockefeller Foundation's Paris office from Copenhagen. One of these was written by Hevesy, who stressed that

> . . . [n]ot only is a cooperation between this institute and that of Prof. Krogh planned on a firm basis, but in fact we started an intense cooperation already [a] few months ago.

He explained that the original emphasis on heavy water resulted from the scarcity of radioactive indicators, and emphasized the need for accelerator equipment by stating that the radioactively weak preparations presently available could only be used to improve the experimental technique, not to obtain biological results.[87]

Another of the three letters from Copenhagen was a statement from Ole Chievitz, which was enclosed with Hevesy's letter. Chievitz referred to

> . . . the negotiations of Professor Bohr and Professor Hevesy with the Carlsberg and Rockefeller Foundations regarding sup-

port for the planned collaboration between the scientists of the institute of theoretical physics in Copenhagen and Danish biologists.

Thus, Chievitz implied that the Carlsberg Foundation, too, saw its funding as contributing to biology. His statement also confirmed Hevesy's increasing cooperation with Danish biologists on a general level. He thus stressed the importance for the Finsen Institute and the Radium Station – Chievitz worked for both of them – of the ongoing negotiations for support. These institutes, Chievitz explained, intended to pursue "the problem of the part played by substances like calcium and phosphorus in organic metabolism." Whereas Bohr's institute would provide the radioactive indicators and "undertake the radioactive analysis of the products obtained from after the application of these sources to the organisms," Chievitz's institutes would carry out the intervening "purely biological part of the work." With his own letter and the enclosed statement from Chievitz, Hevesy sought to reassure the Rockefeller Foundation about his close cooperation with Krogh, as well as to show the broader potential for cooperation involved in the expansion of his indicator technique.[88]

In his response dated 12 March 1935, Krogh, in the first of the three explanatory letters sent to the Paris office of the Rockefeller Foundation, made plain that the present collaboration between Hevesy and himself had as its "starting point the experiments made by Hevesy and his colleagues in Freiburg on the absorption and elimination of 'heavy water.'" He continued:

> I jumped at the possibilities which this substance appeared to offer for the biological study of animal membrane permeabilities and we discussed also some further aspects of the heavy water problem, its possible toxicity and the exchange of H[ydrogen] and D[euterium] atoms in the animal body.

He went on to report that his permeability studies were yielding interesting results. Nevertheless, when he turned specifically to the need for funding, Krogh wrote:

> From the very beginning we have discussed the possibility of utilizing the radioactive isotopes for biologic purposes. It soon became clear that experiments along this line would require

larger amounts of these substances than can be produced by the methods now available here, and this is why Bohr and Hevesy want to obtain means to produce such substances on a larger scale.

Thus, when arguing to obtain support from the Rockefeller Foundation, Krogh, like Hevesy, now stressed the methodological aspects of the cooperative project at the expense of well-defined biological problems. It seems that Krogh and Hevesy had successfully coordinated their arguments.[89]

Krogh also made an effort to explain his "rather abrupt" request for the support of an assistant. Not only had he now found the right man for the job, but the work had also evolved much faster than expected. Hence, an assistant was required to develop the indicator technique with weak preparations before the strong ones could be obtained from the accelerator equipment. Here again, Krogh's reasoning echoed Hevesy's.[90]

Obviously satisfied with the reaction to his complaints, Tisdale wrote to Hevesy that "[t]hese several communications will be of extreme importance to me." He indicated that he would have the final response from the New York office within a month.[91]

Funding for experimental biology at Copenhagen University was granted at the meeting of the trustees of the Rockefeller Foundation on 18 April 1935. It amounted to somewhat more than what had been originally proposed – $54,000 instead of $45,000. As Bohr had requested in his first application, he received $15,000 for the cyclotron. In addition, the Rockefeller Foundation provided Dkr 14,000 [$3,100] per year for assistants and the same amount for materials and equipment to Bohr and Hevesy; finally, it supplied Dkr 3,000 [$650] annually for materials and equipment for Krogh. Because part of the funds was requested in Danish currency, the increased dollar amount of the grant can be ascribed at least in part to a slight rise in the value of the Danish krone from February to April 1935; the amount specifically granted to Krogh also came in addition to the original request. Considering that the other major source of annual expenses for research was the Dkr 24,000 from the Carlsberg Foundation, the new support from the Rockefeller Foundation entailed a considerable improvement of the institute's economic situation.[92]

CONCLUSION

Despite some initial communication problems, from October 1934 Bohr proved himself to be an astute negotiator, making a successful appeal to the Rockefeller Foundation's experimental biology program, which had supplanted the Rockefeller philanthropies' previous best science approach to the funding of basic science. Bohr only made his appeal, however, after several missed opportunities during the foundation's slow reorganization. Bohr's initial reluctance owed itself to his biological interests being too vague and philosophical by comparison with the Rockefeller Foundation's desire to set up concerted experimental projects. In October 1934, Bohr reversed himself completely, requesting support for the kind of concrete research project demanded by the Rockefeller Foundation. Bohr's turnabout is even more striking in that it took place at a time when the Rockefeller Foundation was making its demand for a specific kind of research even more stringent. Moreover, Bohr's lack of appreciation of the foundation's desire to recruit him for its program on the basis of his prestige alone makes his turnabout particularly curious. Having originally been doubtful of the new project-oriented policy for basic science funding, Bohr succeeded in obtaining support for experimental biological research in collaboration with George Hevesy, August Krogh, as well as other biologically oriented persons and institutions in Copenhagen.

Having accepted Bohr's application, the Rockefeller Foundation officers were thrilled by his participation in their program. Thus, a month after support had been granted, Wilbur Tisdale noted in his diary that Bohr

> . . . is so full of enthusiasm for the plans for physico-biological research that he can talk of nothing else. He says that he anticipates that in three years he will be giving all of his time to this work.

Wittingly or unwittingly, Bohr gave the Rockefeller Foundation the impression that he considered experimental biology to be a more important activity at the institute than physics itself.[93]

As already intimated several times in this chapter, however, it is more likely that Bohr's turn to the experimental biology program was motivated by his desire to complete the redirection of physics research at the institute to the experimental and theoretical study of the atomic nucleus. After all, the equipment applied for was not for biology alone; it was exactly the kind of expensive equipment needed for a full-scale transition to experimental nuclear physics research. It seems to have been no coincidence, then, that Bohr's sudden turnabout in negotiating experimental biology support coincided with his desire, beginning in April 1934, to introduce nuclear physics under the leadership of Franck and Hevesy.

Conversely, Bohr's complete reversal toward the Rockefeller Foundation's funding program from April to October 1934 substantiates my contention that he decided to redirect the institute to nuclear physics as a result of Franck and Hevesy's presence there. As they became more self-confident in their negotiations, Bohr and Hevesy increased their emphasis on experimental apparatus as compared to biological problems. This is another indication that the nuclear physics program was closer to their hearts. Finally, in his dealings with the Carlsberg Foundation, Bohr emphasized his desire to turn to nuclear physics without mentioning biology at all. It seems, then, that Tisdale's enthusiasm can be ascribed at least in part to his interpreting Bohr's remarks in the light of his own wishes. Even though in this case Tisdale's statement was particularly extreme, it was not the first time the Rockefeller Foundation officers viewed Bohr's statements from their own perspective.

In order to appreciate the relative roles of experimental biology and nuclear physics at the institute, as well as to ascertain the fate of the Copenhagen spirit, it is necessary to ask what the scientific redirection there really came to entail. Thus, we will now turn to Bohr's continued activities as a policymaker and fund-raiser before the Second World War, as well as the concurrent scientific developments at the institute, especially as they relate to experimental biology and nuclear physics.

5

Consolidation of the transition, 1935 to 1940

After years of no growth or even a reduction in the institute's finances, facilities, and personnel, Bohr improved its material situation substantially by obtaining support from the Rockefeller Foundation's Special Research Aid Fund for European Scholars, and, more importantly, from the same foundation's experimental biology program. As we will see in the first section of this chapter, Bohr continued his activities as a policymaker and fund-raiser during the subsequent years, obtaining additional support from the Carlsberg Foundation and a variety of other sources. In the process, he was able not only to improve the financial situation of the institute, but also to complete the redirection of theoretical and experimental research begun there in 1934.

As we will see in the second section, the new funds enabled Hevesy to develop his experimental biology effort into a full-scale research program. During the next few years, Hevesy's activities, which conformed to the Rockefeller Foundation's demand for project-oriented basic science perfectly, became increasingly independent of the institute's main activities in physics. As will be shown in the last section, Bohr and other physicists in Copenhagen were thus able to concentrate their activities more or less exclusively on the physics of the atomic nucleus. After the redirection had been thus consolidated, physics at the institute continued to be pursued in the Copenhagen spirit.

ACQUISITION OF FUNDING

In obtaining support from the Carlsberg and Rockefeller Foundations in 1935, Bohr had only begun what would amount to a revolution in the funding of the institute. Almost a month before funding had been obtained from the Rockefeller Foundation, for example, the Nordic Insulin Foundation – a nonprofit organization managing, among other institutions, the Nordic Insulin Laboratory and a Copenhagen hospital – informed Bohr that Dkr 15,000 ($3,200) had been granted "for atomic-physical investigations as preparations for physiological experiments." This support was repeated four years later.[1]

More substantially, the Thomas B. Thrige Foundation, established from the profits of the largest electrical firm in the country "to advance the electrical arts in Denmark," in 1935 provided the institute with the electromagnet for the cyclotron. As Warren Weaver of the Rockefeller Foundation – who estimated the magnet to be worth about Dkr 100,000 ($22,000) – noted in his diary, Bohr "was able to convince the trustees of [the Thrige Foundation] that a contribution from them of a great magnet and the power source from it would be consistent with their purposes." According to its internal records, the foundation decided to provide the gift, which in reality would amount to Dkr 45,000 ($9,700), on the very same day it was applied for – 19 August 1935. The Thrige Foundation thus became involved in funding the redirection of the institute long after the Rockefeller and Carlsberg Foundations. In 1936, the Thrige Foundation provided a sum half the size of its previous grant for further extensions. It granted another Dkr 65,000 ($12,450) in 1939.[2]

An amount somewhat less than the total granted from the Thrige Foundation was secured when Hevesy masterminded a gift to Bohr for his fiftieth birthday on 7 October 1935. Sixteen Danish firms and foundations – including the Carlsberg Breweries and the Carlsberg Foundation, the Danish East Asian Company, and the Thrige Foundation – jointly paid for the acquisition of 600 milligrams of radium. In the official congratulatory letter to Bohr, it was observed

> . . . that the investigations of new radioactive elements recently instigated by [Bohr], produced by ordinary elements through bombardment with radium rays, demand considerably larger quantities of radium than the Radium Station in the future can afford.

This was the same argument Bohr had used when he sought support for a cyclotron from the Rockefeller Foundation's experimental biology program. After earning interest, the grant would amount to Dkr 91,371.90 ($20,000) for 600 milligrams of radium and Dkr 16,641.74 ($3,700) for pertinent equipment and for rare metals to be irradiated. When Hevesy proudly announced his feat in a letter to his mentor Ernest Rutherford, Rutherford replied:

> This idea of celebrating people at a relatively youthful age is quite foreign in this country – it is only occasionally we venture on any appreciation of an individual, and then generally wait to see whether he can survive to 70, or still better to 80! However, I think there are some points in the Continental idea to encourage the others!

Undoubtedly, the "radium gift,". as it was called, confirmed Bohr's high standing in the foremost Danish commercial and industrial circles. It also confirmed the turn of the institute toward nuclear physics.[3]

Having thus obtained substantial funding in 1935, Bohr applied in September the following year to the Carlsberg Foundation for Dkr 60,000 ($13,200) in support of the cyclotron. This was the first time he formally presented his plans to install such a device to the Danish foundation; seeking the support as a necessary addition to the Thrige Foundation's gift of an electromagnet, he did not mention that $15,000 of his Rockefeller grant was also intended for this purpose. He explained that the new equipment would be used for "experimental investigations on nuclear reactions," without referring to the experimental biology project. This support was granted two months later, when Bohr also received a smaller amount from a fund, established in 1927 in memory of the Danish attorney Lauritz Zeuthen, "for the instigation of physical investigations on artificially radioactive substances with a particular view to their application in research on biological problems of metabolism." In his usually detailed application, Bohr asked specifically for support for Danish assistants to help in developing Hevesy's indicator technique. Clearly, even among Danish foundations, Bohr was apt to change his arguments for support to accommodate their major interests. The support from the Zeuthen Fund was repeated three years later.[4]

In 1937, Bohr finally set out on his long-overdue journey around the world. In the course of this trip, he visited Warren

Weaver's home in Scarsdale, New York, in February 1937. Explaining that the grants obtained for the cyclotron still would not suffice, Bohr asked for "$12,000 to complete the design, development, construction and testing of the cyclotron." In a letter to his colleague Wilbur Tisdale, Weaver noted: "I think this is the one place in the world where we would perhaps be willing to make an exception for the support of pure atomic physics." Moreover, Raymond Fosdick, now president of the foundation, said in a conversation with Weaver

> . . . that this was the sort of case where he [i.e., Fosdick] is not at all concerned as to whether the item is in [the] program or not. The ability and character of the man are such that the R[ockefeller] F[oundation] cannot miss such an opportunity.

The restrictions imposed by the experimental biology program had lessened considerably since the period up to 1935. The economic situation of the Rockefeller Foundation had improved, and Weaver's Division of Natural Sciences had obtained more autonomy from the Division of Medical Sciences. In 1936 and 1937, the projects supported within the experimental biology program were generally more oriented toward physical science, and also involved larger sums than before. In this situation, Weaver still had reason to consider that Bohr's "request connects in important ways with our program interests."[5]

Indeed, the Rockefeller Foundation's annual report for 1937 presented the Copenhagen cyclotron, the support for which was granted in March, under the general rubric of "Apparatus for Tagging the Atoms" for biological purposes. It was thus grouped together with two other biological research projects making use of the radioactive indicator technique. First, $36,000 was provided for construction and testing of a Van de Graaff generator at the physics department of the University of Minnesota under John T. Tate and for research at that university and the affiliated Mayo Clinic. Second, a biological research program directed by Frédéric Joliot and conducted by four institutions in France – the Collège de France, the Institute for Physical Chemistry at the Sorbonne, the Rothschild Foundation, and the Radium Institute – received $18,000, which included support for the design and installation of a cyclotron. Individuals and institutions closer to Bohr's interest were now allowed under the experimental biology umbrella. Another example of this development was a grant of $10,000 in 1936 to James Franck at the Johns Hopkins University "for his work on photosynthesis and photooxidation processes."[6]

Continuing his world tour, in early April Bohr visited Ernest Orlando Lawrence's Radiation Laboratory in Berkeley, where the cyclotron had been invented and first developed. He reported to Weaver that the visit made him more convinced than ever of "the decisive importance of the new grant." Not only, then, had the Rockefeller Foundation relaxed its demand for relevance, but in addition, Bohr now felt more confident in stating his aims in physics to the foundation officers.[7]

In 1935, Bohr was elected President of the Danish Cancer Society, a private organization established in 1928 to increase the effort to combat cancer in Denmark. One precursor of the Cancer Society was the Radium Foundation, which was established in 1912, and which was responsible for the administration of the Radium Station, the main Danish institution for research and treatment of cancer. Bohr had been a member of the Radium Foundation's national committee from 1921, and joined the Cancer Society in 1929 when that organization took over the responsibilities of the Radium Foundation, which was now disbanded. Bohr therefore had substantial experience when he was elected chairman.[8]

Before 1938, the efforts of both the Radium Foundation and the Cancer Society had been independent of the activities at the institute. In that year, the institute and the Cancer Society began a collaboration, and by the outbreak of the Second World War, the Society had provided Dkr 30,000 ($6,600) "to cover the expenses of the institute in the construction and testing of an x-ray tube for the million volt set-up at the Radium Station" in preparation for an intense effort toward treating cancer. In May 1940, just after the German occupation of Denmark, Bohr recommended a different installation than that originally planned; he suggested that the Radium Station install an electrostatic belt-charged Van de Graaff generator. According to Bohr, this solution would be significantly cheaper and would require less space. Besides, Thomas Lauritsen from Pasadena, an expert in building such generators, was presently visiting the institute. Before learning whether his advice had been heeded, Bohr applied for and received Dkr 40,000 ($7,700) from the Carlsberg Foundation to complete the installation of this equipment at the institute. After receiving the support, Bohr explained his action to the Cancer Society as follows:

> In consideration of . . . the Society's explicit wish to have complete freedom in the final choice of the kind of high voltage installation that is most suitable for the Radium Station's purpose, I have furthermore, on the motivation that the said electrostatic high voltage generator may find extended application in the purely atomic physical investigations at the institute, obtained a grant from the Carlsberg Foundation covering all expenses involved in the construction of this generator.

Bohr's advice to change the plans for an installation at the Radium Station had no effect. Incidentally, the Radium Station was not able to put into operation the high-voltage apparatus that was in fact built. Despite this rare defeat for Bohr, his connection with Danish organizations at this point was such that he had little difficulty obtaining financial support for his institute, whether he argued on the basis of fundamental physics research or medical applications.[9]

In addition to acquiring new support for *construction and equipment,* Bohr also persuaded a number of organizations to cover the increased *operating expenses* at the institute. Whereas the five-year support from the Rockefeller Foundation for operating expenses constituted an immediate increase in the annual budget for costs directly related to research, the nonscientific operating expenses paid by the Danish government rose more gradually. Indeed, the sharp distinction established in the 1920s between private foundations and the government as providers of scientific and nonscientific operating expenses, respectively, continued. For example, in an application to the government in May 1935, Bohr emphasized that all *scientific* operating expenses were taken care of by the Rockefeller Foundation. From 1933–34 to 1939–40, the operating expenses provided by the government increased more than 80 percent – from Dkr 23,200 ($5,000) to Dkr 42,100 ($8,500). In 1935, the government accepted Bohr's request for the establishment of a new permanent assistantship at the institute. Ebbe Rasmussen advanced to this position in April 1936, when Johan Ambrosen took over the annually renewable assistantship previously held by Rasmussen.

This was the first increase in the institute's government-paid staff since Sven Werner obtained the institute's first annually renewable assistantship in 1924.[10]

In a successful application to the Carlsberg Foundation for operating expenses in the fall of 1938, Bohr asked that the annual grant for scientific assistance and equipment be increased from Dkr 24,000 ($5,100) to Dkr 36,000 ($7,700). In contrast to his previous request for an increase in 1933, when he had argued for a rise on the basis of the growing mathematical sophistication of atomic physics, Bohr now sought support for theoretical and experimental work in nuclear physics. Thus, he sought

> . . . the expansion of the investigations which the extraordinary development in the area of atomic nuclear research has brought with it. Not least the thorough examination of the recently developed theoretical conceptions of atomic nuclear reactions presents us with a series of tasks, whose solution must be expected to provide results of decisive significance for further progress.

In contrast to previous approaches to the Carlsberg Foundation, Bohr's application now noted the "close collaboration with a

The institute after the expansion in the 1930s.

number of Danish chemical and biological institutions" based on "the radioactive indicator method." He stressed, however, "that [this collaboration] in a most thorough and mutually supportive way will be connected with the performance of those investigations of a purely physical purpose, for which the said appropriation is sought." Bohr's application testifies to the increasing emphasis on theoretical and experimental nuclear physics at the institute. Moreover, although mentioning the biological effort, Bohr carefully distinguished it from physics research proper.[11]

We can now sum up the developments in the funding of the institute in the latter half of the 1930s. In 1935, after years of virtual stagnation in the institute's funding, Bohr was able to obtain special-purpose grants from the Rockefeller and Carlsberg Foundations for building and equipment for more than Dkr 441,000 ($97,000). Before the German occupation of Denmark in April 1940, Bohr applied successfully for another Dkr 256,000 ($55,500) in such grants from a variety of sources. Moreover, in 1935 the total annual scientific and nonscientific operating expenses increased 38 percent – from about Dkr 78,000 ($17,100) to about Dkr 108,000 ($23,700), virtually because of the Rockefeller Foundation support alone. Operating expenses from other sources did not increase as abruptly. Nevertheless, from this point on they did rise steadily; in 1935, for example, Bohr argued successfully for a permanent increase from the government by referring to the cooperative project in experimental biology. To these amounts should be added the salaries paid by the government, as well as the salaries for Franck and Hevesy supplied by the Rockefeller Foundation. From 1934 to 1939, Danish wholesale prices rose 21 percent, whereas the cost of living index increased only 11 percent. Thus, after having established connections with the Rockefeller Foundation's programs for the refugee scientists and for biology, Bohr had accomplished no less than a revolution in the institute's finances. We will now turn to how he used the funding.[12]

RISE OF THE EXPERIMENTAL BIOLOGY PROJECT

In this section I will show that although Hevesy at first had no specific plans for research in Copenhagen, he eventually devoted his time completely to his experimental biology project supported by the Rockefeller Foundation. By the outbreak of the Second World War, he had developed this activity into a full-fledged, world-renowned research effort, which both in approach and content differed substantially from the work in physics.

When he moved to Copenhagen, Hevesy brought along Erich Hofer, an assistant from Freiburg who had helped him in applying heavy water to biological problems. As early as 1934, the two had published a note on the "Elimination of Water from the Human Body," and in July 1935 they completed a longer article with August Krogh using heavy water in the study of the permeability problem. Hofer soon returned to Germany, however, and neither heavy water, the permeability problem, nor the collaboration with Krogh would figure prominently in Hevesy's future work. Instead, it was Rudolf Schönheimer – another refugee scientist from Germany – working at Columbia University in New York, who independently would lead the effort to employ deuterium as a biological tracer. Schönheimer's work, too, was supported by the Rockefeller Foundation's experimental biology program.[13]

When James Franck had decided definitely to leave Denmark, Bohr asked Franck's assistant Hilde Levi to work under Hevesy. Levi accepted the offer, partly because she had no choice, partly, as she recalls, because it was impossible to say no to Bohr's fatherly way of asking. According to Levi, she accepted in part because of "the way he made it so interesting, it was not sort of a matter of obedience, it was just being very quickly convinced that this was frightfully interesting and you should do that."[14]

Levi's recollections also confirm Bohr's newly aroused enthusiasm for launching an experimental nuclear physics program at the institute. She thus remembers her first work with Hevesy as being devoted exclusively to induced radioactivity. During 1935 and 1936, Hevesy pursued the experiments he had told Fermi about in October 1934 when he did not heed Fermi's request for rare earth samples. Hevesy thus published on the induced radioactivity of scandium and calcium as well as of potassium. He had begun work on the natural radioactivity of the latter element in Freiburg, and from the new data he now drew the correct conclusion regarding the long-standing question as to which potassium isotope is naturally radioactive. Yet Hevesy spent most of his time investigating induced radioactivity in the rare earths of which he had claimed to Fermi that he had too small samples. Prolific as ever, he published eight articles on induced radioactivity during this period.[15]

Only in the summer of 1935, after funding from the Rockefeller Foundation's experimental biology program had been granted, did Hevesy find time and opportunity to prepare his first publication reporting the application of an artificially prepared radioactive indicator in biology. This publication was in the form of a letter to *Nature* coauthored with Ole Chievitz; it was no coincidence that this schoolmate and close friend of Bohr's had recommended the Copenhagen experimental biology project to the Rockefeller Foundation in March as the only person not officially involved. Indeed, Hevesy later recalled that the "only [early] support [for his biology project] came from the surgeon Chiewitz [sic]." Employing the radioactive phosphorus isotope that Hevesy had told H.M. Miller about in January, Chievitz and Hevesy investigated the phosphorus metabolism in rats. Studying the exchange of phosphorus in several organs, they concluded in the letter to *Nature* that their result

> . . . strongly supports the view that the formation of the bones is a dynamic process, the bone continuously taking up phosphorus atoms which are partly or wholly lost again, and are replaced by other phosphorus atoms.

The editor of *Nature* judged the conclusion of the letter as so improbable as to express his reservation publicly.[16]

Only months later, however, Hevesy obtained similar results with the same isotope in a study of plants conducted with the prominent biochemist Kaj Ulrik Linderstrøm-Lang and his assistant Carsten Olsen at the internationally renowned Carlsberg Laboratory in Copenhagen. Growing plants in a culture containing radioactive phosphorus, the three scientists concluded that their results

> . . . clearly indicate that the phosphorus atoms of the leaves are present in a mobile state, and that during the growth of the plant a continuous interchange of phosphorus atoms takes place between the different leaves.

Maintaining that Hevesy's experiments with radioactive lead in plants more than twelve years before had produced a similar result, they asserted that "we have to do with a general property of plant constitution." They announced that they were preparing experiments with other elements to verify this hypothesis. Despite Linderstrøm-Lang's earlier doubts about his biological work, Hevesy had convinced one more Copenhagen scientist and scientific institution that his technique could be applied in biology.[17]

During the year following the publication of Hevesy, Linderstrøm-Lang, and Olsen's letter to *Nature* in January 1936, Hevesy produced no more papers presenting new biological applications of his radioactive indicator technique. However, in three articles explaining the potential of the technique written for *Naturens Verden (The World of Nature)* – the leading Danish semi-popular natural science journal – biological applications played a prominent role. As we have seen, when he first developed the indicator and x-ray techniques, Hevesy had published such survey articles in anticipation of a full-scale attack. Typically, the pattern would now repeat itself.[18]

In 1936, Bohr obtained funding from the Rockefeller Foundation for two conferences on biology to be held at the institute. He started planning the first of these conferences – which was devoted to genetics – on receiving the paper on gene mutations by Delbrück and collaborators described in Chapter 2. The first conference took place in September 1936 without the participation of either Hevesy, who was visiting Budapest, or Krogh, who was in the United States celebrating the tercentennial of Harvard University. The physicist P.A.M. Dirac, who did attend, told a Rockefeller Foundation officer that he "was genuinely enthusiastic and said that he had learned a lot." This conference, in other words, was held in the tradition of Bohr and his younger physics collaborators' interest in biology as guided by the Copenhagen spirit.[19]

However, only when applying for support for a second conference – which was devoted to the "problems of metabolism in animals and plants" and the "outlooks for the radioactive indicator researches" – was Bohr assured support for both conferences. After several postponements, the second conference was held in May 1938. The addresses of Hevesy, Dorothy Needham from Cambridge University, Jakub Karol Parnas from Poland, Otto Meyerhof, who was in the process of emigrating from Heidelberg, and Krogh were all directly related to the Copenhagen experimental biology project supported by the Rockefeller Foundation. This conference was by far the larger of the two. The two conferences confirm that the biological concerns of Bohr and his physicist collaborators, on the one hand, and of Hevesy and the Rockefeller Foundation, on the other, were as distinct as before. Moreover, the second conference highlighted the increasing international interest in Hevesy's project.[20]

By the time of this conference, Hevesy had turned all his energies toward "experimental biology." In 1937, eight out of ten of his publications reported original biological applications of his radioactive indicator technique; the remaining two comprised a tribute to the deceased Rutherford, and a survey of chemical and biological applications of the indicator technique coauthored with his old collaborator Fritz Paneth.[21]

During the first few years, Hevesy developed his biological activities almost exclusively by arousing interest in his technique among the foremost Danish biological and medical scientists. Besides publishing again with Chievitz and Linderstrøm-Lang, he also collaborated with Einar Lundsgaard, who had recently been appointed professor of human physiology at the university's medical faculty. In addition, he resumed his collaboration with Krogh. This time, however, it was Hevesy's indicator technique that defined the framework for their investigations. Thus, they studied the exchange of phosphorus in teeth, bringing in Johannes Juul Holst from the College of Dentistry as a third collaborator. The Danish State Farm at Hillerød north of Copenhagen was solicited for a study involving chick embryos. Thanks to the cooperation of the Diabetes and Finsen Hospitals, and on a greater scale the University Hospital, Hevesy was able to experiment on phosphorus metabolism in the human body.[22]

Hevesy even instituted international collaboration in the biological application of his technique; in 1938, he coauthored two papers with a group at Parnas's Biochemical Institute in Lwów, Poland, and in 1940, he wrote a paper with Ida Smedley-MacLean at the Lister Institute of Preventive Medicine in England, which had been established in 1891 to support biomedical research.[23]

In spite of the variety of scientists and institutions engaged in the work, as a rule problems were defined and initiatives taken by Hevesy. The work followed the pattern anticipated in Chievitz's letter to the Rockefeller Foundation from March 1935. First, radioactive phosphorus was produced at the institute. Second, personnel at a medical or biological institution distributed it to a living organism. Finally, a radioactive sample from this work was investigated at the institute. As the Rockefeller Foundation's support for experimental biology was earmarked for Bohr's and Krogh's institutes, there was never a question of reimbursement to other institutions. Hevesy worked actively and successfully to win genuine enthusiasm for his technique among researchers in other disciplines.[24]

From the outset, Hevesy's only collaborator at the institute was Hilde Levi, whose background was in physics, and who during the first years published with Hevesy only on physical questions. Levi published her first biologically oriented article, which was coauthored with Hevesy, in 1938. By that time, Hevesy had obtained additional collaborators at the institute for his experimental biology project. Nevertheless, once she had turned to biology, Levi, like Hevesy, would never again pursue questions of pure physics. From the end of the Second World War until retirement in 1979, she worked in Copenhagen University's Zoophysiological Laboratory, which Krogh had founded, and which in 1970 moved into the newly constructed August Krogh Institute.[25]

Among foreign visitors, the Czech student Ladislaus Hahn, who arrived in November 1936, and the Dutch student Adriaan Hendrik Willem Aten, Jr., who arrived in January 1937, were particularly helpful. Moreover, in early 1937 the Danish high school teacher and chemist Otto Rebbe became Hevesy's assistant, receiving his salary from the Rockefeller Foundation's five-year grant. Levi also received her salary from this source during the first years of the project; she was funded thereafter by the Rask–Ørsted Foundation. Aten was supported by both foundations, whereas Hahn was paid exclusively by Rask–Ørsted.[26]

As Hevesy gained more experience with his biological research, as well as more biological collaboration at the institute, the work became less dependent upon help from outside institutions. In 1937, two articles by Hahn and Hevesy were the only

Hevesy (left) and Aten chat over coffee in the institute's lunchroom, c. 1938.

contributions to experimental biology in Copenhagen written exclusively by personnel at the institute; people from other institutions were involved in all of the other publications of that year. Subsequently, the great majority of experimental biology publications reported to the Rockefeller Foundation were written by Hevesy together with one or more collaborators at the institute, with no other institutions involved. Although more often than not these publications acknowledged help from other Danish institutions, the institute had clearly taken on a new dimension, which included keeping a few animals for experimentation.[27]

In 1938, Hevesy's group was further strengthened by the arrival of two Americans who would stay for a year. The dentist Wallace David Armstrong worked on the phosphorus exchange in teeth, while the physicist-turned-biologist William Archibald Arnold, staying as a Rockefeller Foundation fellow, contributed to the construction of more sensitive Geiger counters, sought to apply deuterium in biological experiments, and attempted unsuccessfully to produce the isotope carbon 14. Martin Kamen and Samuel Ruben established the existence of this element somewhat later with the cyclotron at Berkeley.[28]

Krogh's laboratory was the only institution in Copenhagen publishing papers in experimental biology that were not coauthored by personnel from the institute. From 1935 through 1940, these publications numbered nine out of the fifty-five papers coming out of the Copenhagen experimental biology project. Here, the emphasis continued to be on the application of deuterium as an indicator in permeability problems.[29]

Although Hevesy had expressed an intention "to follow up the metabolism and circulation of all elements of biological importance in animals and plants by making use of isotopic indicators," phosphorus 32 remained by far the isotope most frequently employed by Hevesy and his group. This isotope was convenient to prepare, even with the radon–beryllium supplied with radon from the Radium Station before Bohr's "radium gift" could be put to use. At least from 1938, Hevesy also received radioactive phosphorus in the mail from Lawrence's cyclotron in Berkeley, and from 1939 the institute was able to produce radioelements with its own cyclotron. From then on, shorter-lived radioisotopes, notably potassium, were employed in addition to phosphorus.[30]

In the course of their efforts, Hevesy and his group gained a reputation as pioneers in the application of radioactive isotopes to biological research. Arnold, for example, recalled that he chose to go to Copenhagen precisely for this reason. In May 1939, Harold Urey expressed enthusiasm about inviting Hevesy to the United States, noting that "he would be the best man in the world to speak on the uses of the radioactive isotopes." By the late 1930s, scientists and laboratories in several countries were beginning to follow suit, and in a review article in 1940 on the "Application of Radioactive Indicators in Biology," Hevesy referred to ninety-five publications, of which no more than thirty-three came from Copenhagen or originated directly in his own work. Since the negative response to his paper with Chievitz in 1935, Hevesy had gained substantial acceptance for the applicability of his radioactive indicator technique in biology.[31]

As we have seen, Hevesy's first major effort at the institute after his arrival in 1934 had been to follow up Bohr's rapidly increasing interest in experimental nuclear physics. Soon, however, Hevesy developed the Copenhagen experimental biology program in his own image, applying his radioactive indicator technique to investigate the "fate in the body" of the elements – especially phosphorus. By the outbreak of the Second World War, Hevesy, supported by the Rockefeller Foundation, devoted all of his time to this project.

Although less diversified, Hevesy's work in Copenhagen followed the pattern of his earlier major research efforts. Rather than letting the conceptual problems of physics direct his choice of the appropriate experimental technique, Hevesy sought opportunities to apply such techniques – preferably his own radioactive indicator technique – in other fields of investigation. Although this approach to science was contrary to Bohr's, it corresponded well with the Rockefeller Foundation's goal of improving biology by introducing the most advanced techniques from physics and chemistry.

Although they shared experimental equipment with the physicists, Hevesy and his group worked on entirely different research problems. The assistants paid by the Rockefeller Foundation's experimental biology program included builders of physics apparatus. Yet Hevesy's collaborators and the physicists generally had little scientific interest in common. The two groups even competed for the same equipment, the physicists disapproving of the severe contamination of the cyclotron each time the biologists employed it to produce radiophosphorus. As we will see in the next section, the work of Hevesy's group, which by the outbreak of war contributed as much as one fifth of the institute's published papers, differed from that of the physicists in spirit as well as in substance. Indeed, the physicists were able to retain the Copenhagen spirit of work in physics.[32]

CONSOLIDATION OF NUCLEAR PHYSICS

During the five years before the outbreak of the Second World War, Bohr completed the redirection at the institute to a full-fledged research program in nuclear physics. Except for sharing experimental equipment, the physicists worked in complete independence of Hevesy's goal-oriented research project. Thus, a clear distinction was maintained between Hevesy's pointed efforts to expand techniques from physics to new fields and Bohr's discursive guidance of physics according to the Copenhagen spirit.

As judged from the extensive Bohr Scientific Correspondence, Bohr's main concerns in physics in 1935 drifted momentarily away from the atomic nucleus, to return once more to the conceptual problems of relativistic quantum physics. Thus Bohr began a collaboration with Rosenfeld on a sequel to their long article of 1933. Typically, then, Bohr's interest continued to be guided by open-ended discussions with colleagues.[33]

In contrast to the late 1920s and early 1930s, however, Bohr was now deeply involved in overseeing the redirection of the institute begun in 1934. Thus, in May 1935 he reported in a letter to Heisenberg that the work with Rosenfeld was impeded by the institute's expansion, especially the "cooperation with biological institutions." Although in the summer of 1935 Bohr did find

Bohr takes a symbolic first step toward expanding the institute, 1935/36.

time to rebut a paper by Einstein and two collaborators in Princeton questioning the completeness of quantum theory, Bohr and Rosenfeld's sequel was not published until 1950. In fact, for many years after 1935, Bohr found little or no time to concern himself with the problem of relativistic quantum physics. At the beginning of this period, his effort to expand and redirect the institute was taking priority over freewheeling discussions of theoretical questions in relativistic quantum physics.[34]

Bohr's redirection of the institute is clearly reflected in the kind and number of research publications produced. After the sharp decline, described in Chapter 1, in the numbers of both visitors and publications after the peak year of 1927, the total number of publications started to pick up again in 1934. Almost half the twenty-four publications of that year dealt with nuclear problems. As we saw in Chapter 2, however, these publications continued research programs already begun by George Gamow and Hans Kopfermann, and did not reflect a planned redirection of scientific research at the institute.[35]

Only the papers on artificial radioactivity published in 1935 expressed Bohr's concerted redirection of research through his established experimentalist friends James Franck and George Hevesy. These publications were dominated by Hevesy's five papers on the artificial radioactivity of rare earths and other elements requested by Fermi the year before. Two additional articles on similar experiments by Frisch proved that the new program was not limited to Hevesy's work. A popular survey in a Danish physics journal on the "properties and constitution of atomic nuclei" by Bohr's young Danish collaborators Torkild Bjerge and Fritz Kalckar indicates further that the scientific emphasis at the institute was shifting. In that year, the total number of publications from the institute also rose to thirty-three.[36]

The arrival of the refugee physicist Otto Robert Frisch in October 1934 proved to be particularly crucial for the activities in nuclear physics at the institute. Frisch provided important expertise in the planning, construction, and application of new scientific apparatus, including equipment for the biology project. Instead of the one year he had originally expected to stay in Copenhagen, Frisch stayed on for almost five years. Supported initially by the Rask–Ørsted Foundation, and then by the Rockefeller Foundation's experimental biology grant, he remained until July 1939, when he went to England to work with Mark Oliphant at the University of Birmingham.[37]

The year 1935 represented a clear turn in the research emphasis at the institute toward nuclear physics. Nevertheless, less than half the publications from the institute in that year were devoted directly to that field. Only in 1936, the year when Bohr published his own first articles devoted exclusively to nuclear physics, was the large majority of papers concerned with nuclear processes. Although Hevesy was still publishing on artificial radioactivity, the approach to nuclear physics was clearly broadening, both in terms of personnel and experimental and theoretical work. In 1936, the number of long-term visitors also increased to seventeen after having remained at a constant level of fourteen for the previous three years. Bohr's scientific redirection of the institute had reached a milestone.[38]

Along with his policy efforts, Bohr's first research article on the atomic nucleus undoubtedly helped boost the redirection. The article was based on a lecture he gave to the Royal Danish Academy of Sciences and Letters toward the end of January 1936. Published in *Nature* soon after, it was to have a tremendous impact not only on the work at the institute, but on the development of theoretical nuclear physics generally.[39]

In his lecture and article, Bohr addressed for the first time in print the much-discussed question of why experiments showed that neutrons, particularly the slow ones, were so readily absorbed by atomic nuclei. In not containing a single mathematical equation, Bohr's publication of his ideas in *Nature* is unique at this stage in the development of physics. Bohr had previously considered the nucleus as a system of mutually noninteracting particles by analogy with electrons orbiting the atom. He now proposed that a considerable interaction might occur among the nuclear constituents – or nucleons – when a neutron impinged on the nucleus. The kinetic energy of the colliding neutron would then be distributed among all the nucleons, and the probability that one of them would acquire sufficient energy to escape would be significantly smaller than for a nucleus considered as a system of noninteracting particles.[40]

Bohr's proposition constitutes at the same time a departure from and a continuation of his previous thinking about the atomic nucleus. On the one hand, as discussed briefly in Chapter 3, he had previously been opposed on theoretical grounds to Fermi's interpretation of his own experiments that neutrons were readily absorbed by nuclei. With his new contribution, Bohr's attitude to this particular question had reversed completely; he now presented a theoretical basis for Fermi's interpretation.

On the other hand, Bohr's conception that the system of nucleons was fundamentally different from the atomic system of the nucleus and its orbiting electrons had clear precedents in his earlier writings. Indeed, as we saw in Chapter 2, for several years Bohr had sought to explain anomalous phenomena such as beta decay by arguing that traditional quantum mechanics did not apply to the atomic nucleus. Instead, he believed that a proper treatment would have to await the formulation of an entirely new relativistic quantum physics. As a result, his statements before 1936 about the atomic nucleus had always been appended to broader presentations of the theoretical problems of quantum physics.

In his 1936 contribution, Bohr maintained his position that the atomic and nuclear systems were different. Now, however, he referred this difference to the dissimilar degree of interaction among the particles constituting, respectively, the outer atom and its nucleus. That is, he abandoned his search for an explanation in terms of a new and overarching quantum theory, a goal that had long provided intellectual fodder in the discussions among theoretical physicists at the institute. Thus, the article made no mention of his earlier search for a theoretical basis incorporating radical proposals such as the nonconservation of energy, which Bohr for years had thought to be indispensable for explaining beta decay. Indeed, only months later, Bohr was to renounce publicly the idea of nonconservation of energy in the atomic nucleus, and accept – after several years of procrastination – Pauli and Fermi's alternative explanation of beta decay, which introduced another new particle – the neutrino.

Typically, Bohr explained his turnabout in the context of experimental results. The American experimentalist Robert Shankland had created a stir by reporting experiments contradicting the conservation of energy and momentum evidenced in the so-called Compton effect, which results from the collision between free electrons and light and provides experimental confirmation that light consists of particles, or "photons." Bohr's

Bohr and collaborator Fritz Kalckar, 1934. [Courtesy of AIP Niels Bohr Library: Weisskopf Collection]

recanting was appended to a letter to *Nature* from the Copenhagen physicist J.C. Jacobsen, which announced experiments contradicting Shankland and redeeming energy conservation. Independently, Walther Bothe and Hermann Maier-Leibnitz at the Kaiser Wilhelm Institute for Medical Research in Heidelberg had arrived at the same result. Jacobsen's and Bothe and Maier-Leibnitz's results confirmed experiments by Bothe and Hans Geiger in 1925, which at that time had forced Bohr to give up his suggestion of energy nonconservation for atomic processes. The new evidence, in addition to Pauli and Fermi's concept of the neutrino, induced Bohr to abandon the idea of energy nonconservation even for nuclear processes. Interestingly, P.A.M. Dirac, who previously had expressed his opposition to Bohr's proposal of energy nonconservation in no uncertain terms, had now become a proponent of the idea himself. The tables had been turned.[41]

True to the Copenhagen spirit, Bohr solicited his younger colleagues for comments about his new concept of the atomic nucleus. Thus, during the month of February he sent the manuscript for his note to appear in *Nature* to Werner Heisenberg, Oskar Klein, Max Delbrück, William Houston, and George Gamow. As usual, Bohr's solicitations produced prompt and forthright responses. Whereas Klein goodhumoredly found Bohr's approach too mechanical, only P.A.M. Dirac, who replied several months later, was opposed to Bohr's theory. Thus, Bohr's substantial effort in matters of policy and funding had not stymied the Copenhagen spirit of uninhibited scientific communication between him and his younger colleagues. In this case, though, the discussion took place only *after* Bohr had written his manuscript. There is no indication of an intimate and extensive collaboration – such as, for example, the collaboration with Rosenfeld described in Chapter 2 – leading to his 1936 statement. But when Bohr subsequently elaborated his ideas in collaboration with Fritz Kalckar, their collaboration followed the old pattern exactly.[42]

This continuity in work style is confirmed by an incident that also sheds further light on the Copenhagen spirit itself. After the editor of the German journal *Die Naturwissenschaften* had invited Bohr to submit a German version of his 1936 article in *Nature,* Max Delbrück offered in a letter from Berlin to take care of the translation. Bohr accepted the offer, and soon after received the translation by Delbrück and his collaborator Hermann Reddemann in the mail. Although he expressed general satisfaction with the translation, Bohr typically suggested a few changes, including the recombination of a long sentence taken apart by the translators. Delbrück replied by postcard with undisguised anger, withdrawing his name from the translation, as he considered it "hopeless to convince [Bohr] of the inadequacy of [his] use of the German language." Bohr did not answer Delbrück's outburst himself, but let Rosenfeld explain that "the readership will have to reconcile themselves to the fact that there is no 'via regia' to Bohr's thoughts." In spite of Rosenfeld's intervention, the disagreement was not resolved, and the German version named only Reddemann as translator.[43]

This incident confirms both Bohr's uncompromising attitude toward the written word and his need for collaboration to achieve it. Although, as Rosenfeld's reply clearly indicates, Delbrück's reaction was unexpectedly harsh, it does show that the relationship between Bohr and his younger collaborators was close enough to allow even rather severe outbursts of disagreement. In contrast to the relationship with Jordan described in Chapter 2, Bohr's friendship with Delbrück continued to thrive. In fact, under more relaxed circumstances Delbrück's cordial mockery of Bohr's involved language could be a source of considerable amusement in Copenhagen. In 1935, for example, when he sent Bohr his translation into German of Bohr's reply to the criticism of Einstein and his collaborators in Princeton, Delbrück, mimicking Bohr's style of writing, formulated the accompanying letter of several pages as a single sentence. From his reply, it is clear that Bohr appreciated the prank. From these incidents, as well, it seems that the Copenhagen spirit continued to thrive after the mid-1930s.[44]

The 1936 annual informal physics conference at the institute is yet another example of a traditional vehicle of the Copenhagen spirit that was retained after the transition to nuclear physics. This conference, originally planned for September 1935, was postponed until the summer of 1936 because of the activities involved in expanding the institute. In addition to the traditional discussion of theoretical problems, Bohr now looked forward to showing the expanded premises to the participants of the conference, which, in line with the institute's redirection, would be devoted to nuclear and biological problems.[45]

Thus, the first biological conference at the institute sponsored by the Rockefeller Foundation was organized as part of the 1936 physics conference. Moreover, the conference immediately preceded the Copenhagen Unity of Science congress where Bohr denounced misguided interpretations of his complementarity idea. As a result, more visitors than usual attended the 1936 informal conference. At the conference, nuclear questions, in particular, were thoroughly discussed. Undoubtedly, the participants debated Bohr's proposition about the atomic nucleus from only months before in the usual open and informal manner.[46]

This proposition, soon to be known as the "compound nucleus" model, would guide nuclear physics for several years to come. Indeed, according to Hans Bethe, a prominent participant in this development, Bohr's authority was such that for almost

two decades this model was to dominate theoretical nuclear physics at the expense of other possible concepts of the atomic nucleus. Bohr's influence in the physics community at the time was truly enormous.[47]

In the following years, the theoretical and experimental work at the institute and Bohr's own approach to nuclear problems went hand in hand. The Nineteenth Scandinavian Meeting of Natural Scientists, held in Helsinki in 1936, provides striking evidence of the extent to which both had changed. While Bohr gave a general lecture on the "properties of atomic nuclei," as many as five of the younger Danish collaborators at the institute – theoreticians Niels Arley, Fritz Kalckar, and Christian Møller, and experimentalists Torkild Bjerge and Ebbe Rasmussen – spoke on various aspects of nuclear physics. Only one physicist from Denmark (not connected to the institute) devoted his talk to a different subject, whereas none of the talks of the twenty physicists from Finland, Norway, and Sweden addressed nuclear problems. Unlike any other leader in Scandinavian

Informal conference at the institute, June 1936.

First row: Wolfgang Pauli, Pascual Jordan, Werner Heisenberg, Max Born, Lise Meitner, Otto Stern, James Franck, George Hevesy. *Second row,* seated: Erwin David, Mark Oliphant, P. Kapur, Carl Friedrich von Weizsäcker, Friedrich Hund, unidentified, Hans Jensen, Fritz London, Otto Robert Frisch, John R. Dunning. *Third row,* seated: Hans Kopfermann, Owen Richardson, Ugo Fano, unidentified, Ladislaus Hahn, unidentified, Erwin Fues, Walter Heitler, Milton S. Plesset. *Fourth row,* seated: Ivar Waller, Rudolf Peierls, unidentified, Hans Euler, Homi Jehangir Bhabha, unidentified, Llewellyn Hilleth Thomas, Niels Arley, Fritz Kalckar, unidentified. *Fifth row,* seated: Edward Teller, Victor Weisskopf, Max Delbrück, Martin Strauss, McKay, Hendrik Anthony Kramers, H. B. G. Casimir, Eugene Rabinowitch. *Sixth row,* seated: Hilde Levi, three unidentified persons, Torkild Bjerge, Buch Andersen, J. K. Bøggild. *In the back:* Ebbe Rasmussen (seated), George Placzek, Christian Møller, C. H. Manneback, J. C. Jacobsen, K. J. Brostrøm, unidentified, Johan Ambrosen, unidentified, Sørensen, Sven Høffer-Jensen, Hugo Asmussen. *Standing on the side:* Niels Bohr, Léon Rosenfeld, Edoardo Amaldi, Gian Carlo Wick, Arthur von Hippel, unidentified, Jørgen Koch.

physics, Bohr had been able to gather around him a homegrown core of nuclear physicists.[48]

During his trip around the world in 1937, Bohr had several opportunities to lecture to colleagues on the problems of physics that concerned him most. Whereas in the late 1920s and early 1930s he had devoted such lectures to the problems of relativistic quantum physics, discussing the atomic nucleus briefly toward the end, he now concentrated on the problems of the nucleus as such. Bohr continued to give these lectures promoting nuclear physics after his return to Denmark. This is another clear indication that his priorities in physics had changed.[49]

Even before the large-scale experimental equipment for nuclear investigations had been completed, a considerable experimental effort was made at the institute to understand the nucleus. Whereas Hevesy, concentrating on experimental biology, published his last papers on artificial radioactivity in 1936, Frisch continued his work actively. In 1937 and 1938, for example, Frisch, in collaboration with Hans von Halban and Jørgen Koch, published a series of articles reporting their experiments on slow neutrons.[50]

In 1937 and 1938, Bohr, as well as the nuclear physics community generally, suffered two severe losses. In 1937, Bohr's mentor Ernest Rutherford, whom Bohr in a memorial lecture would call the "founder of nuclear science," died. Then, early the next year, Fritz Kalckar, Bohr's young Danish collaborator in working out the theory of the compound nucleus, died suddenly and unexpectedly.[51]

Despite these losses, Bohr, like the physicists at his institute more generally, continued to contribute to the development of nuclear physics. Although he followed the general developments in physics with his usual energy and interest, in his own research during these years Bohr concentrated increasingly on nuclear physics. In early 1938, for example, he published a note applying his compound nucleus model in order to explain the nuclear photoeffect, in which the nucleus absorbs a photon, emitting a proton, a neutron, or an alpha particle. This effect had been proposed by Maurice Goldhaber in the summer of 1934, and was confirmed experimentally only months later by Goldhaber and James Chadwick in Cambridge, England. For his new paper,

Bohr once more solicited comments from several colleagues. Although he was forced to abandon some of his original statements in a publication later in the year, Bohr's work constituted an important contribution to a new and thriving field. In addition to the publications, several preserved manuscripts on various aspects of nuclear physics during this period testify to Bohr's activity in this area.[52]

While Bohr and other theoreticians at the institute concentrated on nuclear theory, and Frisch and others did nuclear experiments, other physicists completed the installation of the large-scale apparatus for nuclear research funded by the Rockefeller Foundation, the Carlsberg Foundation, and the Thrige Foundation. Thus, the construction of the high-voltage equipment – about which Hevesy had negotiated in London even before his move to Copenhagen in 1934 – was conceived, and in its early stages also directed, by the German experimental physicist Arthur von Hippel. Hippel, who had arrived in Copenhagen on the last day of 1934, also took care of the relationship with the German firm Koch and Sterzel, which on Hippel's suggestion in the end was chosen over Metropolitan-Vickers to supply the equipment. Not Jewish himself, Hippel was married to James Franck's daughter. He thus deemed it essential to find permanent work outside Germany, and in September 1936, after less than two years in Copenhagen, the Hippels left for the United States on the same boat as James Franck – both to find permanent occupation and to live close to her parents. The spectroscopist Ebbe Rasmussen thereupon took over responsibility for installing the high-voltage equipment. The American physicist Charles Lauritsen – father of Thomas and a close friend of Bohr's, with experience from Caltech in developing high-voltage equipment – helped over the first setbacks. The equipment was put to use, first, in the spring of 1938 to provide x-rays for cancer therapy, and then, about New Year 1939, as a particle accelerator for nuclear studies.[53]

In contrast to the high-voltage equipment, which to a great extent was tested and built by Koch and Sterzel, the cyclotron was mostly built on the premises of the institute. In March 1937, when he visited Berkeley on his trip around the world, Bohr agreed with Ernest Orlando Lawrence that one of Lawrence's

Frisch conducting experiments in nuclear physics, 1936.

Bohr poses after installation of the high-voltage equipment in 1938.

people should assist in building the Copenhagen cyclotron. The Rockefeller Foundation suggested at first that Bohr should share Lawrence's cyclotron builder with Frédéric Joliot, who was just beginning his own Rockefeller-funded experimental biology project. After Lawrence's objection that his men should be treated as scholars participating in scientific activity, and not just as traveling technicians, the Rockefeller Foundation decided to pay for two people from Lawrence's laboratory. While Hugh Campbell Paxton went to Joliot, Lawrence Jackson Laslett, whom Lawrence considered "definitely the best man in his laboratory," went to Bohr.[54]

Laslett came to Copenhagen in September 1937. Before his arrival he had traveled in the United States to see several of the cyclotrons already functioning there. When he arrived, the cyclotron magnet and a few other components were already in place. However, substantial modifications proved to be necessary, and Laslett's activities in Copenhagen consisted mostly of designing and constructing the cyclotron, which started working on 1 November 1938. Later that month, Laslett rushed to the United States because of his mother's illness. Only days before his departure he had the opportunity to see the cyclotron produce neutrons. A few days later, Bohr reported to Wilbur Tisdale that the Copenhagen physicists had been able to obtain "deuteron beams of 5 Mill[ion] Volts, producing disintegration effects of a scale which could only have been obtained by one kilogramm of radium." The Copenhagen installation was the second cyclotron, after the one at the Cavendish, which produced its first beam in August 1938, to operate in Western Europe. They were preceded by Yoshio Nishina's cyclotron in Tokyo, and the cyclotron at the Radium Institute in Leningrad, designed and constructed by L.V. Mysovskii. Both of these machines started operating in the summer of 1937 as the first cyclotrons outside the United States. At the beginning of the Second World War, Thomas Lauritsen assisted in constructing the Van de Graaff generator at the institute provided by the Carlsberg Foundation.[55]

Just around the time when the high-voltage equipment and the cyclotron were put to use, a discovery was made that definitely sealed the transition at the institute to theoretical and experimental nuclear physics. An important contribution to this discovery was made by Otto Robert Frisch, who was still working at the institute. Taking place about New Year 1939, the discovery was not an outcome of the application of the powerful new apparatus. Instead, it stemmed from the intense discussion of results from other laboratories. Soon, however, the new equipment was put to active use to exploit the new discovery.

L. J. Laslett of Ernest Lawrence's Berkeley laboratory, who was sponsored by the Rockefeller Foundation to design and build the Copenhagen cyclotron in 1938.

J. C. Jacobsen, the institute's resident experimentalist, and Niels Bohr at the cyclotron in 1938.

The prehistory of the discovery goes back to Fermi's 1934 program of creating artificial radioactivity by bombarding each chemical element with neutrons. In this effort, the heaviest elements soon attracted particular interest. In June of that year, Fermi reported that he might have produced elements heavier than any previously known, and in his Nobel Prize lecture four and a half years later he named two such "transuraniums." A little earlier, however, the chemist Otto Hahn in Berlin, in collaboration with Fritz Strassmann, had found evidence that when uranium, the heaviest known naturally occurring element, was bombarded with neutrons, radium, four places *below* uranium in the periodic table, and not a *heavier* element, was created. Hahn's finding went against the accepted creed and therefore met with decided opposition. In particular, Lise Meitner, working with Hahn in Berlin, and Bohr argued on theoretical grounds that an element removed as much as four places from uranium in the periodic table could not be created by neutron bombardment.[56]

In 1938, Meitner, an Austrian Jew, was finally forced to flee Germany, eventually to take up work at Manne Siegbahn's Nobel Institute in Stockholm. On her arrival there, her nephew Otto Robert Frisch was still working at the institute. Frisch, according to his own account, was invited to join Meitner in visiting friends near Göteborg, Sweden. Just before they met in the last days of the year, Meitner had received a letter from Hahn reporting that the reaction product he had first identified as radium was instead its homologue barium, element number 56. This element was substantially farther below uranium in the periodic table than radium. Hence, the result of Hahn's correction was even more radical than his original finding.[57]

Recognizing her long-time collaborator's chemical competence and the care he would take before drawing such an unexpected conclusion, Meitner found it difficult to doubt Hahn's result. As Frisch has described it, during an excited walk the two discussed the previously unheard of possibility that the uranium nucleus was split by the neutron into two pieces of approximately the same weight. As the historian of physics Roger Stuewer has written:

> . . . inspired by an awareness of the liquid-drop model of the nucleus [see Chapter 2], Frisch and Meitner calculated that the repulsive force of the high surface charge of the uranium nucleus almost completely cancels the attractive force of its surface tension, so that a bombarding neutron might induce vibrations and hence instability. Furthermore, they found that if the uranium nucleus were split into two roughly equal parts, a transformation of rest mass into kinetic energy would occur which, according to Einstein's relationship [between mass and energy], would amount to about 200 Mev, an enormous figure.

If Meitner and Frisch's explanation was correct, Fermi in 1934 and later Hahn and Meitner, the Joliot-Curies, and others had repeatedly split the uranium atom without realizing it. Even more remarkably, because the impinging neutron split the nucleus instead of attaching to it, the transuraniums predicted by Fermi, which had come to be seen as an established fact in the physics community, would be proven nonexistent.[58]

Frisch found it easy to convince Bohr, who was only astonished that he had not thought about this possibility before. As Stuewer has pointed out, however, Bohr's relative slowness in accepting Frisch's suggestion is not surprising. In his 1936 paper on the compound nucleus, Bohr had presented a very different view of how the nucleus might break up, suggesting that absorption and reemission of neutrons resembled, respectively, "the adhesion of a vapor molecule to the surface of a liquid or solid body" and "the evaporation of such substances at high temperatures." Furthermore, he had previously been inclined to see the nucleus as an elastic solid rather than as a liquid drop, even though George Gamow had already published his first proposal of a liquid drop model in 1929 while visiting the institute in Copenhagen.[59]

Bohr and Frisch's first conversation on the possibility of splitting the atomic nucleus took place only four days before Bohr departed on 7 January 1939 for an extended visit to the United States. Promising Frisch not to reveal this momentous interpretation of Hahn's experiment before it was published, Bohr urged him to prepare a paper for publication as soon as possible. Immediately upon the arrival of Bohr and Rosenfeld in the United States, however, the news was broken, and an unprecedented race began to verify the splitting of the atomic nucleus experimentally.[60]

The communication between Frisch and Bohr in this situation provides unique insight into how the Copenhagen spirit could face up to exceptional circumstances. Feeling responsible for having broken the news of Meitner and Frisch's proposal before the two had published it, and eager to make certain that they obtain priority for their discovery, Bohr dispatched a number of telegrams to Copenhagen urging immediate publication of the Meitner–Frisch proposal, as well as an experimental test. At every opportunity, he underscored Meitner and Frisch's priority to American physicists and news media. As the news from Copenhagen continued to be incomplete, Bohr's frustration increased. When Frisch finally provided a full account, he excused his slow response to Bohr by writing that he had not appreciated the urgency of the situation. The unyielding race to be first seemed foreign to Frisch's conduct of science, which conformed more to the Copenhagen spirit of thorough and dispassionate discussion. As one participant, Hilde Levi, has observed in connection with Frisch and Meitner's discovery, "it was not a competitive affair" either for them or other scientists working at the institute; "it was just simply when we got something very exciting here we would tell Father Bohr." Bohr's frustration proved to be superfluous, however, as Meitner and Frisch did become the first to publish both an interpretation of Hahn's finding on the basis of Bohr's compound nucleus model and an experimental verification of nuclear splitting.[61]

Incidentally, these activities provided a rare occasion for a direct influence on nuclear physics by the institute's experimental biology group. Asking the American W.A. Arnold whether there was a biological term for the process of a bacterium dividing into two, Frisch received the reply, "binary fission." Such was the introduction of a term into physics that soon became associated with that field rather than with biology.[62]

Just as the announcement of fission produced a race among physicists – particularly in the United States – to verify the process experimentally, so it influenced Bohr's main effort during his American visit. Thus, during the spring of 1939, he collaborated in Princeton with John Archibald Wheeler – a younger American physicist who had visited the institute in the mid-1930s – to develop a paper on "The Mechanism of Nuclear Fission." In the course of this work, the two physicists were able to fit the fission process in elaborate detail into the framework of Bohr's compound nucleus model. This collaboration was carried on in a true Copenhagen spirit, with young Wheeler serving as Bohr's sounding board and helper.[63]

In his first experimental verification of the fission process, Frisch employed as a neutron source one of the four radium–beryllium mixtures produced from the radium gift. In their next report of experiments on fission, however, Meitner and Frisch were able to make use of the institute's high-voltage equipment to produce neutrons. The installation of the new equipment, then, proved particularly timely for further experimental investigation of the fission process.[64]

Indeed, during the subsequent period the study of fission provided the major research program at the institute, involving a particularly close relationship between theory and experiment. While Bohr himself published on the theoretical aspects, the Danish experimentalists, in collaboration with Thomas Lauritsen, used the new equipment – which, in addition to the high-voltage apparatus was soon to include the cyclotron – for fission studies. In a rare coauthorship in 1940, Bohr even contributed to the experimental literature.[65]

In 1937, the numbers of publications and foreign visitors at the institute peaked for a second time. The two numbers now reached forty and twenty-one, respectively, somewhat below their previous peaks ten years before. Unlike the rapid decline after 1927, the slowdown in activity ten years later was due neither to a declining research program nor to less funding. Instead, it reflected the deteriorating political situation in Europe, where war broke out in 1939 and Denmark was occupied the following year. Thomas Lauritsen was the last temporary visitor to leave the institute, and then Europe; he barely made it. The institute was thus forced to proceed at a substantially lower level of research activity, and in the fall of 1943 Bohr himself was compelled to escape from Denmark to neutral Sweden, bringing with him his family and Jewish collaborators. Around New Year 1944, even the institute itself was occupied for a brief period by the Germans.[66]

The second decline in scientific activities at the institute was therefore due entirely to circumstances outside the pursuit and policy of research as guided by Bohr. In contrast to the first such decline after 1927, it thus did not reflect a need for a scientific reorientation when the situation returned to normal upon Bohr's return at the end of the war. On the contrary, the institute continued to thrive as a center for theoretical and experimental nuclear physics, with Bohr successfully obtaining new collaborators and ever more advanced equipment. Even some specific nuclear physics projects that had been discontinued because of the war, were taken up again when the situation returned to normal. Thus, Bohr's scientific redirection of the institute in the mid-1930s would last beyond the interwar period. In fact, nuclear physics remains a major preoccupation of the institute to this day.[67]

CONCLUSION

From 1935 until the Second World War, Bohr consolidated the scientific redirection of the institute. A substantial part of this effort consisted of active fund raising. Bohr, taking into account the special interests of each agency approached, based his arguments for support either on George Hevesy's new effort in experimental biology, or on the new theoretical and experimental emphasis on nuclear physics. By the outbreak of war, Bohr had improved the economic situation of the institute substantially.

The biological activity directed by Hevesy profited greatly from Bohr's efforts to obtain funding. Consisting at first mainly of Hevesy's enterprising effort to convince other Copenhagen institutions of the usefulness of his radioactive indicator technique in biology, more and more of the project took place within the institute, where Hevesy obtained assistants and apparatus, and even animals on which to conduct his experiments. By the outbreak of war, Hevesy's project had gained considerable international renown in its own right.

The pursuit of Bohr's own discipline of physics benefited even more from his renewed funding efforts. Whereas before the redirection, Bohr and his collaborators at the institute had devoted their main effort to general discussions of relativistic quantum theory, by 1936 the vast majority of papers resulting from their work concentrated on the thriving new field of nuclear physics. In the process, the approach to research as well as its subject matter changed: There was an increasing emphasis on experiment. Indeed, by conscientiously changing the attention at the institute to a growing subfield of physics, Bohr was able to reinstate the unity between theory and experiment that a decade and a half before had provided his motivation for establishing the institute in the first place. In this way, Bohr finally remedied the lack of experimental emphasis that he had regretted publicly already in late 1931 when he thanked the Royal Danish Academy of Sciences and Letters for the opportunity to move into the Carlsberg mansion.

Yet, unlike Hevesy's effort in experimental biology, the transition in physics did not involve a full-scale transformation from freewheeling theoretical discussions of general problems to a carefully directed research activity based on experiment. Discussions – for example, of Bohr's own compound nucleus model or his contribution to the nuclear photoeffect – were still pursued between Bohr and his younger colleagues in the open-ended Copenhagen spirit. In particular, the theoretical and experimental work on nuclear fission beginning in late 1938 cannot be argued to be part of the planned redirection to nuclear physics. Although Bohr's decision to introduce a new research program provided a useful context for Meitner and Frisch's explanation of nuclear fission, the process itself was entirely unexpected. By redirecting the institute, Bohr had accomplished a reunion between theoretical and experimental research as well as retained the Copenhagen spirit; by the late 1930s, the two enjoyed a mutually supportive coexistence.

It may not be surprising, then, that the physicists saw their own activities as increasingly detached from Hevesy's experimental biology project, which in contrast to the physicists' efforts represented a directed research activity. In fact, with the notable exception of Bohr himself, many physicists came to see Hevesy's campaign for his radioactive indicator technique, and, in particular, the use by his group of the accelerator equipment, as an unfortunate but necessary deviation from the narrow path of physics; for them, the experimental biology effort was simply an excuse for obtaining financial support for the more valid and valuable search for knowledge of the atomic nucleus. Thus, although the Copenhagen experimental biology project received continued support after the first five-year period except for a brief interruption at the end of the war, the cultural divide between the physicists fostering the Copenhagen spirit and Hevesy's group toiling with applied problems widened. Eventually, the physicists would win out. Thus, Hevesy moved his biological work to Stockholm after the war, where he came to spend the rest of his long career. According to Hilde Levi, the institute continued to provide isotopes for biological research until 1952 or 1953, when the release of isotopes ceased to be considered a security risk by the United States government. After Hevesy's departure, the experimental biology effort in Copenhagen supported by the Rockefeller Foundation would be centered at Krogh's institute, now under the leadership of Krogh's successor Paul Brandt Rehberg. The move of the effort into a biological environment owed itself not only to the enmity of the physicists; it also reflected the biologists' own desire for independence. The marriage between theoretical physics and experimental biology at the institute thus proved to be ephemeral. Yet it was instrumental in the scientific redirection there in the mid-1930s.[68]

Conclusion

The main thesis of this book can be summed up in one sentence: The transition to nuclear physics at the institute in the 1930s depended on Bohr's response to and action on changes in funding opportunities for international basic natural science, particularly as represented by the Rockefeller philanthropies. What does this statement mean, and what are its implications?

I have argued, first, that the transition would not have occurred as abruptly or the way it did without the Rockefeller Foundation's Special Research Aid Fund for European Scholars and its experimental biology program. This means not simply that these developments provided Bohr with the financial opportunity to redirect the institute in a way that he would have sought in any case, but could not otherwise afford. More fundamentally, it means that the new developments in funding opportunities were crucial in making Bohr consider a concerted scientific redirection of the institute at all.

It is to bolster this argument that I devoted so much space in Chapter 2 to delineate the purely scientific concerns and activities at the institute up to the redirection to a full-scale theoretical and experimental research program in nuclear physics. Although following eagerly the experimental developments of the "miraculous year" – 1932 – of nuclear physics, Bohr did not see them as demanding a concerted redirection. Instead, he saw them as providing fodder for freewheeling discussions of fundamental theoretical problems. Until early 1934, such discussions continued to be the central preoccupation of Bohr and his collaborators. Radical changes in physics theory did not change this situation. For example, Fermi's theory of beta decay, first proposed in late 1933, has come to be seen as the final step toward ridding the nucleus of the impetuous electron and hence clearing the way for nuclear physics as a field independent of the problematical relativistic quantum physics. Nevertheless, it did not move Bohr to redirect research at the institute. The atomic nucleus, as well as philosophically oriented questions of biology, continued to be just two of many subjects for general discussion guided by Bohr's complementarity argument.

Bohr's success at establishing his preferred discursive approach to research depended on the material conditions under which it was pursued. Because contemporary funding policy for basic science emphasized quality and prestige, Bohr encountered no difficulty in obtaining personnel and equipment to pursue his own choice of research. This policy included bringing the best young scientists to the best research centers for limited periods to confer with him. This was precisely the mode of work Bohr wanted, and it is not surprising that he exploited the opportunity for discussion with the cream of younger physicists. The dominant contemporary policy toward basic science – represented most conspicuously by the Rockefeller-funded International Education Board (IEB) – and the division of scientific labor it encouraged, were ideal for Bohr's promotion of the Copenhagen spirit.

When the situation of German scholars deteriorated after Hitler came to power in January 1933, Bohr continued to show his concern for the younger physicists, and provided many of them with a temporary haven at the institute. Thus, Bohr's own preferred course of action on the refugee problem entailed a continuation of the open-ended discussions with younger physicist colleagues. This response blended perfectly with the approach to the refugee problem of the European relief agencies, which provided temporary help to younger and unestablished physicists rather than securing the future of established professors. Even during an emergency situation, then, Bohr was able to use contemporary funding opportunities to sustain the Copenhagen spirit.

Only when confronted with the Rockefeller Foundation's different approach to the refugee problem did Bohr consider breaking with the ensconced division of labor between himself and the younger temporary visitors from abroad. Contrary to the policies of European foundations and organizations to help the refugees, the American Rockefeller Foundation sought out the most senior and established of the refugee scientists from Germany. When he saw the opportunity to obtain his old friends and experimentalist professors James Franck and George Hevesy, Bohr did not hesitate to grasp it. In contrast to the young and unestablished physicists commonly visiting the institute, these senior physicists required independence in their research. This, more than particular developments in physics itself, was the background for Bohr's decision to introduce experimental research in nuclear physics at the institute. Bohr persuaded his friends not only to conduct research in experimental nuclear physics; he also made them look into the possibilities of obtaining new equipment for that purpose. Toward the end of 1934, Bohr reported to colleagues in other institutes about the new experimental venture in nuclear physics at his institute led by Franck and Hevesy.

The conclusion that the presence of Franck and Hevesy precipitated Bohr's decision to redirect research at the institute toward nuclear physics is substantiated by several facts. First, it was the experimental discoveries of the Joliot-Curies in Paris and of Fermi in Rome – and not theoretical developments such as the evolution of the neutrino concept – that provoked Bohr to act. Bohr had followed the theoretical and experimental advances in nuclear physics closely all along. Only when compelled to find a research project for his friends, however, did he deem such advances to have practical implications for work at the institute.

Second, Bohr's reaction to the Rockefeller Foundation's experimental biology program, which supplanted the funding policy of the IEB, confirms my interpretation of both the timing and motivation for Bohr's redirection of the institute to nuclear physics. On several occasions until mid-April 1934, the foundation had enticed Bohr to present a project fitting into its new scheme of an experimental biology based on developments in the sciences of mathematics, physics, and chemistry, which were perceived as more advanced. Bohr's first response had been hesitant, and neither he nor the foundation officers were able to find a suitable compromise between Bohr's theoretical and philosophical concerns in biology and the Rockefeller Foundation's own more practical interest.

By October 1934, Bohr's attitude had completely changed. He

then presented a plan corresponding exactly with the Rockefeller Foundation's ambitions. In collaboration with George Hevesy and the Danish physiologist August Krogh, Bohr proposed to the foundation that the radioactive indicator technique developed by Hevesy be fruitfully applied to biological problems. For this purpose, Bohr needed the same experimental equipment that was required for a full-scale study of the atomic nucleus.

The Rockefeller Foundation's previous – and unsuccessful – approach to Bohr had coincided with Franck's arrival at the institute. Thus, the new proposal to the Rockefeller Foundation's experimental biology program developed in Bohr's mind only as Franck and Hevesy's experimental nuclear physics was getting off the ground. This supports my contention that Bohr decided to turn the institute to nuclear physics as a result of the presence of his two established colleagues. Moreover, Hevesy's central role in Bohr's biological plan suggests a close connection in Bohr's mind between the presence of his older colleagues, his decision to pursue nuclear physics experimentally, and his proposed project in experimental biology.

That extrascientific developments were crucial in spurring Bohr to change the direction of research at the institute does not mean that he reacted passively to them. On the contrary, once having made his decision, he was more enthusiastic than either of his friends about the new work. Although needing the presence of his friends to stir him, once stirred, it was Bohr who in the end decided vigorously to go on with the redirection. His enthusiasm for launching theoretical and experimental nuclear physics at the institute was genuine.

Moreover, Bohr's determined efforts to obtain funding – from the very first presentation of the proposal until support was granted in March 1935 – indicate that he was moved to appeal to the experimental biology program by his new enthusiasm for a concerted effort in nuclear physics. During this period he negotiated even more ably and successfully than he had a decade before. Just as his approach to the IEB in the mid-1920s had brought him expensive experimental spectroscopical equipment to study the outer part of the atom, so his consultations with the Rockefeller Foundation officers in 1934 and 1935 resulted in the acquisition of a cyclotron as well as substantial support for scientific operating expenses. In the following years, Bohr continued his efforts, turning to several agencies in order to secure the transition to theoretical and experimental nuclear physics as well as to obtain support for experimental biology. By the outbreak of the Second World War, the redirection was successfully completed. In fact, as already noted, the emphasis at the institute on nuclear physics has continued until this day.

As noted in the Introduction, a controversial issue in the history and sociology of scientific knowledge is whether factors external to science itself are responsible not only for the rate of growth, but also the direction, and even the conceptual content, of science. Which implications can be drawn from the conclusion of this book about this issue?

There can be little doubt that the changing funding opportunities for basic science were decisive for the timing of Bohr's redirection of the institute to theoretical and experimental nuclear physics. Were they also responsible for the particular choice of field? Certainly not in the most simplistic sense: No funding agency sought to exert pressure on Bohr to turn to nuclear physics. Nevertheless, the change in the division of labor at the institute brought about by the presence of Franck and Hevesy did persuade Bohr to work toward a new unity between theory and experiment there. If open-ended discussion of theoretical questions on the one hand, and research based on a unity between theory and experiment on the other, are understood as different scientific "directions," one might argue that the new funding opportunities did induce Bohr to change the direction of research at the institute. That the new unity was to be based on nuclear physics, however, was due to developments within physics itself.

If the funding opportunities were not responsible for the choice of field, they can hardly be expected to explain the content of Bohr's ideas about the atomic nucleus once the choice had been made. Again, the story is not quite so simple. In spite of the abundance of Bohr's letters and manuscripts from this period, the conceptual origins of his new ideas about the atomic nucleus from early 1936 have proven to be notoriously difficult to track down. The preceding account, in fact, suggests a broader approach that includes Bohr's policy, as well as pursuit, of science at the institute.[1]

As we saw in Chapter 5, Bohr's concern in his famous 1936

lecture, which introduced the compound nucleus, contrasted with his earlier approach to nuclear problems; he no longer sought to use anomalies in the experimental study of the nucleus as a means of obtaining a deeper general understanding of quantum theory. On the contrary, he now tried more constructively to postulate a concrete model for the nucleus in order to explain experimental results. This motivation was vastly different from his attempts in the preceding years to base energy nonconservation in the atomic nucleus – as well as the uniqueness of biological phenomena – on his complementarity argument. In fact, not only did he become less general and philosophical in his approach; he no longer saw energy nonconservation as a viable option at all.

Whereas Bohr designed his earlier speculations about the atomic nucleus as examples of the problems involved in a generalization of quantum theory, then, he formulated his compound nucleus model of early 1936 to explain specific experimental results. As such, it reflected his decision to reintroduce a unity between theory and experiment at the institute. Because this decision arose from his response to the change in the Rockefeller philanthropies' funding policy for basic science, even the *content* of Bohr's nuclear theorizing may not have been entirely independent of the "external factors" considered in this book.

Obviously, it is too simplistic to conclude that the change of funding policy "caused" Bohr's concept of the nucleus, and more specifically the compound nucleus model. Nevertheless, a broadening of perspective from the push of scientific influences might be helpful; this study indicates that the pull of Bohr's motivation for and effort toward a concerted redirection of research also played a role. A more general approach along such lines might help not only in enriching the history of Bohr and the institute, but even contribute to an understanding of the origins of new conceptual contributions such as Bohr's new model of the atomic nucleus from 1936.

In addition to pointing to a relationship between the "internal" event of the transition to nuclear physics at the institute and an "external" change in funding opportunities, this book also argues for an *interdisciplinary* approach to the history of science. Although the main field of investigation at the institute was physics, the redirection to nuclear physics depended upon the projects and involvements of the physical chemist Hevesy and the physiologist Krogh. Of course, there can be no doubt that Bohr used the arguments supplied by Hevesy's and Krogh's projects in order to secure the transition at the institute to nuclear physics. Yet he had to transcend the boundaries of physics in order to do so. Moreover, Bohr's biological interest – not only in its philosophical aspects, but also in Hevesy's concrete project once it got off the ground – was genuine, with Bohr providing encouragement and participating in seminars throughout the period. Thus, Bohr's introduction at the institute of experimental biology was not just a strategy for developing nuclear physics, but a real concern in its own right. A fully interdisciplinary approach seems required to understand changes even in a discipline as established and apparently autonomous as physics.

How can the physicists' reminiscences of the institute between the world wars, described in the Prologue, be reconciled with the thesis of this book? Can their recollection of a research atmosphere completely unhampered by extrascientific considerations be squared with the indisputable importance of funding opportunities argued here?

I have shown that the Copenhagen spirit recalled by the physicists was not primarily responsible for the scientific redirection of the institute in the 1930s. Indeed, it constituted a conservative rather than a progressive force in the institute's development. Thus, the Copenhagen spirit of disinterested theoretical discussion had replaced the unity between theory and experiment initiated at the institute's establishment, and it took the presence of Franck and Hevesy – which was contrary to the traditional division of labor between Bohr and his younger colleagues – to trigger Bohr to reintroduce this unity on a new basis.

Yet, in spite of some of the reminiscing physicists, there need be no contradiction between Bohr's exploitation of funding opportunities and the Copenhagen physicists' pursuit of pure physics for its own sake. Whereas Bohr was strongly involved in both activities, the younger visiting physicists were sheltered from extrascientific concerns. First, coming to Copenhagen for only a limited time, they had no stake in participating in the administration of science there, and were most likely to be totally uninterested in matters of policy. Second, Bohr discouraged such activity on the part of his junior colleagues, knowing that they would profit most from an intensive attention to science during their short stay. The Copenhagen physicists even felt inhibited *discussing* such matters with their spiritual father. Thus, in order to explain in an interview why she was not certain about the source of her stipend at the institute, Hilde Levi candidly exclaimed, "But you never asked Bohr where he got the money from."[2]

Nevertheless, Bohr's natural inclination was to promote and participate in the Copenhagen spirit, which, after all, was an expression of his unique teaching and research qualities. Only when persuaded by other factors – such as the need at the end of the century's first decade to establish a new institute, the requirement for expansion in the mid-1920s, or the problem of the refugee physicists in the 1930s – did Bohr turn into an active policymaker and fund-raiser.

In pointing out the importance of philanthropy in the scientific redirection of the institute in the 1930s, then, I do not imply that the content and practice of science there can be reduced to economic considerations. In particular, the Copenhagen spirit recalled by the physicists constituted a real and important basis for scientific work. The peculiar working atmosphere at the institute provided the cornerstone for its accomplishments not only *before* Bohr's decision about a scientific redirection; even *after* the transition had been accomplished, work in nuclear physics – as earlier in quantum theory more generally – was pursued in the same spirit. On this level, then, the physicists' recollections of disinterested and informal discussions of physics with the great master Bohr are accurate.

Bohr's genius as a policymaker lay in completely separating his extrascientific efforts from his collaboration with the young physicists visiting the institute. This separation, which was helped immensely by the existing funding policies for international basic science, lends reality to the physicists' version of the history of the institute between the world wars. A full account of that history, however, has to consider both sides of the divide.

Finally, let me reiterate the limitation of my approach mentioned in the Introduction. Bohr and his institute is a natural and well-considered choice, and the limitation to one institute allows, furthermore, for close attention to historical detail. In short, the emphasis on one director of one research institute for basic science has made it possible to concentrate in some detail on the interrelationship between the policy of science and the pursuit of science. Nevertheless, this limitation also prohibits a treatment of such interrelationships at other institutes, as well as a discussion of the interactions and collaborations between such institutes. To what extent, for example, may elements of the Copenhagen spirit be found at other institutes? Some general tenets of the Copenhagen spirit, such as the emphasis on disinterestedness, was part of the ideology of any contemporary basic science. Moreover, the atmosphere at the institute, and especially the informal conferences, was seen as a model by many, particularly those who were to direct new research centers after a long stint in Copenhagen.[3]

Other scientists in other places, too, were slow to appreciate the nucleus as an entity in its own right for theoretical and experimental investigation. What were the circumstances for the transformation of *their* concept of nuclear physics? These directors and institutes, of course, lived in the same world as Bohr and experienced, to a greater or lesser extent, the same developments in funding opportunities and other factors affecting their science. But did these developments affect *their* science, and if so, how? Such questions will have to be answered before a fuller understanding can be achieved of the change in physics research to nuclear physics more generally. This book offers some ideas on how to approach such questions historically. It suggests, in particular, that concentrated studies of the interrelationship between an institute director's science and policy activities may provide useful perspectives. In turn, several such studies may provide a basis for a more synthetic approach.

In spite of the obvious limitations of this book, I hope to have supplied at least one of the missing links between the history of science before the Second World War, which too often has concentrated exclusively on scientific developments in their own right, and the history of the later period, when the relationship between science and society has become too obvious to ignore. If so, I have achieved one important purpose of my book.

Notes on sources

The notes for each chapter give detailed information on the published and unpublished sources on which this book is based. For the benefit of the reader, this section provides in one place the basic data on the material used.

UNPUBLISHED MATERIAL

The unpublished material is of two kinds. First, and most substantively, I have sought letters, manuscripts, and other kinds of primary sources in several archives. Second, I have sought information directly from the historical subjects themselves, through correspondence and interviews.

Archives

I have used the resources of the following archives:

Cambridge University Archives
The Carlsberg Breweries Archives, Copenhagen
The Carlsberg Foundation Archives, Copenhagen
The Danish State Archive, Copenhagen
Hebrew University Archives, Jerusalem
Institutt for medisinsk genetikk at the University of Oslo
The Joseph Regenstein Library at the University of Chicago
Library of Congress, Washington, D.C.
The Lilly Library at Indiana University
The Niels Bohr Archive, Copenhagen
The Niels Bohr Library, American Institute of Physics, New York City
The Rockefeller Archive Center, Pocantico Hills, New York
The Royal Library, Copenhagen
The Staatsbibliothek, West Berlin

The Department of Rare Books and Special Collections of the University of Rochester Library

The following are the archives that proved to be the most useful.

The Niels Bohr Archive, Copenhagen
This archive has provided most of the material for the book. The Bohr Scientific Correspondence (BSC), the Bohr Scientific Manuscripts (MSS), and oral history interviews with physicists – all of which form part of the Archive for the History of Quantum Physics (AHQP) – have been invaluable. Although the AHQP is available in several other places, access to these sources in the Niels Bohr Archive (NBA) is superior for two reasons. First, in the few cases where the microfilm was unreadable, I was able to consult the original letter or manuscript. Second, the NBA has unique chronological listings of all of Bohr's letters, which for a project like mine are more helpful than listings by correspondents.

The NBA also houses collections that cannot be found anywhere else. Most useful for this study has been the Bohr General Correspondence (BGC), which comprises material that can be classified neither as "scientific" nor "private." Part of this collection has been microfilmed as a "Special File" (BGC-S). This special file includes Bohr's dealings with funding agencies – which have been consulted extensively for the book. In this connection, the handwritten Budget Book of the institute has also been extremely useful.

I have also made considerable use of the George Hevesy Scientific Correspondence (HSC), which was compiled and arranged by Hilde Levi during my stay at the NBA. In this collection, the letters in Hevesy's own files are supplemented with photocopies of Hevesy's own letters, obtained from other archives. The HSC has thus given me access to material originally in collections such as the Fritz Paneth Nachlass in the Max Planck Gesellschaft, the Johannes Stark Nachlass in the Staatsbibliothek – both in West Berlin – and the Ernest Rutherford Collection at Cambridge University.

Also useful have been the collection of manuscripts submitted to Bohr and, especially, *Universitetets Institut for teoretisk Fysik, Afhandlinger* – a collection of publications from the institute from 1918 to 1959. Although not fully complete, this collection constitutes an invaluable reference tool for research at the institute during the period.

While doing research at the NBA, I profited from being close to Hilde Levi's effort to collect and classify photographs for the archive. The assortment of photos in this book is thanks to that fortunate experience. Audiovisual material – which also came to light during the preparations for this book – also taught me much about Bohr and his institute, even though in ways too subtle to be reflected explicitly in the text.

Finally, because the NBA is physically part of Bohr's institute itself, working there gave me a unique opportunity for close contact with some of the subjects in this book. This fact is reflected in my list of interviews, and whatever authenticity this book has results from that experience.

Rockefeller Archive Center, Pocantico Hills, New York
This immensely rich archive has been crucial for my work, not least in bringing to my attention the importance of funding and funding policy for the scientific development of Bohr's institute; the nature of this connection came as a complete surprise at the time. The Rockefeller Archive Center (RAC) contains detailed files on the International Education Board's (IEB) and the Rockefeller Foundation's funding of the institute. In addition, I have made use of information about the foundation's "program and policy," its involvement in the refugee problem during the Nazi period, and the IEB's and the Rockefeller Foundation's fellowship programs. Material dealing specifically with support of other science institutions has also been useful. Particularly candid comments and information are found in the comprehensive diaries of the Rockefeller philanthropies' officers.

The Carlsberg Foundation Archives, Copenhagen
I have made extensive use of this well-arranged archive, even though the use of its collections for my purposes is somewhat limited by two circumstances. First, much of the written communication is duplicated in the BGC at the NBA. Second, most of the contact between Bohr and the foundation took the form of informal conversations that were never recorded.

Niels Bohr Library, American Institute of Physics, New York City

My workplace while I developed my dissertation into a publishable book manuscript, the Niels Bohr Library (NBL) is a convenient repository for the microfilmed AHQP. In addition, I have made extensive use of the large collection of oral history interviews with people working at the institute, which unlike the AHQP interviews extend into the 1930s. The NBL also contains one of the largest and most used photograph collections of physicists, from which this book has also benefited.

Other archives

I have also consulted the James Franck Papers in the Joseph Regenstein Library at the University of Chicago; the Nachlass Born in the Staatsbibliothek, West Berlin; the Otto Robert Frisch Papers at Cambridge University; the August and Marie Krogh Papers in the Royal Library, Copenhagen; the Aage Friis Papers in the Danish State Archive, Copenhagen; the Albert Einstein Papers at Hebrew University, Jerusalem; the J. Robert Oppenheimer Papers in the Library of Congress, Washington, D.C.; the archives of the Carlsberg Breweries, Copenhagen; the Otto Lous Mohr Papers at the Institutt for medisinsk genetikk of the University of Oslo; the Hermann Muller Papers at the Lilly Library of Indiana University; and the Rush Rhees Papers in the Department of Rare Books and Special Collections of the University of Rochester Library.

Correspondence and interviews

My correspondence and interviews for this book have been deposited in the NBA. They span a period of nine years.

Correspondence

The bulk of my correspondence was carried out in 1980, before the focus of the book had been clearly defined. At that time, I wrote to fifty-one scientists, asking whether they would be willing to respond to questions about their experiences at the institute between the two world wars. Thirty-three responded, of whom twenty-one subsequently provided detailed answers to a long list of questions relating to my interest in the institute's history. Many letters contain information that is broader than or different from the focus of the book. The following is a list of the most substantial of the responses.

W.D. Armstrong, 16 Apr 1980
W.A. Arnold, 24 Aug 1980
E.W. Beth, 7 Jul 1980
F. Bloch, 21 May 1980
M. Delbrück, 21 Apr 1980, 13 May 1980
D. ter Haar, 5 Jun 1982
J. Kistemaker, 21 Apr 1980, 29 May 1980
R. de Laer Kronig, 2 Nov 1980, 1 Jan 1981
L.J. Laslett, 29 Apr 1980, 15 Jul 1980
H. Levi (tape recording), 7 Dec 1987
R. Bruce Lindsay, 30 Apr 1980, 18 Jun 1980
L. Nordheim, 25 Apr 1980
L. Pauling, 29 Apr 1980, 28 May 1980
P. Brandt Rehberg, 5 May 1981
S. Rozental, 10 Aug 1980
B. Schneider [widow of E.E. Schneider], 9 Feb 1981
L. Simons, 9 Jan 1981
E. Teller, 11 Dec 1980
L.H. Thomas, 23 May 1980
G.E. Uhlenbeck, 23 Jun 1980
V. Weisskopf, 30 Oct 1980, 24 Feb 1983, 6 Feb 1989, 17 Feb 1989, 17 Mar 1989
C.F. von Weizsäcker, 18 Dec 1980
H. Wergeland, 27 Oct 1980
J.A. Wheeler, 30 Aug 1980

A few more letters, usually containing more specific information in connection with the research for this book, have been deposited in the same collection in the NBA.

Interviews

In the course of my research, I have conducted the following interviews, some of which were tape-recorded (r), and one of which was transcribed (t). The recorded interviews were conducted either in Danish (D) or English (E). Notes for most of the interviews, as well as tapes and transcript, are deposited in the NBA. The interview with Wheeler was part of my oral interview project for the American Institute of Physics, and was therefore first deposited in that institution's Niels Bohr Library. NBI = Niels Bohr Institute; MIT = Massachusetts Institute of Technology; Inst. = Institute.

H. Bethe	NBI, 1 Apr, 8 Apr, 14 Apr 1981
G. Breitscheid (r,D)	Copenhagen, 18 Jan 1983
N.O. Lassen	NBI, 12 May 1981
H. Levi	NBI, 28 Oct 1980
H. Levi (r,D)	NBI, 10 Sep, 16 Sep, 22 Sep, 24 Sep 1981

L.J. Mullins	University of Maryland, 12 May 1980
R. Peierls	NBI, 11 Sep 1980
S. Rozental	NBI, 9 Oct 1980
B. Schmidt-Nielsen	Bar Harbor, Maine, Jun 1981
H.E. Ussing (r,D)	August Krogh Inst., Copenhagen, 6 May 1981
V. Weisskopf (r,E)	NBI, 7 Oct 1980
V. Weisskopf (r,E)	MIT, 5 Jun 1981
V. Weisskopf (r,E)	Cornell University, 25 Oct 1984
J. Wheeler (t,E)	Princeton University, 4 May 1988

PUBLISHED MATERIAL

The published material used for the research on this book is of three main kinds: reference works, contemporary scientific publications, and historical monographs and articles. Historical articles are often found in anthologies, which I have listed in a separate section; information on specific articles in these anthologies is provided in the section on monographs and articles.

Reference works

American Men of Science: A Biographical Directory, Jaques Cattell (ed.) (New York: Science Press), particularly the sixth and seventh editions, published in 1938 and 1944, respectively.

Banking and Monetary Statistics, 1914–1941 (Washington, D.C.: The Board of Governors of the Federal Reserve System, 1976).

Biologisk Selskabs Forhandlinger i Vinter-Halvaaret 1897–98 (Copenhagen: Biologisk Selskab, 1898).

Carlsbergfondets Understøttelser 1876–1936 (Copenhagen: Bianco Luno, 1937).

Copenhagen University Yearbook, published in Danish as *Aarbog for Københavns Universitet, Kommunitetet og Den Polytekniske Lereanstalt, Danmarks Tekniske Højskole* (Copenhagen: A/S J.H. Schultz Bogtrykkeri), several years.

Dansk Biografisk Leksikon (Copenhagen: Gyldendal), in particular, the third edition published in 1981.

Dictionary of Scientific Biography [15 volumes], Charles Gillispie (gen. ed.) (New York: Charles Scribner's Sons, 1970–1978).

Landsforeningen til Kræftens Bekæmpelse: Aarsberetning 1929.

European Historical Statistics 1705–1975 (Second Revised Edition), B.R. Mitchell (ed.), (New York: Facts on File, 1981).

Nobel Lectures, including presentation speeches and laureates' biographies: Physics 1922–1941 (Amsterdam, London, New York: Elsevier Publishing Company, 1965).

Nobel Lectures, including presentation speeches and laureates' biographies: Chemistry, 1922–1941 (Amsterdam, etc.: Elsevier Publishing Company, 1966).

Nobel Lectures, including presentation speeches and laureates' biographies: Chemistry, 1942–1962 (Amsterdam, etc.: Elsevier Publishing Company, 1964).

Oversigt over Det Kongelige Danske Videnskabernes Selskabs Forhandlinger (Copenhagen: Andr. Fre. Høst & Søn), in particular, *Juni 1924–Maj 1925* (1925).

Rask–Ørsted Fondet: Beretning for 1919–1920, 1 (Copenhagen: Rask–Ørsted Fondet, 1920).

Rockefeller Foundation: Annual Report (New York: Rockefeller Foundation, undated), several years.

Sources for the History of Quantum Physics: An Inventory and Report, Thomas S. Kuhn, John L. Heilbron, Paul Forman, and Lini Allen (eds.) (Philadelphia: The American Philosophical Society, 1967).

Collections of scientists' papers

This section contains a list of published collections of scientists' published or unpublished papers used in this book. For a complete list of scientific papers, which may or may not have been republished in such collections, I refer to the notes to the individual chapters.

The *Niels Bohr Collected Works,* the first volume of which appeared in 1972 and which is still being published, has been particularly useful. It claims completeness only for Bohr's published works, using unpublished manuscripts mainly for illustration. It is published by the North-Holland Publishing Company, which has its main office in Amsterdam. The first three volumes, under the general editorship of Léon Rosenfeld, are: J. Rud Nielsen (ed.), *Early Work (1905–1911)* [published 1972]; Ulrich Hoyer (ed.), *Work on Atomic Physics (1912–1917)* [1981]; J. Rud Nielsen (ed.), *The Correspondence Principle (1918–1923)* [1976]. Volume four is J. Rud Nielsen (ed.), *The Periodic System (1920–1923)* [1977]. For subsequent volumes, Erik Rüdinger has served as general editor: Klaus Stolzenburg (ed.), *The Emergence of Quantum Mechanics (Mainly 1924–1926)* [volume five, 1984]; Jørgen Kalckar (ed.), *Foundations of Quantum Physics I (1926–1932)* [volume six, 1985]; Jens Thorsen (ed.), *The Penetration of Charged Particles through Matter (1912–1954)* [volume eight, 1987]; Rudolf Peierls (ed.), *Nuclear Physics (1929–1952)* [volume nine, 1986]. Volumes seven, ten, and eleven are yet to be published. Volume seven, to be edited by Jørgen Kalckar, will be titled *Foundations of Quantum Physics II.* Volume ten, under the editorship of David Favrholdt, will contain Bohr's generally oriented articles on biology, psychology, and culture. Finally, volume eleven will deal with Bohr's efforts toward internationalism both within and outside physics, and his general contributions to Danish society.

The Philosophical Writings of Niels Bohr, Volume 1: Atomic Theory and the Description of Nature; Volume 2: Essays 1933–1957 on Atomic Physics and Human Knowledge; Volume 3: Essays 1958–1962 on Atomic Physics and Human Knowledge (Woodbridge, Connecticut: Ox Bow Press, 1987) constitute handy reprints of Bohr's contributions to philosophy.

Collections of papers of other scientists (alphabetized by the scientists in question) include:

Enrico Fermi, *Collected Papers, volume 1: Italy 1921–1938* (Chicago: University of Chicago Press, 1961).

George Hevesy, *Selected Papers of George Hevesy* (London, etc.: Pergamon Press, 1967).

George Hevesy, *Adventures in Radioisotope Research: Collected Papers* [2 volumes, consecutively paginated, 2nd volume beginning on p. 517] (New York, etc.: Pergamon Press, 1972).

Frédéric and Irène Joliot-Curie, *OEvres Scientifiques Complètes* (Paris: Presses Universitaires de France, 1961).

Herbert G. Dingle and G.R. Martin (eds.), *Chemistry and Beyond: A selection from the writings of the late Professor F.A. Paneth* (New York, London, Sydney: Interscience Publishers, 1964).

Wolfgang Pauli, *Aufsätze und Vorträge über Physik und Erkenntnistheorie*, Victor F. Weisskopf (ed.) (Braunschweig: Friedr. Vieweg & Sohn, 1961).

Ralph de Laer Kronig and Victor F. Weisskopf (eds.), *Collected Scientific Papers by Wolfgang Pauli in Two Volumes, Vol. 2* (New York, London, Sydney: Interscience, 1964).

Armin Hermann, Karl von Meyenn, and Victor F. Weisskopf (eds.), *Wolfgang Pauli, Scientific Correspondence with Bohr, Einstein, Heisenberg, a.o., Volume I: 1919–1929* (New York, Heidelberg, Berlin: Springer-Verlag, 1979).

Karl von Meyenn, Armin Hermann, and Victor F. Weisskopf (eds.), *Wolfgang Pauli, Scientific Correspondence with Bohr, Einstein, Heisenberg a.o., Volume II: 1930–1939* (Berlin, etc.: Springer-Verlag, 1985).

Robert S. Cohen and John J. Stachel (eds.), *Selected Papers of Léon Rosenfeld* [Robert S. Cohen and Marx W. Wartowsky (eds.), *Boston Studies in the Philosophy of Science, Volume XXI*] (Dordrecht, Boston, London: D. Reidel Publishing Company, 1979).

Ernest Rutherford, *The Collected Papers of Lord Rutherford: Volume Three, Cambridge* (London: George Allen and Unwin Ltd., 1965).

Finally, I have made use of anthologies of original scientific articles by several authors.

Romer, Alfred (ed.), *Radiochemistry and the Discovery of Isotopes* [*Classics of Science, Volume VI*] (New York: Dover, 1970).

Julian Schwinger (ed.), *Selected Papers on Quantum Electrodynamics* (New York: Dover, 1958).

Strachan, Charles, *The Theory of Beta Decay* (Oxford, etc.: Pergamon Press, 1961).

Anthologies of historical articles

Birks, J.B. (ed.), *Rutherford at Manchester* (London: Heywood & Company Ltd., 1962).

de Boer, Jorrit, Erik Dal, and Ole Ulfbeck (eds.), *The Lesson of Quantum Theory: Niels Bohr Centenary Symposium October 3–7, 1985* (Amsterdam: North-Holland Physics Publishing, 1986).

Fleming, Donald and Bernard Bailyn (eds.), *The Intellectual Migration: Europe and America, 1930–1960 [Perspectives in American History Volume 2, 1968]* (Cambridge, Massachusetts: Harvard University Press, 1969).

French, A.P. and P.J. Kennedy, *Niels Bohr: A Centenary Volume* (Cambridge and London: Cambridge University Press, 1985).

Hendry, John (ed.), *Cambridge Physics in the Thirties* (Bristol: Adam Hilger, Ltd., 1984).

Holter H. and K. Max Møller (eds.), *The Carlsberg Laboratory 1876–1976* (Copenhagen: Rhodos, 1976).

Holton, Gerald, *Thematic Origins of Scientific Thought: Kepler to Einstein* (Cambridge, Massachusetts: Harvard University Press, 1973).

Holton, Gerald, *The Scientific Imagination: Case Studies* (Cambridge, England: Cambridge University Press, 1978).

International Congress for the History of Science, two proceedings: *Dixième Congrès international d'histoire des sciences, 1962* [vol. I] (Paris: Hermann, 1964); *XIVth International Congress of the History of Science* [Tokyo and Kyoto 19–27 August 1974]: *Proceedings No. 2* (Tokyo: Science Council of Japan, 1975).

Københavns Universitet 1479–1979 [14 volumes] *bind XII: Det matematisk-naturvidenskabelige Fakultet, 1. del* (Copenhagen: G.E.C. Gads Forlag, 1983).

Københavns Universitet 1479–1979, bind XIII: Det matematisk-naturvidenskabelige Fakultet, 2. del (Copenhagen: G.E.C. Gads Forlag, 1979).

Kuhn, Thomas S., *The Essential Tension: Selected Studies in Scientific Tradition and Change* (Chicago and London: University of Chicago Press, 1977).

Niels Bohr – Et Mindeskrift [Fysisk Tidsskrift 60 (1962)] (Copenhagen: Selskabet for Naturlærens udbredelse, 1963).

Price, William C., Seymour S. Chissick, and Tom Ravensdale (eds.), *Mechanics: The first fifty years* (London: Butterworths, 1973).

Reingold, Nathan (ed.), *The Sciences in the American Context: New Perspectives* (Washington, D.C.: Smithsonian Institution Press, 1979).

Rozental, Stefan (ed.), *Niels Bohr: His life and work as seen by his friends and colleagues* (Amsterdam: North-Holland Publishing Company, 1968).

Shea, William R. (ed.), *Otto Hahn and the Rise of Nuclear Physics* (Dordrecht, Boston, Lancaster: D. Reidel Publishing Company, 1983).

Spiegel-Rösing, Ina and Derek de Solla Price (eds.), *Science, Technology and Society: A Cross-Disciplinary Perspective* (London and Beverley Hills: SAGE Publications, 1977).

Stuewer, Roger H., (ed.), *Nuclear Physics in Retrospect: Proceedings of a Symposium on the 1930s* (Minneapolis: University of Minnesota Press, 1979).

Ullmann-Margalit, Edna (ed.), *The Kaleidoscope of Science: The Israel Colloquium: Studies in History, Philosophy, and Sociology of Science, Volume 1* [Robert S. Cohen and Marx W. Wartofsky (eds.), *Boston Studies in the Philosophy of Science, Volume 94*] (Dordrecht, etc.: D. Reidel Publishing Company, 1986).

Weiner, Charles (ed.), *History of Twentieth Century Physics* [*Proceedings of the International School of Physics "Enrico Fermi," Course LVII, Varenna on Lake Como, Villa Monastero, 31st July – 12th August 1972*] (New York and London: Academic Press, 1977).

Weisskopf, Victor F., *Physics in the Twentieth Century: Selected Essays* (Cambridge, Massachusetts and London, England: Massachusetts Institute of Technology Press, 1972).

Historical monographs and articles

Aaserud, Finn, *The Redirection of the Niels Bohr Institute in the 1930s: Response to Changing Conditions for Basic Science Enterprise*, PhD dissertation at the Johns Hopkins University 1984 (microfilms international 8510398).

Aaserud, Finn, "Niels Bohr as Fund Raiser" in *Physics Today 38* (Oct 1985), 38–46.

Abir-Am, Pnina, "The Discourse of Physical Power and Biological Knowledge in the 1930s: A Reappraisal of the Rockefeller Foundation's 'Policy' in Molecular Biology" in *Social Studies of Science 12* (1982), 341–382.

Adler, David Jens, "Childhood and Youth" in Rozental (ed.), *Niels Bohr* (see previous section on anthologies), pp. 11–37.

Allen, Garland E., "J.S. Haldane: The Development of the Idea of Control Mechanisms in Respiration" in *Journal of the History of Medicine and Allied Sciences 22* (1967), 392–412.

Allen, Garland E., *Thomas Hunt Morgan: The Man and His Science* (Princeton: Princeton University Press, 1978).

Allen, O.A., "Hugo Fricke and the Development of Radiation Chemistry: A Perspective View" in *Radiation Research 17* (1962), 255–261.

Allibone, T.E., "Metropolitan-Vickers Electrical Company and the Cavendish Laboratory" in Hendry (ed.), *Cambridge Physics* (see previous section on anthologies), pp. 150–173.

Anderson, C.D., "Early Work on the Positron and Muon" in *American Journal of Physics 29* (1961), 825–830.

Badash, Lawrence, "Nuclear Physics in Rutherford's Laboratory before the Discovery of the Neutron" in *American Journal of Physics 51* (1983), 884–889.

Bethe, Hans A., "The Happy Thirties" in Stuewer (ed.), *Nuclear Physics in Retrospect* (see previous section on anthologies), pp. 11–26.

Beyerchen, Alan D., *Scientists under Hitler: Politics and the Physics Community in the Third Reich* (New Haven and London: Yale University Press, 1977).

Beyerchen, Alan D., *James Franck and the Social Responsibility of the Scientist*, forthcoming book.

Blædel, Niels, *Harmoni og Enhed: Niels Bohr – En Biografi* (Copenhagen: Rhodos, 1985); translated into English as Niels Blædel, *Harmony and Unity: The Life of Niels Bohr* (Madison, Wisconsin: Science Tech Publishers; Berlin, etc.: Springer-Verlag, 1988).

Blumberg, Stanley A. and Gwinn Owens, *Energy and Conflict: The Life and Times of Edward Teller* (New York: G.P. Putnam's Sons, 1976).

Bohr, Niels, "Mindeord over Harald Høffding" in *Oversigt over Det Kongelige Danske Videnskabernes Selskabs Forhandlinger 1931–1932*, pp. 131–136.

Bohr, Niels, "Ole Chievitz" in *Ord och Bild 55* (1947), 49–53.

Bohr, Niels, "The Rutherford Memorial Lecture 1958: Reminiscences of the Founder of Nuclear Science and of Some Developments Based on his Work" in *Proceedings of the Physical Society 78* (1961), 1083–1115; reprinted in Birks (ed.), *Rutherford at Manchester* (see previous section on anthologies), pp. 114–167, and in Bohr, *Philosophical Writings 3* (see previous section on publications of scientists' papers), pp. 30–73.

Born, Max, *My Life: Recollections of a Nobel Laureate* (London: Taylor & Francis Ltd., 1978).

Brickwedde, F.G., "Harold Urey and the Discovery of Deuterium" in *Physics Today 35* (Sep 1982), 34–39.

Bromberg, Joan, "The Impact of the Neutron: Bohr and Heisenberg" in *Historical Studies in the Physical Sciences 3* (1971), 307–341.

Bromberg, Joan, "The Concept of Particle Creation Before and After Quantum Mechanics" in *Historical Studies in the Physical Sciences 7* (1975), 161–191.

Broszat, Martin, *The Hitler State: The Foundation and Development of the Internal Structure of the Third Reich* (London and New York: Longman, 1981).

Brown, Laurie M., "The Idea of the Neutrino" in *Physics Today 31* (Sep 1978), 23–28.

Brown, Laurie M., "Yukawa's Prediction of the Meson" in *Centaurus 25* (1981), 71–132.

Brown, Laurie M. and Donald F. Moyer, "Lady or Tiger? – The Meitner-Hupfeld Effect and Heisenberg's Neutron Theory" in *American Journal of Physics 52* (Feb 1984), 130–136.

Busch, Alexander, *Die Geschichte des Privatdozenten: Eine soziologische Studie zur Grossbetrieblichen Entwicklung der deutschen Universitäten* (Stuttgart: Ferdinand Enke Verlag, 1959).

Cahan, David, "Werner Siemens and the Origins of the Physikalisch-Technische Reichsanstalt, 1872–1887" in *Historical Studies in the Physical Sciences 12* (1982), 253–285.

Cahan, David, *An Institute for an Empire: The Physikalisch-Technische Reichsanstalt 1871–1918* (Cambridge University Press: Cambridge, New York, Melbourne, Sydney, 1989).

Casimir, Hendrik B.G., "Recollections from the Years 1929–1931" in Rozental (ed.), *Niels Bohr* (see previous section on anthologies), pp. 109–113.

Casimir, Hendrik B.G., *Haphazard Reality: Half a Century of Science* (New York, etc.: Harper & Row, 1983).

Cassidy, David C., "Cosmic Ray Showers, High Energy Physics, and Quantum Field Theories: Programmatic Interactions in the 1930s" in *Historical Studies in the Physical Sciences 12* (1981), 1–39.

Cassidy, David C., biography of Heisenberg, forthcoming book.

Chadwick, James, "Some Personal Notes on the Search for the Neutron" in *Dixième Congrès* (see previous section on anthologies), pp. 159–162; reprinted in Hendry (ed.), *Cambridge Physics* (see previous section on anthologies), pp. 42–45.

Christiansen, J.A., "Julius Petersen, Einar Biilmann og J.N. Brønsted" in *Kemien i Danmark, III: Danske Kemikere* [En forelæsningsrække ved Folkeuniversitetet, København 1964] (Copenhagen: Nyt Nordisk Forlag, 1968), pp. 69–94.

Clark, Ronald W., *Einstein: The Life and Times* (New York: World Publishing Company, 1971).

Clemmensen, C.A., *Radiumfondet, Oprettet til Minde om Kong Frederik VIII, 1912–1929: Et Afsnit af Kampen mod Kræften i Danmark* (Copenhagen: Radiumfondet, 1931).

Coben, Stanley, "The Scientific Establishment and the Transmission of Quantum Mechanics to the United States, 1919–32" in *American Historical Review 76* (1971), 442–466.

Cockcroft, John D., "George de Hevesy 1885–1966" in *Biographical Memoirs of Fellows of the Royal Society 13* (1967), 125–166.

Cornell, Thomas David, *Merle A. Tuve and His Program of Nuclear Studies at the Department of Terrestrial Magnetism: The Early Career of a Modern American Physicist*, PhD dissertation at the Johns Hopkins University 1986 (microfilms international 8609316).

Courant, Richard, "Fifty Years of Friendship" in Stefan Rozental (ed.), *Niels Bohr* (see previous section on anthologies), pp. 301–309.

Culotta, Charles A., "Tissue Oxidation and Theoretical Physiology: Bernard, Ludwig, and Pflüger" in *Bulletin of the History Medicine 44* (1970), 109–140.

Davies, Shannon, *American Physicists Abroad: Copenhagen, 1920–1940*, PhD dissertation at the University of Texas at Austin, 1985 (microfilms international 8609491).

De Maria, Michelangelo and Arturo Russo, "The Discovery of the Positron" in *Rivista di Storia della Scienza 2* (1985), 237–286.

Dickson, David, *The New Politics of Science* (New York: Pantheon Books, 1984), republished, with a new preface, by Chicago University Press in 1988.

Dingle, Herbert G., and G.R. Martin, "Introduction" in *idem*. (eds.), *Chemistry and Beyond* (see previous section on publications of scientists' papers), pp. ix–xxi.

Dirac, P.A.M., "Recollections of an Exciting Era" in Weiner (ed.), *History of Twentieth Century Physics* (see previous section on anthologies), pp. 109–146.

Dresden, M., *H.A. Kramers: Between Tradition and Revolution* (New York, etc.: Springer-Verlag, 1987).

Dupree, A. Hunter, *Science in the Federal Government: A History of Policies and Activities* (Baltimore and London: Johns Hopkins University Press, 1986), originally published by Harvard in 1957.

Favrholdt, David, "Niels Bohr and Danish Philosophy" in *Danish Yearbook of Philosophy 13* (1976), 206–220.

Favrholdt, David, "On Høffding and Bohr: A Reply to Jan Faye" in *Danish Yearbook of Philosophy 16* (1979), 73–77.

Favrholdt, David, "The Cultural Background of the Young Niels Bohr" in *Rivista di Storia della Scienza 2* (1985), 445–461.

Faye, Jan, "The Influence of Harald Høffding's Philosophy on Niels Bohr's Interpretation of Quantum Mechanics" in *Danish Yearbook of Philosophy 16* (1979), 37–72.

Faye, Jan, "The Bohr-Høffding Relationship Reconsidered" in *Studies in History and Philosophy of Science 19* (1988), 321–346.

Feigl, Herbert, "The Wiener Kreis in America" in Fleming and Bailyn (eds.), *The Intellectual Migration* (see previous section on anthologies), pp. 630–673.

Fermi, Laura, *Illustrious Immigrants: The Intellectual Migration from Europe 1930–41* (Chicago and London: University of Chicago Press, 1968).

Feuer, Lewis S., *Einstein and the Generations of Science* (New York: Basic Books, 1974).

Flexner, Abraham, *Funds and Foundations: Their Policies Past and Present* (New York: Harper & Brothers Publications, 1952).

Folse, Henry J., *The Philosophy of Niels Bohr: The Framework of Complementarity* (Amsterdam, Oxford, New York, Tokyo: North-Holland, 1985).

Forman, Paul, *The Environment and Practice of Atomic Physics in Weimar Germany: A Study in the History of Science*, PhD dissertation at the University of California at Berkeley, 1967 (microfilms international 6810322).

Forman, Paul, "The Financial Support and Political Alignment of Physicists in Weimar Germany" in *Minerva 12* (1974), 39–66.

Fosdick, Raymond B., *The Story of the Rockefeller Foundation* (New York: Harper and Brothers, 1952).

Fosdick, Raymond B. (based on an unfinished manuscript prepared by the late Henry F. Pringle and Katharine Douglas Pringle), *Adventure in Giving: The Story of the General Education Board, A Foundation Established by John D. Rockefeller* (New York: Harper & Row, 1962).

Fridericia, L.S., "Bohr, Christian Harald Lauritz Peter Emil" in *Dansk biografisk Leksikon III* (1934) (see previous section on reference works), pp. 371–374.

Friis, Aage, "De tyske politiske Emigranter i Danmark 1933–46" in *Politiken* 8 and 10 May 1946, pp. 9–10, 8–9.

Frisch, Otto Robert, "The Interest Is Focussing on the Atomic Nucleus" in Rozental (ed.), *Niels Bohr* (see previous section on anthologies), pp. 137–148.

Frisch, Otto Robert, *What Little I Remember* (Cambridge, England, etc.: Cambridge University Press, 1979).

Galison, Peter, "The Discovery of the Muon and the Failed Revolution in Quantum Electrodynamics" in *Centaurus 26* (1983), 262–316.

Galison, Peter, *How Experiments End* (Chicago and London: University of Chicago Press, 1987).

Gamow, George, *My World Line: An Informal Autobiography* (New York: Viking Press, 1970).

Glamann, Kristof, *Carlsbergfondet* (Copenhagen: Rhodos, 1976).

Graham, Loren R., *Between Science and Values* (New York: Columbia University Press, 1981).

Gray, George W., *Education on an International Scale: A History of The International Education Board 1923–1938* (New York: Harcourt Brace and Company, 1941).

Haldane, J.S., "Chapter I. Historical Introduction" in *idem., Respiration* (Oxford: Oxford University Press; New Haven: Yale University Press, 1922), pp. 1–14.

Heilbron, John L., "The Earliest Missionaries of the Copenhagen Spirit" in *Revue d'histoire des sciences et leurs applications 38* (1985), 194–230.

Heilbron, John L., *The Dilemmas of an Upright Man: Max Planck as Spokesman for German Science* (Berkeley, etc.: University of California Press, 1986).

Heilbron, John L., "The First European Cyclotrons" in *Rivista di storia della scienza 3* (1986), 1–44.

Heilbron, John L. and Thomas S. Kuhn, "The Genesis of the Bohr Atom" in *Historical Studies in the Physical Sciences 1* (1969), 211–290.

Heilbron, John L., Robert W. Seidel, and Bruce R. Wheaton, *Lawrence and His Laboratory: Nuclear Science at Berkeley 1931–1961* (Berkeley: Office of the History of Science and Technology, University of California, 1981).

Heilbron, John L. and Robert W. Seidel, *A History of the Lawrence Berkeley Laboratory*, vol. 1: *Lawrence and His Laboratory* (Berkeley: University of California Press, in press).

Heisenberg, Werner, *Der Teil und das Ganze: Gespräche im Umkreis der Atomphysik* (Munich: R. Piper & Co. Verlag, 1969); English translation, *Physics and Beyond: En-*

counters and Conversations [Ruth Nanda Anshen (ed.), *World Perspectives*] (London: George Allen and Unwin Ltd., 1971).

Hendry, John, "Bohr–Kramers–Slater: A Virtual Theory of Virtual Oscillators and Its Role in the History of Quantum Mechanics" in *Centaurus 25* (1981), 189–221.

Hendry, John, "The History of Complementarity: Niels Bohr and the Problem of Visualization" in *Rivista di Storia della Scienza 2* (1984), 392–407.

Hendry, John, *The Creation of Quantum Mechanics and the Bohr-Pauli Dialogue* (Dordrecht, Boston, Lancaster: D. Reidel Publishing Company, 1984).

Henriques, V., "Chr. Bohrs videnskabelige Gerning" in *Oversigt over Det Kongelige Danske Videnskabernes Selskabs Forhandlinger 1911*, pp. 395–405.

Hevesy, George, "Freiherr Auer von Welsbach" in *Akademische Mitteilungen aus Freiburg 4* (1929), 17–18.

Hevesy, George, "A Scientific Career" in *Perspectives in Biology and Medicine 1* (1958), 345–365; reprinted in *idem., Adventures in Radioisotope Research* (see previous section on publications of scientists' papers), pp. 11–30.

Hevesy, George, "Gamle Dage" in *Niels Bohr – Et Mindeskrift* (see previous section on anthologies), pp. 26–30.

Hewlett, Richard G. and Francis Duncan, *Atomic Shield, 1947/1952: A History of the United States Atomic Energy Commission, Volume 2* (University Park and London: Pennsylvania State University Press, 1969).

Hippel, Arthur R. von, *Life in Times of Turbulent Transitions* (Anchorage, Alaska: Stone Age Press, 1988).

Høffding, Harald, "Mindetale over Christian Bohr" in *Tilskueren 1911*, pp. 209–212.

Høffding, Harald, *Erindringer* (Copenhagen: Gyldendalske Boghandel, Nordisk Forlag, 1928).

Hoffmann, Dieter, "Zur Teilnahme deutscher Physiker an den Kopenhagener Physikerkonferenzen nach 1933 sowie am 2. Kongress für Einheit der Wissenschaft, Kopenhagen 1936" in *NTM – Schriftenreihe für Geschichte der Naturwissenschaften, Technik und Medizin 25* (1988), 49–55.

Holter, H., "K.U. Linderstrøm-Lang" in Holter and Max Møller (eds.), *The Carlsberg Laboratory* (see previous section on anthologies), pp. 88–117.

Holton, Gerald, "Striking Gold in Science: Fermi's Group and the Recapture of Italy's Place in Physics" in *Minerva 12* (1974), 158–198; reprinted as "Fermi's Group and the Recapture of Italy's Place in Physics" in *idem., The Scientific Imagination* (see previous section on anthologies), pp. 155–198.

Holton, Gerald, "The Roots of Complementarity" in *idem., Thematic Origins* (see previous section on anthologies), pp. 115–161.

Jackman, Jarrell C. and Carla M. Borden (eds.), *The Muses Flee Hitler: Cultural Transfer and Adaptation 1930–1945* (Washington, D.C.: Smithsonian Institution Press, 1983).

Jammer, Max, *The Conceptual Development of Quantum Mechanics* (New York: McGraw-Hill, 1966).

Jammer, Max, *The Philosophy of Quantum Mechanics: The Interpretation of Quantum Mechanics in Historical Perspective* (New York, etc.: John Wiley & Sons, 1974).

Jensen, Carsten, "A History of the Beta Spectrum and Its Interpretation, 1911–1934," PhD dissertation, Copenhagen University, forthcoming.

Jørgensen, C. Barker, "Dyrefysiologi og gymnastikteori" in *Københavns Universitet bind XIII, 2. del* (see previous section on anthologies), pp. 447–488.

Jørgensen, Erik Stiig, "Friis, Aage" in *Dansk Biografisk Leksikon* [third edition] *IV* (1980) (see previous section on reference works), pp. 649–653.

Jungnickel, Christa and Russell McCormmach, *Intellectual Mastery of Nature: Theoretical Physics from Ohm to Einstein, Volume 1: The Torch of Mathematics, 1800–1870; Volume 2: The Now Mighty Theoretical Physics, 1870–1925* (Chicago and London: University of Chicago Press, 1986).

Kamen, Martin D., *Radiant Science, Dark Politics: A Memoir of the Nuclear Age* (Berkeley, Los Angeles, London: University of California Press, 1985).

Kargon, Robert H., *The Rise of Robert Millikan: Portrait of a Life in American Science* (Ithaca and London: Cornell University Press, 1982).

Kay, Lily E., "Conceptual Models and Analytical Tools: The Biology of the Physicist Max Delbrück, 1931–1946" in *Journal of the History of Biology 18* (1985), 207–246.

Kay, Lily E., "The Secret of Life: Niels Bohr's Influence on the Biology Program of Max Delbrück" in *Rivista di storia della scienza 2* (1985), 487–510.

Kevles, Daniel J., "Towards the *Annus Mirabilis:* Nuclear Physics Before 1932" in *The Physics Teacher 10* (1972), 175–181.

Kevles, Daniel J., *The Physicists: The History of a Scientific Community in Modern America* (New York: Alfred A. Knopf, 1978).

Klein, Oskar, "Glimpses of Niels Bohr as Scientist and Thinker" in Rozental (ed.), *Niels Bohr* (see previous section on anthologies), pp. 74–93.

Kohler, Robert E., "The Management of Science: The Experience of Warren Weaver and the Rockefeller Foundation Programme in Molecular Biology" in *Minerva 14* (1976), 279–306; republished in a slightly revised version as *idem.*, "Warren Weaver and the Rockefeller Foundation Program in Molecular Biology: A Case Study in the Management of Science" in Reingold (ed.), *New Perspectives* (see previous section on anthologies), pp. 249–293.

Kohler, Robert E., "Rudolf Schoenheimer, Isotopic Tracers, and Biochemistry in the 1930's" in *Historical Studies in the Physical Sciences 8* (1977), 257–298.

Kohler, Robert E., "A Policy for the Advancement of Science: The Rockefeller Foundation, 1924–29" in *Minerva 16* (1978), 480–515.

Kohler, Robert E., "Science, Foundations, and American Universities in the 1920s" in *Osiris 3* (1987), 135–164.

Kohler, Robert E., *Managers of Science: Foundations and the Natural Sciences 1900–1950*, forthcoming book.

Kopp, Carolyn, "Max Delbrück – How It Was" in *Engineering & Science* (California Institute of Technology), first installment *43* (Mar-Apr 1980), 21–26, second installment *43* (May-Jun 1980), 21–27.

Krafft, Fritz, *Im Schatten der Sensation: Leben und Wirken von Fritz Strassmann* (Weinheim, etc.: Verlag Chemie, 1981).

Kraft, Victor, *The Vienna Circle: The Origin of Neo-Positivism, A Chapter in the History of Recent Philosophy* (New York: Philosophical Library, 1953).

Kragh, Helge, "Niels Bohr's Second Atomic Theory" in *Historical Studies in the Physical Sciences 10* (1979), 123–186.

Kragh, Helge, "Anatomy of a Priority Conflict: The Case of Element 72" in *Centaurus 23* (1980), 275–301.

Kragh, Helge, "The Genesis of Dirac's Relativistic Theory of Electrons" in *Archive for History of Exact Sciences 24* (1981), 31–67.

Krogh, August, "Visual Thinking: An Autobiographical Note" in *Organon 2* (1938), 87–94.

Kuhn, H.G., "James Franck 1882–1964" in *Biographical Memoirs of Fellows of the Royal Society 11* (1965), 53–74.

Kuhn, Thomas S., *The Structure of Scientific Revolutions* (Chicago and London: University of Chicago Press, 1962, enlarged edition 1970).

Kuhn, Thomas S., "Mathematical versus Experimental Traditions in the Development of Physical Science" in *Journal of Interdisciplinary History 7* (1976), 1–31; reprinted in *idem., The Essential Tension* (see previous section on anthologies), pp. 31–65.

Larsen, Poul, "Carsten Olsen," in Holter and Møller (eds.), *The Carlsberg Laboratory* (see previous section on anthologies), pp. 130–138.

Lemberg, I.Kh., V.O. Najdenov, and V.Ya. Frenkel, "The Cyclotron of the A.F. Ioffe Physico-Technical Institute of the Academy of the Sciences of the USSR (on the fortieth anniversary of its startup)" in *Soviet Physics Uspekhi 30* (1987), 993–1006.

Lemmerich, Jost, *Max Born, James Franck, Physiker im ihrer Zeit: Der Luxus des Gewissens* (Berlin: Staatsbibliothek Preussischer Kulturbesitz, 1982).

Levi, Hilde, "George de Hevesy, 1 August 1885 – 5 July 1966" in *Nuclear Physics A98* (1967), 1–24.

Levi, Hilde, *George de Hevesy: Life and Work* (Bristol: Adam Hilger; Copenhagen: Rhodos, 1985).

Lewis, W. Bennett, "Early Detectors and Counters" in *Nuclear Instruments and Methods 162* (1979), 9–14.

Livingston, M. Stanley, *Particle Accelerators: A Brief History* (Cambridge, Massachusetts: Harvard University Press, 1969).

McCagg, William O., Jr., *Jewish Nobles and Geniuses in Modern Hungary* (Boulder: East European Quarterly, 1972).

McLeod, Roy, "Changing Perspectives in the Social History of Science" in Spiegel-Rösing and Price (eds.), *Science, Technology and Society* (see previous section on anthologies), pp. 149–195.

Mehra, Jagdish and Helmut Rechenberg, *The Historical Development of Quantum Theory*. The following books have been published in this series. *Volume 1* has been published in two parts, each under different cover, *The Quantum Theory of Planck, Einstein, Bohr and Sommerfeld: Its Foundations and the Rise of Its Difficulties 1900–1925, Part 1* and *Part 2*. Then follow: *Volume 2, The Discovery of Quantum Mechanics 1925; Volume 3, The Formulation of Matrix Mechanics and Its Modifications 1925–1926; Volume 4, Part 1, The Fundamental Equations of Quantum Mechanics 1925–1926; Volume 4, Part 2, The Reception of the New Quantum Mechanics 1925–1926; Volume 5, Erwin Schrödinger and the Rise of Wave Mechanics – Part 1, Schrödinger in Vienna and Zurich, 1887–1925, Part 2, The Creation of Wave Mechanics: Early Response and Applications, 1925–1926* (New York, Heidelberg, Berlin: Springer-Verlag, Volumes 1–4 1982, Volume 5 1987).

Mendelsohn, K., *The World of Walther Nernst: The Rise and Fall of German Science, 1864–1941* (Pittsburgh: University of Pittsburgh Press, 1973).

Meyer-Abich, Klaus Michael, *Korrespondenz, Individualität und Komplementarität: Eine Studie zur Geistesgeschichte der Quantentheorie in den Beitragen Niels Bohrs* (Wiesbaden: F. Steiner, 1965).

Møller, Christian, "Nogle Erindringer fra Livet paa Bohr's Institut i sidste Halvdel af Tyverne" in *Niels Bohr – Et Mindeskrift* (see previous section on anthologies), pp. 54–64.

Møller-Christensen, Vilhelm (ed.), *Finsen Instituttet 1896–23. oktober–1946* (Copenhagen: Det Berlingske Bogtrykkeri, 1946).

Moore, Ruth, *Niels Bohr: The Man, His Science, and the World They Changed* (New York: Alfred A. Knopf, 1966).

Moyer, Donald Franklin, "Origins of Dirac's electron, 1925–1928" in *American Journal of Physics 49* (Oct 1981), 944–949; *idem.*, "Evaluations of Dirac's electron, 1928–1932" in *ibid.* (Nov 1981), 1055–1062; *idem.*, "Vindications of Dirac's electron, 1932–1934" in *ibid.* (Dec 1981), 1120–1125.

Nielsen, J. Rud, "Introduction" in *Niels Bohr Collected Works, Volume 3* (see previous section on publications of scientists' papers), pp. 3–45.

Olby, Robert, *The Path to the Double Helix* (London: Macmillan, 1974).

Pais, Abraham, "*Subtle Is the Lord . . .*": *The Science and Life of Albert Einstein* (New York and Oxford: Clarendon Press and Oxford University Press, 1982).

Pais, Abraham, *Inward Bound: Of Matter and Forces in the Physical World* (Oxford and New York: Clarendon Press and Oxford University Press, 1986).

Pais, Abraham, Bohr biography to be published by Random House in 1990.

Paneth, Fritz, "Zum 70. Geburtstag Auer von Welsbach" in *Die Naturwissenschaften 16* (1928), 1037–1038; translated into English as "Auer von Welsbach" in Dingle and Martin (eds.), *Chemistry and Beyond* (see previous section on anthologies), pp. 73–76.

Pastor, Peter, *Hungary Between Wilson and Lenin: The Hungarian Revolution of 1918–1919 and the Big Three* (Boulder: East European Quarterly, 1976).

Pauli, Wolfgang, "Zur älteren und neueren Geschichte des Neutrinos," in Weisskopf (ed.), *Aufsätze und Vorträge* (see previous section on publications of scientists' papers), pp. 156–180; reprinted in Kronig and Weisskopf (eds.), *Collected Scientific Papers by Wolfgang Pauli II* (see previous section on publications of scientists' papers), pp. 1313–1337.

Pedersen, Johannes, "Niels Bohr and the Royal Danish Academy of Sciences and Letters" in Rozental (ed.), *Niels Bohr* (see previous section on anthologies), pp. 266–280.

Peierls, Rudolf, "Introduction" in *Niels Bohr Collected Works, Volume 9* (see previous section on publications of scientists' papers), pp. 3–83.

Petersen, Aage, "The Philosophy of Niels Bohr" in *Bulletin of the Atomic Scientists 19* (Sep 1963), 8–14.

Pihl, Mogens, "Fysik" in *Københavns Universitet bind XII, 1. del* (see previous section on anthologies), pp. 365–462.

Purcell, Edward M., "Nuclear Physics without the Neutron; Clues and Contradictions" in *International Congress 10* (see previous section on anthologies), pp. 121–133.

R[abinowitch], E[ugene], "James Franck, 1882–1964, Leo Szilard, 1898–1964" in *Bulletin of the Atomic Scientists 20* (Oct 1964), 16–20.

Rehberg, P. Brandt, "August Krogh, 15.11.1874 – 15.11.1974" in *Dansk medicinhistorisk Aarbog 1974*, pp. 7–28, with a summary in English on pp. 27–28.

Reingold, Nathan, "The Case of the Disappearing Laboratory" in *American Quarterly 29* (1977), 77–101.

Rhodes, Richard, *The Making of the Atomic Bomb* (New York, etc.: Simon & Schuster, Inc., 1986).

Robertson, Peter, *The Early Years: The Niels Bohr Institute 1921–1930* (Copenhagen: Akademisk Forlag, 1979).

Romer, Alfred, "The Science of Radioactivity, 1896–1913: Rays, Particles, Transmutations, Nuclei and Isotopes" in *idem.* (ed.), *Radiochemistry and the Discovery of Isotopes* (see previous section on publications of scientists' papers), pp. 3–60.

Röseberg, Ulrich, *Niels Bohr: Leben und Werk eines Atomphysikers 1885–1962* (East Berlin: Akademie-Verlag, 1985).

Rosenfeld, Léon, *Niels Bohr: An Essay Dedicated to Him on His Sixtieth Birthday 1945* (Amsterdam: North-Holland Publishing Company, 1949, 2nd edition 1961); reprinted in Cohen and Stachel (eds.), *Selected Papers of Léon Rosenfeld* (see previous section on publications of scientists' papers), pp. 313–326.

Rosenfeld, Léon, "Niels Bohr in the Thirties: Consolidation and extension of the conception of complementarity" in Rozental (ed.), *Niels Bohr* (see previous section on anthologies), pp. 114–136.

Rosenfeld, Léon, "Niels Bohr's Contribution to Epistemology" in *Physics Today 16* (Oct 1963), 47–54; reprinted in Cohen and Stachel (eds.), *Selected Papers of Léon Rosenfeld* (see previous section on publications of scientists' papers), pp. 522–535.

Rosenfeld, Léon, "Nogle minder om Niels Bohr" in *Niels Bohr – Et Mindeskrift* (see previous section on anthologies), pp. 65–75.

Roy-Poulsen, Niels Ove, Niels Ove Lassen, and Mikael Jensen, Kohler, *History of the Copenhagen Cyclotron*, forthcoming book.

Rozental, Stefan, "The Forties and the Fifties" in *idem.* (ed.), *Niels Bohr* (see previous section on anthologies), pp. 149–190.

Rozental, Stefan, *NB: Erindringer om Niels Bohr* (Copenhagen: Gyldendal, 1985).

Schröer, Heinz, *Carl Ludwig: Begründer der messenden Experimentalphysiologie 1816–1895* [Heinz Degen (ed.), *Grosse Naturforscher, Band 33*] (Stuttgart: Wissenschaftliche Verlagsgesellschaft m.b.H., 1967).

Segrè, Emilio, *Enrico Fermi: Physicist* (Chicago and London: University of Chicago Press, 1970).

Slater, John C., "The Development of Quantum Mechanics in the Period 1924–1926" in Price, Chissick, and Ravensdale (eds.), *Wave Mechanics* (see previous section on anthologies), pp. 19–25.

Slater, John C., *Solid-State and Molecular Theory: A Scientific Biography* (New York, etc.: John Wiley & Sons, 1975).

Snorrason, Egill, "Krogh, Schack August Steenberg" in *Dictionary of Scientific Biography* (see previous section on reference works), *VII* [1973], pp. 501–504.

Snorrason, Egill (L.S. Fridericia), "Lundsgaard, Einar" in *Dansk Biografisk Leksikon* [third edition] *IX* (see previous section on reference works), pp. 196–197.

Sopka, Katherine R., *Quantum Physics in America 1920–1935* (New York: Arno Press, 1980); republished as *idem.*, *Quantum Physics in America: The years through 1935 [The History of Modern Physics 1800–1950, Volume 10]* (New York: Tomash Publishers and American Institute of Physics, 1988).

Spärck, R., "August Krogh 15. November 1874 – 3. September 1949" in *Videnskabelige Meddelelser fra Dansk naturhistorisk Forening 3* (1949), V–XXX.

Spence, R., "George Charles de Hevesy" in *Chemistry in Britain 3* (1967), 527–532.

Strömgren, Bengt, "Niels Bohr and the Royal Danish Academy of Sciences and Letters" in de Boer, Dal, and Ulfbeck (eds.), *The Lesson of Quantum Theory* (see previous section on anthologies), pp. 3–12.

Stuewer, Roger H., "The Nuclear Electron Hypothesis" in Shea (ed.), *Otto Hahn* (see previous section on anthologies), pp. 19–67.

Stuewer, Roger H., "Nuclear Physicists in the New World: The Emigrés of the 1930s in America" in *Berichte zur Wissenschaftsgeschichte 7* (1984), 23–40.

Stuewer, Roger H., "Bringing the News of Fission to America" in *Physics Today 38* Oct 1985, 48–56.

Stuewer, Roger H., "Niels Bohr and Nuclear Physics" in French and Kennedy (eds.), *Niels Bohr* (see previous section on anthologies), pp. 197–220.

Stuewer, Roger H., "The Naming of the Deuteron" in *American Journal of Physics 54* (1986), 206–218.

Stuewer, Roger H., "Gamow's Theory of Alpha-Decay" in Ullmann-Margalit (ed.), *The Kaleidoscope of Science* (see previous section on anthologies), pp. 147–186.

Stuewer, Roger H., "The Origins of the Liquid-Drop Model of the Nucleus," forthcoming article in *Proceedings* of conference on "50 Years of Nuclear Fission" (West Berlin, 30–31 Mar 1989).

Sturdy, Steve, "Biology as Social Theory: John Scott Haldane and Physiological Regulation" in *British Journal for the History of Science 21* (1988), 315–340.

Szabadváry, F., "George Hevesy" in *Journal of Radioanalytical Chemistry 1* (1968), 97–102.

Walker, Mark, *German National Socialism and the Quest for Nuclear Power 1939–1949* (Cambridge: Cambridge University Press, 1989).

Weart, Spencer, "The Physics Business in America, 1919–1940: A Statistical Reconnaissance" in Reingold (ed.), *New Perspectives* (see previous section on anthologies), pp. 295–358.

Weart, Spencer, "The Discovery of Fission and a Nuclear Physics Paradigm" in Shea (ed.), *Otto Hahn* (see previous section on anthologies), pp. 91–133.

Weaver, Warren, *Scene of Change: A Lifetime in American Science* (New York: Charles Scribner's Sons, 1970).

Weinberg, Steven, "The Search for Unity: Notes for a History of Quantum Field Theory" in *Daedalus 106* (1977), 17–35.

Weiner, Charles, "A New Site for the Seminar: The Refugees and American Physics in the Thirties" in Fleming and Bailyn (eds.), *The Intellectual Migration* (see previous section on anthologies), pp. 190–234.

Weiner, Charles, "1932 – Moving into the New Physics" in *Physics Today 25* (May 1972), 40–49.

Weiner, Charles, "Cyclotrons and Internationalism: Japan, Denmark and the United States, 1935–1945" in *XIVth International Congress* (see previous section on anthologies), pp. 353–365.

Weisskopf, Victor, "Niels Bohr, A Memorial Tribute" in *Physics Today 16* (Oct 1963), 58–64.

Weisskopf, Victor, "My Life as a Physicist" in *idem., Physics in the Twentieth Century* (see previous section on anthologies), pp. 1–21.

Weisskopf, Victor, "Niels Bohr, the Quantum and the World" [1967] in *idem., Physics in the Twentieth Century* (see previous section on anthologies), pp. 52–65.

Wheeler, John Archibald, "Niels Bohr and Nuclear Physics" in *Physics Today 16* (Oct 1963), 36–45.

Wheeler, John Archibald, "Some Men and Moments in the History of Nuclear Physics: The Interplay of Colleagues and Motivations" in Stuewer (ed.), *Nuclear Physics in Retrospect* (see previous section on anthologies), pp. 217–282.

Witt-Hansen, Johs., "Leibniz, Høffding, and the 'Ekliptika Circle'" in *Danish Yearbook of Philosophy 17* (1980), 31–58.

Notes

ACKNOWLEDGEMENTS

1. Finn Aaserud, *The Redirection of the Niels Bohr Institute in the 1930s: Response to Changing Conditions for Basic Science Enterprise*, PhD dissertation at the Johns Hopkins University 1967 (microfilms international 8510398); see also *idem.*, "Niels Bohr as Fund Raiser" in *Physics Today 38* (Oct 1985), 38–46.

INTRODUCTION

1. Numerous books and articles deal with the increasing integration of science into contemporary society. An illuminating overview, written by a journalist, is David Dickson, *The New Politics of Science* (New York: Pantheon Books, 1984), republished, with a new preface, by Chicago University Press in 1988.
2. The classic account of science and government in the United States up to the Second World War is A. Hunter Dupree, *Science in the Federal Government: A History of Policies and Activities* (Baltimore and London: Johns Hopkins University Press, 1986), originally published by Harvard in 1957. Dupree has prepared a brief essay on subsequent developments for the new edition of the book (pp. vii–xviii). The changing role of physics in the United States during and after the Second World War is discussed in a broad historical context in Daniel J. Kevles, *The Physicists: The History of a Scientific Community in Modern America* (New York: Alfred A. Knopf, 1978), pp. 287ff. Kevles's book also contains an extensive "Essay on Sources" (pp. 435–464), which is a useful guide to the rich literature.
3. A concise overview of the different developments of scientific fields is Thomas S. Kuhn, "Mathematical versus Experimental Traditions in the Development of Physical Science" in *Journal of Interdisciplinary History 7* (1976), 1–31; reprinted in *idem., The Essential Tension: Selected Studies in Scientific Tradition and Change* (Chicago and London: University of Chicago Press, 1977), pp. 31–65.
4. The by now classic attempt to restrict the interconnection between scientific and extrascientific developments to periods of scientific upheaval is Thomas S. Kuhn, *The Structure of Scientific Revolutions* (Chicago and London: University of Chicago Press, 1962, enlarged edition 1970). For a general historiographical discussion of the extent of such interconnections, see, for example, Roy McLeod, "Changing Perspectives in the Social History of Science" in Ina Spiegel-Rösing and Derek de Solla Price (eds.), *Science, Technology and Society: A Cross-Disciplinary Perspective* (London and Beverley Hills: SAGE Publications, 1977), pp. 149–195.

5. On the Rockefeller Foundation, see, in particular, Robert E. Kohler, "A Policy for the Advancement of Science: The Rockefeller Foundation, 1924–29" in *Minerva 16* (1978), 480–515; *idem.*, "The Management of Science: The Experience of Warren Weaver and the Rockefeller Foundation Programme in Molecular Biology" in *Minerva 14* (1976), 279–306, a slightly revised version of which has been published as *idem.*, "Warren Weaver and the Rockefeller Foundation Program in Molecular Biology: A Case Study in the Management of Science" in Nathan Reingold (ed.), *The Sciences in the American Context: New Perspectives* (Washington, D.C.: Smithsonian Institution Press, 1979), pp. 249–293. For a more general account, see *idem.*, "Science, Foundations, and American Universities in the 1920s" in *Osiris 3* (1987), 135–164. Kohler is presently completing his book, *Managers of Science: Foundations and the Natural Sciences 1900–1950*. I am grateful to Kohler for letting me read the parts of his manuscript that are particularly relevant for this book. For an insightful approach to American physics between the wars in a broad economic context, see Spencer R. Weart, "The Physics Business in America, 1919–1940: A Statistical Reconnaissance" in Reingold (ed.), *New Perspectives*, pp. 295–358.

6. The history of nuclear physics between the two world wars is presently being investigated in detail by Roger H. Stuewer. See, in particular, the collection of physicists' reminiscences, Roger H. Stuewer (ed.), *Nuclear Physics in Retrospect: Proceedings of a Symposium on the 1930s* (Minneapolis: University of Minnesota Press, 1979), and his own articles "The Nuclear Electron Hypothesis" in William R. Shea (ed.), *Otto Hahn and the Rise of Nuclear Physics* (Dordrecht, Boston, Lancaster: D. Reidel Publishing Company, 1983), pp. 19–67; "Gamow's Theory of Alpha-Decay" in Edna Ullmann-Margalit (ed.), *The Kaleidoscope of Science: The Israel Colloquium: Studies in History, Philosophy, and Sociology of Science, Volume 1* [Robert S. Cohen and Marx W. Wartofsky (eds.), *Boston Studies in the Philosophy of Science, Volume 94*] (Dordrecht, etc.: D. Reidel Publishing Company, 1986), pp. 147–186. Stuewer is presently developing his historical investigators into a full-length monograph. For other references to the many writings on this topic, see the notes for the main body of this book.

7. Bohr's inclination to emphasize the collective efforts in physics at the expense of his own is particularly evident in a taped conversation with J. Robert Oppenheimer from 1957, retained in the Niels Bohr Archive, Copenhagen.

8. Some physicists' reminiscences of Bohr and his institute are collected in Stefan Rozental (ed.), *Niels Bohr: His Life and Work as Seen by His Friends and Colleagues* (Amsterdam: North-Holland Publishing Company, 1968). The first ten years of the institute's history is covered in Peter Robertson, *The Early Years: The Niels Bohr Institute 1921–1930* (Copenhagen: Akademisk Forlag, 1979). See also Ruth Moore, *Niels Bohr: The Man, His Science, and the World They Changed* (New York: Alfred A. Knopf, 1966); Ulrich Röseberg, *Niels Bohr: Leben und Werk eines Atomphysikers 1885–1962* (East Berlin: Akademie-Verlag, 1985); A. P. French and P. J. Kennedy, *Niels Bohr: A Centenary Volume* (Cambridge and London: Cambridge University Press, 1985); Niels Blædel, *Harmony and Unity: The Life of Niels Bohr* (Madison, Wisconsin: Science Tech Publishers; Berlin, etc.: Springer-Verlag, 1988). Abraham Pais is currently completing a substantial biography of Bohr to be published by Random House in 1990.

9. These numbers are based on the bound volumes of publications from the institute between 1918 to 1959, retained at the Niels Bohr Institute, Copenhagen: *Universitetets Institut for teoretisk Fysik: Afhandlinger.*

10. *Ibid.*

PROLOGUE

1. W. Heisenberg, *Die physikalischen Prinzipien der Quantentheorie* (Leipzig: S. Hirzel, 1930), p. VI. In the unpaginated Preface to the English translation, *The Physical Principles of the Quantum Theory* (New York: Dover, 1930), the German term is not translated.

2. On the Copenhagen spirit as describing the atmosphere at the institute, see Léon Rosenfeld, *Niels Bohr: An Essay Dedicated to Him on His Sixtieth Birthday 1945* (Amsterdam: North-Holland Publishing Company, 1949, 2nd edition 1961); reprinted in Robert S. Cohen and John S. Stachel (eds.), *Selected Papers of Léon Rosenfeld [Boston Studies in the Philosophy of Science*, vol. 21] (Dordrecht, Boston, London: D. Reidel Publishing Company, 1979), pp. 313–326, on p. 313; Victor Weisskopf, "Niels Bohr, the Quantum and the World" [1967] in *idem.*, *Physics in the Twentieth Century: Selected Essays* (Cambridge and London: Massachusetts Institute of Technology Press, 1972), pp. 52–65, on p. 55; Peter Robertson, *The Early Years: The Niels Bohr Institute 1921–1930* (Copenhagen: Akademisk Forlag, 1979), pp. 152–155. In an article – John L. Heilbron, "The earliest missionaries of the Copenhagen spirit" in *Revue d'histoire des sciences et leurs applications 38* (1985), 194–230 – Heilbron defines the Copenhagen spirit instead in terms of attempts of the circle around Bohr to base a general epistemology on Bohr's interpretation of quantum mechanics. The physicists' freedom and independence is discussed in my interview with Victor Weisskopf on 7 Oct 1980, and with Stefan Rozental on 9 Oct 1980, both conducted at the Niels Bohr Institute, Copenhagen. Weisskopf remembers Bohr and his institute in his obituary "Niels Bohr, A Memorial Tribute" in the Bohr memorial issue of *Physics Today 16* (Oct 1963), 58–64. He has covered the same ground in "Quantum and the World." Rozental recounts his first twenty years at the institute in "The Forties and the Fifties" in Stefan Rozental (ed.), *Niels Bohr: His Life and Work as Seen by His Friends and Colleagues* (Amsterdam: North-Holland Publishing Company, 1968), pp. 149–190. See also Rozental's more recent autobiographical account in Danish, *NB: Erindringer om Niels Bohr* (Copenhagen: Gyldendal, 1985).

3. Otto Robert Frisch, "The Interest Is Focussing on the Atomic Nucleus" in Rozental (ed.), *Niels Bohr* (note 2), pp. 137–148, on p. 138. For a similar reaction to the informal atmosphere at the institute, see Weisskopf, "Quantum and the World" (note 2), p. 55. The quotation is from Otto Robert Frisch, *What Little I Remember* (Cambridge, etc.: Cambridge University Press, 1979), p. 101.

4. On this occasion, Bohr had been lured to the movies by Gamow, Landau, and Casimir. The latter reports the incident in the form of a poem (in German) in Hendrik B. G. Casimir, "Recollections from the Years 1929–1931" in Rozental (ed.), *Niels Bohr* (note 2), pp. 109–113, on p. 113. A translation of the poem into English by Casimir is contained in Hendrik Casimir, *Haphazard Reality: Half a Century of Science* (New York, etc.: Harper & Row, 1983), pp. 98–99. That Gamow took the role of the villain is also reported in a letter of 18 Dec 1980 from Carl Friedrich von Weizsäcker to me. Weizsäcker, however, gives a slightly different account of Bohr's theory, thus quoting Bohr: "Der Gute hat nicht die ganz kurze Hemmung des schlechten Gewissens, deshalb schiesst er schneller." For our purposes it is the incident itself, not Bohr's particular theory of heroes and villains, that matters.

5. Christian Møller, "Nogle Erindringer fra Livet paa Bohr's Institut i sidste Halvdel af Tyverne" in *Niels Bohr – Et Mindeskrift [Fysisk Tidsskrift 60* (1962)] (Copenhagen: Selskabet for Naturlærens udbredelse, 1963), pp. 54–64, on pp. 62–63. Weizsäcker, in his letter to me of 18 Dec 1980

(note 4), gives another reason for Bohr's consternation: He did not want the Ping-Pong activity to be seen from the street. The quotation of Frisch is from his *What Little* (note 3), p. 86. See the comments on Frisch's view in Casimir, *Reality* (note 4), pp. 125–126.

6. See, for example, the two unmistakably similar accounts of first encounters with Bohr: Møller, "Nogle Erindringer" (note 5), pp. 56–57, and Léon Rosenfeld, "Nogle minder om Niels Bohr" in *Mindeskrift* (note 5), pp. 65–75, on pp. 68–69. See also, George Gamow, *My World Line: An Informal Autobiography* (New York: Viking Press, 1970), pp. 85–89; Werner Heisenberg, *Der Teil und das Ganze: Gespräche im Umkreis der Atomphysik* (Munich: R. Piper & Co. Verlag, 1969), pp. 59–65, 150–162, English translation, *idem., Physics and Beyond: Encounters and Conversations* [Ruth Nanda Anshen (ed.), *World Perspectives*] (London: George Allen and Unwin Ltd., 1971), pp. 38–42; P.A.M. Dirac, "Recollections of an Exciting Era" in Charles Weiner (ed.), *History of Twentieth Century Physics* [*Proceedings of the International School of Physics "Enrico Fermi," Course LVII, Varenna on Lake Como, Villa Monastero, 31st July – 12th August 1972*] (New York and London: Academic Press, 1977), pp. 109–146, on p. 134.

7. The residents of the Carlsberg mansion, chosen by a majority of members of the Royal Danish Academy of Sciences and Letters, have been the philosopher Harald Høffding (1914–1931), the physicist Niels Bohr (1931–1962), the archeologist Johannes Brøndsted (1962–1965), the astronomer Bengt Strömgren (1965–1987), and the sinologist Søren Christian Egerod (1987). See Kristof Glamann, *Carlsbergfondet* (Copenhagen: Rhodos, 1976), pp. 165–167, quotation on p. 165. Frisch's statement is taken from his "Interest" (note 3), p. 138. This quotation is used to recount the atmosphere at Carlsberg in Ruth Moore, *Niels Bohr: The Man, His Science, and the World They Changed* (New York: Alfred A. Knopf, 1966), p. 192. See also Frisch's later account in his *What Little* (note 3), p. 92. The quotation of Rosenfeld is from his *Essay* (note 2), p. 313.

8. Bohr's way of delivering public speeches has been reported in several places. See, for example, the careful comments to this effect by his close friend, colleague, and disciple Oskar Klein, "Glimpses of Niels Bohr as Scientist and Thinker" in Rozental (ed.), *Niels Bohr* (note 2), pp. 74–93, on p. 81. The identification of Bohr's helpers are taken from Léon Rosenfeld, "Niels Bohr in the Thirties: Consolidation and Extension of the Conception of Complementarity" in *ibid.*, pp. 114–136, on pp. 116–120; Weisskopf, "Quantum and the World" (note 2), pp. 61–62; my interview with Weisskopf 7 Oct 1980 (note 2); Rozental, "Forties and Fifties" (note 2), pp. 161–166; and my interview with Rozental 9 Oct 1980 (note 2). Weisskopf has clarified his period as helper in correspondence with me; my letter to Weisskopf 7 Mar 1989, Weisskopf to me 17 Mar 1989. The following description of the work as Bohr's helper draws on Rosenfeld's account, as well as the publications by and personal interviews with Weisskopf and Rozental.

9. On the importance of language in Bohr's philosophical views, see Aage Petersen, "The Philosophy of Niels Bohr" in *Bulletin of the Atomic Scientists 19* (1963), 8–14, pp. 10–11.

10. Personal conversation in March 1984 with Hilde Levi, who arrived at the institute in 1934.

11. My interview with Weisskopf 7 Oct 1980 (note 2); see also transcript of T. S. Kuhn and J. L. Heilbron's interview with Weisskopf 10 Jul 1963, p. 15. The latter interview is part of the Archive for the History of Quantum Physics (AHQP) collection deposited in the Niels Bohr Archive in Copenhagen, in the Niels Bohr Library of the American Institute of Physics in New York, and other places. For general, though dated, information

on this source see, Thomas S. Kuhn, John L. Heilbron, Paul Forman, and Lini Allen, *Sources for the History of Quantum Physics: An Inventory and Report* (Philadelphia: The American Philosophical Society, 1967). Stefan Rozental made the same point in his interview with me 9 Oct 1980 (note 2).

12. T. S. Kuhn and J. H. Van Vleck's interview with John Clarke Slater 3 Oct 1963, AHQP (note 11), pp. 30–34, quotations on pp. 33, 30. See also John C. Slater, "The Development of Quantum Mechanics in the Period 1924–1926" in William C. Price, Seymour S. Chissick, and Tom Ravensdale (eds.), *Wave Mechanics the First Fifty Years* (London: Butterworths, 1973), pp. 19–25, as well as the slightly different account in *idem., Solid-State and Molecular Theory: A Scientific Biography* (New York, etc.: John Wiley & Sons, 1975), pp. 11–19. On Kramers, see M. Dresden, *H. A. Kramers: Between Tradition and Revolution* (New York, etc.: Springer-Verlag, 1987), where the Bohr-Kramers-Slater incident is discussed on pp. 163–171.

13. Linus Pauling to me 29 Apr and 28 May 1980. Weisskopf interview 7 Oct 1980 (note 2).

14. I have corresponded with and interviewed people connected with the institute between the two world wars. See Notes on Sources.

CHAPTER 1

1. On the development of the physics discipline at Copenhagen University, see Mogens Pihl, "Fysik" in *Københavns Universitet 1479–1979* [14 volumes] *bind XII: Det matematisk-naturvidenskabelige Fakultet, 1. del* (Copenhagen: G. E. C. Gads Forlag, 1983), pp. 365–462. The emergence and early development of theoretical physics as a discipline are described in detail in Christa Jungnickel and Russell McCormmach, *Intellectucal Mastery of Nature: Theoretical Physics from Ohm to Einstein, Volume 1: The Torch of Mathematics, 1800–1870; Volume 2: The Now Mighty Theoretical Physics, 1870–1925* (Chicago and London: University of Chicago Press, 1986). A handwritten draft and carbon copy of the letter from Bohr to the Science Faculty at Copenhagen University 18 Apr 1917 are retained in "Science Faculty," Bohr General Correspondence, Special File, film 7, section 7 (BGC-S (7,7)), in the Niels Bohr Archive, Copenhagen. The letter is reprinted in "Instituttet for teoretisk Fysik" in *Aarbog for Københavns Universitet, Kommunitetet og Den polytekniske Læreanstalt indeholdende Meddelelser for de akademiske Aar 1915–1920 – III. Del: Universitetets videnskabelige Institutter [Copenhagen University Yearbook, 1915–1920, III]* (Copenhagen: Copenhagen University, 1923), pp. 316–329, on pp. 316–318. The full letter has been translated into English in Peter Robertson, *The Early Years: The Niels Bohr Institute 1921–1930* (Copenhagen: Akademisk Forlag, 1979), pp. 20–22, quotation on p. 21.

2. Niels Bohr, "Speech at the Dedication of the Institute for Theoretical Physics (3 March 1921)" in Léon Rosenfeld (gen. ed.), *Niels Bohr Collected Works, Volume 3:* J. Rud Nielsen (ed.), *The Correspondence Principle (1918–1923)* (Amsterdam, etc.: North-Holland Publishing Company, 1976), pp. 248–293 (Danish original), 293–301 (English translation).

3. "Instituttet" (note 1). For an example of Bohr's elaborate applications, see, in particular, the handwritten application for the spectrograph, Bohr to the Carlsberg Foundation 31 Oct 1919 in the Carlsberg Foundation Archive, Copenhagen. The quotation is from a history of the foundation written by its present director, Kristof Glamann, *Carlsbergfondet* (Copenhagen: Rhodos, 1976), p. 21.

4. Robertson, *Early Years* (note 1), pp. 20ff.; "Instituttet" (note 1), pp. 318–329. The institute's handwritten "Budget Books," retained in the

Niels Bohr Archive, Copenhagen, contains useful information on the appropriations from private sources and their uses, thus constituting a compact guide to the institute's finances. For the increased allowance from the Carlsberg Foundation, see *Carlsbergfondets Understøttelser 1876–1936* (Copenhagen: Bianco Luno, 1937), p. 68. B. R. Mitchell, *European Historical Statistics 1705–1975* (Second Revised Edition) (New York: Facts on File, 1981) provides Danish "Wholesale Indices" and "Cost of Living Indices," respectively on pp. 774 and 781; the quoted increases are taken from the former table.

5. Rutherford to Bohr 17 Nov 1918, 11 Jan 1919, in the Bohr Scientific Correspondence, film 6, section 3 (BSC (6,3)); this collection is deposited in the Niels Bohr Archive, Copenhagen, which also contains a useful chronological listing of the letters. Microfilm copies of the BSC are deposited in the Niels Bohr Library of the American Institute of Physics in New York, and elsewhere. Planck to Bohr 23 Oct 1920, BSC (5,5). On Einstein's position in Berlin, see Abraham Pais, *"Subtle Is the Lord . . .": The Science and Life of Albert Einstein* (New York and Oxford: Clarendon Press and Oxford University Press, 1982), pp. 239–240. On Bohr's reaction to the offers, see his handwritten drafts of letters to Rutherford 15 Dec 1918, and to Planck undated (but between 23 Oct and 30 Oct 1920), BSC (6,3), (5,5). Bohr received the Nobel Prize for "his services in the investigation of the structure of atoms and of the radiation emanating from them"; *Nobel Lectures, Including Presentation Speeches and Laureates' Biographies: Physics, 1922–1941* (Amsterdam, London, New York: Elsevier Publishing Company, 1965); see his Nobel Prize Lecture, "The Structure of the Atom" in *ibid.*, pp. 7–43, which has been printed in its original Danish and in English translation in *Niels Bohr Collected Works, Volume 3* (note 2), pp. 427–465, 467–482.

6. Robertson, *Early Years* (note 1), pp. 76ff. Jeans to Bohr 17 Jul 1923, BSC (12,3), Rutherford to Bohr 19 Jul 1923, BSC (15,3) (note 5); photocopies of the two letters have been retained under "Royal Society Professorship 1923" in the Bohr General Correspondence (BGC) in the Niels Bohr Archive, Copenhagen. Typewritten copies are also deposited as enclosures to Henriques and Hjelmslev's letter to the Carlsberg Foundation 3 Sep 1923, Carlsberg Foundation Archive, Copenhagen. The exchange rate is taken from *Banking and Monetary Statistics, 1914–1941* (Washington, D.C.: The Board of Governors of the Federal Reserve System, 1976), p. 681.

7. Jeans to Bohr 17 Jul 1923 (note 6); Rutherford to Bohr 19 Jul 1923, BSC (12,3) (note 5).

8. Handwritten drafts of Bohr to Jeans 3 Aug 1923, BSC (12,3) (note 5), and of Bohr to Rutherford 3 Aug 1923, BSC (15,3); Rutherford to Bohr 14 Aug 1923; handwritten drafts of Bohr to Jeans 22 Aug 1923, BSC (12,3), and of Bohr to Rutherford 22 Aug 1923, BSC (15,3); Jeans to Bohr 29 Aug 1923, BSC (12,3), Rutherford to Bohr 30 Aug 1923, BSC (15,3); carbon copy of Bohr to Jeans 9 Sep 1923, BSC (12,3). Photocopies of all letters are in "Royal Society Professorship," BGC (note 6).

9. On Henriques and Hjelmslev, see Glamann, *Carlsbergfondet* (note 3), pp. 174, 178. The quotation is from Henriques and Hjelmslev to the Carlsberg Foundation 3 Sep 1923 (note 6). The increase in Bohr's salary was announced in Carlsberg Foundation to Bohr 30 Mar 1924, "Carlsberg Foundation" in BGC-S (1,1) (note 1); see also the description in *Oversigt over Det Kongelige Danske Videnskabernes Selskabs Forhandlinger Juni 1924 – Maj 1925* (Copenhagen: Andr. Fre. Høst & Søn, 1925), p. 119. The exchange rate is taken from *Banking* (note 6), p. 669.

10. The Ministry's unwillingness was reported in a letter from the Konsistorium [University Governing Board] to Bohr 17 Jul 1923, "University Board of Directors" BGC-S (8,2) (note 1). For the turnabout regarding

Bohr's and Kramer's teaching obligations, see "Dr. phil. H. A. Kramers Ansættelse som Lektor i teoretisk Fysik" in *Copenhagen University Yearbook* (note 1), *1923–1924* (1924), pp. 32–34. The appointment of Werner was reported in "Institutet for teoretisk Fysik. Oprettelse af en ny videnskabelig assistentstilling" in *ibid. 1923–1924*, pp. 126–127.

11. R. B. Owens to Bohr 6 Feb, wherefrom quotation; carbon copy of Bohr to Owens 17 Mar 1924, BSC (14,2) (note 5). Photocopies are in "Royal Society Professorship," BGC (note 6).

12. This division of funding between governmental and private sources is clear from the rich material on the funding of the institute. See, in particular, the several volumes of the *Copenhagen University Yearbook* (note 1).

13. On Rose, see George W. Gray, *Education on an International Scale: A History of the International Education Board, 1923–38* (New York: Harcourt Brace and Company, 1941), pp. vi, 4–6. The book was reprinted in 1978 by Greenwood Press Publishers in Westport, Connecticut. On the General Education Board, see Raymond B. Fosdick (based on an unfinished manuscript prepared by the late Henry F. Pringle and Katharine Douglas Pringle), *Adventure in Giving: The Story of the General Education Board, A Foundation Established by John D. Rockefeller* (New York: Harper & Row, 1962). For a thorough account of both Rose and Rockefeller philanthropies, see the forthcoming Robert E. Kohler, *Managers of Science: Foundations and the Natural Sciences 1900–1950.*

14. Gray, *Education* (note 13), pp. 7–11 (quotation on p. 10), 16–23, 22. On the NRC fellowships, see Daniel J. Kevles, *The Physicists: The History of a Scientific Community in Modern America* (New York: Alfred A. Knopf, 1978), pp. 149–150; Nathan Reingold, "The Case of the Disappearing Laboratory" in *American Quarterly 29* (1977), 77–101, pp. 96–98. The close relationship between the NRC and the Rockefeller Foundation leadership is pointed out in Robert H. Kargon, *The Rise of Robert Millikan: Portrait of a Life in American Science* (Ithaca and London: Cornell University Press, 1982), p. 105.

15. Gray, *Education* (note 13), pp. 23–25. The quotation is from p. vii, while the Paris office is described on p. 19. For informative details of support, see the budgets in "Rockefeller Foundation Agenda for Special Meeting, Apr. 11, 1933," Rockefeller Foundation Archives, Record Group 3, Series 900, Box 22, Folder 169 (RF RG 3, 900, 22, 169), Rockefeller Archive Center (RAC), Pocantico Hills, New York.

16. Fosdick, *Adventure* (note 13), quotation on p. vii. On the IEB's preference for broadness, see Robert E. Kohler, "A Policy for the Advancement of Science: The Rockefeller Foundation, 1924–29" in *Minerva 16* (1978), 480–515, p. 489.

17. Bohr argued for an expansion of the institute in his application to the IEB for support, the original of which, dated 27 Jun 1923, is filed in "Denmark 3: Institute for Theoretical Physics, 1923–1927," International Education Board archives, Series 1, Subseries 2, Box 28, Folder 403 (IEB 1, 2, 28, 403), RAC (note 15). The prehistory of the IEB's first support of the institute is also described in Robertson, *Early Years* (note 1), pp. 90-92. For the price situation, see Mitchell, *Statistics* (note 4), pp. 774, 781.

18. The assistance from prominent Danish friends can be seen from the letters Lundsgaard to Bohr 18 Mar 1923 (wrongly dated, correct date most probably 18 Apr 1923), Berlème to Bohr 8 May 1923, Bohr to Lundsgaard 27 Jun 1923, "Rockefeller Foundation," BGC-S (5,1) (note 1), which also contains copies of Bohr's draft of the application submitted to Lundsgaard, and of the final application to the IEB, both dated 27 Jun 1923. On the Rockefeller Institute and on Flexner, see Fosdick, *Adventure* (note 13), pp. ix, 8, and pp. 238–239, respectively.

19. Bohr reported the purpose of his trip to the United States in a letter to Rose 21 Nov 1923, "Rockefeller Foundation," BGC-S (5,1) (note 1). Gray, *Education* (note 13), pp. 25 and 11–13, respectively, identifies the grant as the first given to an institution, and describes Rose's trip to Europe.

20. The expansion of the institute in general, and the land and maintenance costs in particular, is described in "Institutet for teoretisk Fysik. Institutets Udvidelse" in *Copenhagen University Yearbook, 1923–1924* (note 10), pp. 127–131; on the new assistantship, see "Lærere og andre videnskabelige Tjenestemænd samt Censorer. Afgang og Udnævnelser m.v. Det matematisk-naturvidenskabelige Fakultet" in *ibid. 1924–1925* (1925), pp. 10–11, on p. 11, and Robertson, *Early Years* (note 1), pp. 93–95.

21. The development of prices in Denmark is provided in Mitchell, *Statistics* (note 4,), pp. 774, 781. Bohr's application to the Carlsberg Foundation of 7 Jan 1925 is in the Carlsberg Foundation Archive, Copenhagen, with carbon copy in "Carlsberg Foundation," BGC-S (1,1); the foundation's positive reply of 7 Jan 1925 is in *ibid.;* see also the institute's Budget Books (note 4). The difficulty of increasing the original IEB grant is discussed in Bohr to Rose 29 Nov and Rose to Bohr 17 Dec 1923, in IEB 1, 2, 28, 403 (note 17) and in "Rockefeller Foundation," BGC-S (5,1) (note 1). On the second grant from the IEB, see Augustus Trowbridge to Bohr 25 Sep 1925, Bohr to Trowbridge 6 Nov 1925, IEB 1, 2, 28, 403 (note 17), of which only the latter has been retained in "Rockefeller Foundation," BGC-S (5,3) (note 1). Bohr's second approach to the Carlsberg Foundation of 12 Nov 1926 is in the Carlsberg Foundation Archive, Copenhagen, with carbon copy in "Carlsberg Foundation," BGC-S (1,1) (note 1); the positive reply of 9 Dec 1926 is in *ibid.* An account of these developments, including the completion of the expansion, is given in Robertson, *Early Years*, p. 107.

22. An early expression of Bohr's belief in international scientific cooperation is his speech on receiving the Ørsted Medal: Niels Bohr, "Grundlaget for den moderne Atomforskning" in *Fysisk Tidsskrift 23* (1925), 10–17, reprinted and translated into English ("The Foundations of Modern Atomic Research") in Erik Rüdinger (gen. ed.), *Niels Bohr Collected Works, Volume 5*, Klaus Stolzenburg (ed.), *The Emergence of Quantum Mechanics (Mainly 1924–1926)* (Amsterdam, New York, Oxford, Tokyo: North-Holland Physics Publishing, 1984), pp. 125–135, 136–142, in particular on p. 141. The quotation is from *Rask–Ørsted Fondet: Beretning for 1919–20, 1;* copies of the annual reports of that foundation's activities are deposited in the Carlsberg Foundation Archive, Copenhagen. On the visitors supported by the foundation, see the institute's Budget Books (note 4). The information on the Rask–Ørsted Foundation's support of the informal conference was kindly provided in a letter from Hilde Levi to me 21 Jan 1989.

23. The total number of visitors at the institute is taken from "Visitors from abroad who for longer periods have worked at the Institute for Theoretical Physics, University of Copenhagen 1918–1948 (arranged according to their native countries)," prepared for application purposes circa 1950 and retained in the Niels Bohr Archive, Copenhagen. A copy of the list is deposited in the J. Robert Oppenheimer Papers, container 21, Library of Congress, Washington, D.C. More detailed information on the visitors is in the institute's "Guest Book," which has been microfilmed as part of the Archive for the History of Quantum Physics (AHQP), film 35, section 2. The AHQP is retained in the American Philosophical Society, Philadelphia, the Niels Bohr Archive, Copenhagen, the Niels Bohr Library of the American Institute of Physics in New York, and other places. The microfilm copy, some dates in which are impossible to read, can

be checked against the original of the Guest Book contained in the Niels Bohr Archive. Useful, but dated, information on the AHQP is provided in Thomas S. Kuhn, John L. Heilbron, Paul Forman, and Lini Allen, *Sources for the History of Quantum Physics: An Inventory and Report* (Philadelphia: The American Philosophical Society, 1967). The visitors supported by the Rockefeller Foundation can be identified in *The Rockefeller Foundation: Directory of Fellowships and Scholarships 1917–1970* (New York: Rockefeller Foundation, 1972), pp. 133, 153, 117, 107. Gamow's stay in Copenhagen is mistakenly not mentioned in this source. See also Robertson, *Early Years* (note 1), pp. 156–159.

24. Gray, *Education* (note 13), pp. 25–30, 37–44.

25. For a brief account of the experimental research at the institute in the early 1920s, see Robertson, *Early Years* (note 1), pp. 41–49. An invaluable source for the work at the institute from 1918 to 1959 is the bound volumes of publications from the institute during that period, retained in the Niels Bohr Archive, Copenhagen: *Universitetets Institut for teoretisk Fysik: Afhandlinger*. The discovery of hafnium is described in Helge Kragh, "Anatomy of a Priority Conflict: The Case of Element 72" in *Centaurus 23* (1980), 275–301. Bohr broke the news of the discovery in Bohr, "Structure" (note 5), p. 42.

26. The classic treatment of the theoretical development of quantum theory is Max Jammer, *The Conceptual Development of Quantum Mechanics* (New York, etc.: McGraw-Hill, 1966); see additional references in Chapter 2. The role of the institute has been dealt with in several places; see, for example, Robertson, *Early Years* (note 1), which also delineates the period of consolidation on pp. 139–148.

27. The quotation is taken from *ibid.*, p. 97. The most complete biography of Krogh is P. Brandt Rehberg, "August Krogh, 15.11.1874–15.11.1974" in *Dansk medicinhistorisk Aarbog 1974*, pp. 7–28, with a summary in English on pp. 27–28. See also R. Spärck, "August Krogh 15. November 1874 – 3. September 1949" in *Videnskabelige Meddelelser fra Dansk naturhistorisk Forening 3* (1949), V-XXX. For an appreciation in English of Krogh's life and work, see E. Snorrason, "Krogh, Schack August Steenberg" in Charles Gillispie (gen. ed.), *Dictionary of Scientific Biography* [15 volumes] (New York: Charles Scribner's Sons, 1970–1978), *VII* [1973], pp. 501–504. Krogh's physiologist daughter, Bodil Schmidt-Nielsen, is currently working on a biography of her father and her mother, Marie Krogh, who was also a noted physiologist. Although scarce on historical facts, August Krogh, "Visual Thinking: an Autobiographical Note" in *Organon 2* (1938), 87–94 gives a fascinating glimpse of Krogh's scientific style. On the institutional context of Krogh's work see, C. Barker Jørgensen, "Dyrefysiologi og gymnastikteori" in *Københavns Universitet 1479–1979* [fourteen volumes], *bind XIII: Det matematisk-naturvidenskabelige Fakultet, 2. del* (Copenhagen: G.E.C. Gads Forlag, 1979), pp. 447–488. Krogh's Silliman Lectures were printed as August Krogh, *The Anatomy and Physiology of Capillaries* (New Haven: Yale University Press, 1922). The BSC (note 5) contains no letters either to or from Krogh. Only four such letters, all strictly formal, exist in the August and Marie Krogh Archive, deposited in the Royal Library of Copenhagen, which contains a substantial part of Krogh's scientific correspondence. In my interview in Bar Harbor, Maine, during the early summer of 1981, Bodil Schmidt-Nielsen remembered the relationship between Bohr and her father as friendly; yet she could not recall that Bohr ever visited their house, which was otherwise very open to visitors. I am grateful to Dr. Schmidt-Nielsen for her hospitality at her home in Maine, and for her permission to let me see the Krogh Archive.

28. The original application was Krogh to Vincent 16 Apr 1923, "Denmark 4: Krogh Institute of Physiology 1923–1925," IEB 1, 2, 28, 404 (note 17).

The Krogh Archive (note 27) contains eight drafts of letters from Krogh to Pearce, and eight letters from Pearce to Krogh, all written between 17 Apr 1923 and 19 Jan 1925. On the establishment of the Rockefeller Foundation's Division of Medical Science, see Fosdick, *Adventure* (note 13), p. 105. For Pearce's scheme to fund Copenhagen physiology, see the typewritten excerpt of his letter to Rose 2 Mar 1924, IEB 1, 2, 28, 404. The finalization of the plan is described in "Opførelse og Indretning af en ny Laboratoriebygning m.v. til de fysiologiske Videnskaber for Midler skænkede af 'The Rockefeller Foundation' og 'International Education Board', New York" in *Copenhagen University Yearbook* (note 1), *1927–1928* (1929), pp. 150–168.

29. Brønsted's background is treated briefly in "Det fysisk-kemiske Institut" in *ibid., 1931–1932* (1932), pp. 189–199, on p. 189; a more detailed account is J. A. Christiansen, "Julius Petersen, Einar Biilmann og J. N. Brønsted" in *Kemien i Danmark, III: Danske Kemikere* [En forelæsningsrække ved Folkeuniversitetet, København 1964] (Copenhagen: Nyt Nordisk Forlag, 1968), pp. 69–94, on pp. 85–94. The evidence for Brønsted's first inquiry is a sheet from the Augustus Trowbridge Diary 25 Sep 1925, "Denmark 6: University of Copenhagen, Laboratory of Physical Chemistry 1925–32," IEB 1, 2, 28, 406 (note 17). The earlier correspondence between Brønsted and Trowbridge, which I have been unable to find, is cited in the margin of the sheet. The full Trowbridge Diary, in which the visit to Copenhagen 24 to 27 Sep 1925 occupies 13 double-spaced typewritten pages, is located in Record Group 12, RAC (note 15). Brønsted put forward his request to Bohr in a letter of 14 Jan 1927, "Rockefeller Foundation," BGC-S (5,5) (note 1); see also Rose to Trowbridge 14 Mar 1927, IEB 1, 2, 28, 406 (note 29). Bohr replied in a letter of 4 Feb 1927, "Rockefeller Foundation," BGC-S (5,5).

30. On Trowbridge's appointment and earlier career, see Gray, *Education* (note 13), p. 19. For a more thorough treatment, see Kohler, *Managers* (note 13). Rose made his request for a list in his letter to Trowbridge 3 Mar 1927, IEB 1, 2, 28, 406 (note 29). Trowbridge's extensive reply, dated 6 Apr 1927, is in the Trowbridge Diary (note 29).

31. Like his previous letter, Trowbridge's follow-up letter to Rose was dated 6 Apr 1927, IEB 1, 2, 28, 406 (note 29). The information on Rose's interview with Flexner is from "(Excerpt from Record of Doctor Rose's Interviews)" [paginated 326–327], *ibid.* For the issues considered most important during the IEB's deliberations, see, for example, Rose to Trowbridge 14 Mar 1927, *ibid,* and Charles E. Mendenhall to Trowbridge, quoted in Trowbridge to Rose 6 Apr 1927. The establishment of Brønsted's institute is described in "Det fysisk-kemiske Institut" (note 29), p. 197.

32. "Det matematiske Institut. Institutets Oprettelse" in *Copenhagen University Yearbook* (note 1), *1933–1934* (1935), pp. 221–229.

33. *Ibid.,* pp. 226–227. On the close bonds between the two brothers, see [David Jens Adler,] "Childhood and Youth" in Stefan Rozental (ed.), *Niels Bohr: His Life and Work as Seen by His Friends and Colleagues* (Amsterdam: North-Holland Publishing Company, 1968), pp. 11–37, on pp. 23–31; Richard Courant, "Fifty Years of Friendship" in *ibid.,* pp. 301–309; and Stefan Rozental, "The Forties and the Fifties" in *ibid.,* pp. 149–190, on p. 151.

34. Bohr to the Carlsberg Foundation 15 Apr 1933, Carlsberg Foundation Archive, Copenhagen; carbon copy in "Carlsberg Foundation," BGC-S (1, 2) (note 1).

35. On the development of Danish prices, see Mitchell, *Statistics* (note 4), pp. 774, 781. On the stagnancy in support after the mid-1920s, see the institute's Budget Books (note 4) and Robertson, *Early Years* (note 1), p. 94.

36. The number of visitors is taken from "Visitors" (note 23); the number of publications from *Afhandlinger* (note 25).

CHAPTER 2

1. On the general acceptance of the Copenhagen interpretation, see Max Jammer, *The Philosophy of Quantum Mechanics: The Interpretation of Quantum Mechanics in Historical Perspective* (New York, etc.: John Wiley & Sons, 1974), especially pp. 247–251.
2. For general literature on the miraculous year, see Daniel J. Kevles, "Towards the *Annus Mirabilis:* Nuclear Physics Before 1932" in *The Physics Teacher 10* (1972), 175–181; Charles Weiner, "1932 – Moving into the New Physics" in *Physics Today 25* (May 1972), 40–49; Roger H. Stuewer (ed.), *Nuclear Physics in Retrospect: Proceedings of a Symposium on the 1930s* (Minneapolis: University of Minnesota Press, 1979).
3. On Bohr's work in Manchester, see J. L. Heilbron and Thomas S. Kuhn, "The Genesis of the Bohr Atom" in *Historical Studies in the Physical Sciences 1* (1969), 211–290. Bohr's early explanation of beta decay is described in Abraham Pais, *Inward Bound: Of Matter and Forces in the Physical World* (Oxford and New York: Clarendon Press and Oxford University Press, 1986), pp. 223–224.
4. On Bohr's quantum physical explanation of the periodic table, see Helge Kragh, "Niels Bohr's Second Atomic Theory" in *Historical Studies in the Physical Sciences 10* (1979), 123–186. The development of quantum mechanics has received much attention by historians; the classical account is Max Jammer, *The Conceptual Development of Quantum Mechanics* (New York, etc.: McGraw-Hill, 1966); a more recent treatment is John Hendry, *The Creation of Quantum Mechanics and the Bohr-Pauli Dialogue* (Dordrecht, Boston, Lancaster: D. Reidel Publishing Company, 1984). An enormous enterprise, still in process, is Jagdish Mehra and Helmut Rechenberg, *The Historical Development of Quantum Theory*. The following books have been published in this series. *Volume 1* has been published in two parts, each under different cover, *The Quantum Theory of Planck, Einstein, Bohr and Sommerfeld: Its Foundations and the Rise of Its Difficulties 1900–1925, Part 1* and *Part 2*. Then follow: *Volume 2, The Discovery of Quantum Mechanics 1925; Volume 3, The Formulation of Matrix Mechanics and Its Modifications 1925–1926; Volume 4, Part 1, The Fundamental Equations of Quantum Mechanics 1925–1926; Volume 4, Part 2, The Reception of the New Quantum Mechanics 1925–1926; Volume 5, Erwin Schrödinger and the Rise of Wave Mechanics – Part 1, Schrödinger in Vienna and Zurich, 1887–1925, Part 2, The Creation of Wave Mechanics: Early Response and Applications, 1925–1926* (New York, Heidelberg, Berlin: Springer-Verlag, Volumes 1–4 1982, Volume 5 1987). This work hardly considers the vast unpublished resources available.
5. See, for example, John Hendry, "The History of Complementarity: Niels Bohr and the Problem of Visualization" in *Rivista di Storia della Scienza 2* (1984), 392–407, which does not, however, provide references; for an account with references, see Hendry, *Creation* (note 4), pp. 111–128, 161–165. David Cassidy, of the Einstein Papers and SUNY Stony Brook, is presently completing a substantial biography of Heisenberg.
6. The microfilmed Bohr Scientific Manuscripts (MSS) – in the Niels Bohr Archive, Copenhagen, the Niels Bohr Library of the American Institute of Physics in New York, and other places – contain 193 pages of preparation for the Como lecture; see film 11, section 4 (MSS (11, 4)). The Como lecture itself was published as "The Quantum Postulate and the Recent Development of Atomic Theory" in *Atti del Congresso Internazionale*

dei Fisici *11–20 Settembre 1927, vol. 2* (Bologna: Nicola Zanichelli, 1928), pp. 565–588, and republished in a heavily revised form in *Nature* (Supplement) (14 Apr 1928), 580–590, a slightly modified version of which was printed in *idem., Atomic Theory and the Description of Nature* (Cambridge, England: Cambridge University Press, 1934), pp. 52–91; this selection of writings has recently been reprinted as *The Philosophical Writings of Niels Bohr, Volume 1: Atomic Theory and the Description of Nature* (Woodbridge, Connecticut: Ox Bow Press, 1987). Subsequent volumes in this series have been republished, by the same publisher in the same year, as *The Philosophical Writings of Niels Bohr, Volume 2: Essays 1933–1957 on Atomic Physics and Human Knowledge; Volume 3: Essays 1958–1962 on Atomic Physics and Human Knowledge.* The first two versions of the Como Lecture are reproduced in Erik Rüdinger (gen. ed.), *Niels Bohr Collected Works, Volume 6,* Jørgen Kalckar (ed.), *Foundations of Quantum Physics I (1926–1932)* (Amsterdam, New York, Oxford, Tokyo: North-Holland, 1985), pp. 113–136, 148–590.

7. The quotation is from Bohr, "Quantum Postulate" (*Nature* version) (note 6), p. 590.

8. *Ibid.* Bohr expanded on the complementarity argument in several articles; see the recent republication of his collections of philosophical articles referred to in note 6. For a brief account of Bohr's philosophical views, see Aage Petersen, "The Philosophy of Niels Bohr" in *Bulletin of the Atomic Scientists 19* (Sep 1963), 8–14. A fuller and more recent account is given in Henry J. Folse, *The Philosophy of Niels Bohr: The Framework of Complementarity* (Amsterdam, Oxford, New York, Tokyo: North-Holland, 1985). Referring to Bohr's willingness to expand his complementarity argument beyond physics, Loren R. Graham, *Between Science and Values* (New York: Columbia University Press, 1981), pp. 46–62, classifies Bohr as an "expansionist."

9. David C. Cassidy, "Cosmic Ray Showers, High Energy Physics, and Quantum Field Theories: Programmatic Interactions in the 1930s" in *Historical Studies in the Physical Sciences 12* (1981), 1–39.

10. P. A. M. Dirac, "The Quantum Theory of the Electron" in *Proceedings of the Royal Society of London A117* (1 Feb 1928), 610–624. The origins and early development of the Dirac equation have been described in Helge Kragh, "The Genesis of Dirac's Relativistic Theory of Electrons" in *Archive for History of Exact Sciences 24* (1981), 31–67; and in Donald Franklin Moyer, "Origins of Dirac's electron, 1925–1928" in *American Journal of Physics 49* (Oct 1981), 944–949; *idem.*, "Evaluations of Dirac's electron, 1928–1932" in *ibid.* (Nov 1981), 1055–1062; *idem.*, "Vindications of Dirac's Electron, 1932–1934" in *ibid.* (Dec 1981), 1120–1125. See also Pais, *Inward Bound* (note 3), pp. 286–292.

11. The new interpretation was presented in P. A. M. Dirac, "A Theory of Electrons and Protons" in *Proceedings of the Royal Society of London A126* (1 Jan 1930), 360–365; paper received by journal 6 Dec 1929. For Dirac's own recollection of the event, see *idem.*, "Recollections of an Exciting Era" in Charles Weiner (ed.), *History of Twentieth Century Physics [Proceedings of the International School of Physics "Enrico Fermi," Course LVII, Varenna on Lake Como, Villa Monastero, 31st July – 12th August 1972]* (New York and London: Academic Press, 1977), pp. 109–146, on pp. 141–145. Dirac provided a particularly lucid presentation of his hole theory in his letter to Bohr 26 Nov 1929. This letter is part of the Bohr Scientific Correspondence, with originals at the Niels Bohr Archive in Copenhagen. Microfilm copies of the BSC are also deposited in the Niels Bohr Library of the American Institute of Physics in New York and other places. Dirac's letter is microfilmed on film 9, section 4 (BSC (9, 4)). The quotation is from the abstract of P. A. M. Dirac's lecture, "The Proton," given Monday 8 Sep 1930, as

reproduced in *British Association for the Advancement of Science, Report of the Ninety-Eighth Meeting: Bristol – 1930, September 3–10* (London: Office for the British Association, 1931), p. 303. See Pais, *Inward Bound* (note 3), pp. 346–351.

12. P. A. M. Dirac, "The Quantum Theory of Emission and Absorption of Radiation" in *Proceedings of the Royal Society of London A114* (1 Mar 1917), 243–265, reprinted in Julian Schwinger (ed.), *Selected Papers on Quantum Electrodynamics* (New York: Dover, 1958), pp. 1–23; this paper is discussed by Joan Bromberg, "Dirac's Quantum Electrodynamics and the Wave-Particle Equivalence" in Weiner (ed.), *History* (note 11), pp. 147–157. W. Heisenberg and W. Pauli, "Zur Quantendynamik der Wellenfelder" in *Zeitschrift für Physik 56* (8 Jul 1929), 1–61; idem., "Zur Quantentheorie der Wellenfelder. II" in *ibid. 59* (2 Jan 1930), 168–190. For the historical context of the work of both Dirac and of Heisenberg and Pauli, see Steven Weinberg, "The Search for Unity: Notes for a History of Quantum Field Theory" in *Daedalus 106* (1977), 17–35, pp. 22–23, and the more technical account in Pais, *Inward Bound* (note 3), pp. 334–346.

13. Heisenberg noted his sentiment in his letter to Bohr 23 Jul 1928, BSC (11, 2) (note 11). On Heisenberg's subsequent work, see Joan Bromberg, "The Impact of the Neutron: Bohr and Heisenberg" in *Historical Studies in the Physical Sciences 3* (1971), 307–341, pp. 323–329. Bohr reported Pauli's attitude in his letter to Heisenberg 18 Feb 1931, BSC (20, 2). For the origins of the annual informal conferences, see Peter Robertson, *The Early Years: The Niels Bohr Institute 1921–1930* (Copenhagen: Akademisk Forlag, 1979), pp. 136–137.

14. On Bohr's "enthusiastic resignation," see John L. Heilbron, "The earliest missionaries of the Copenhagen spirit" in *Revue d'histoire des sciences et leurs applications 38* (1985), 194–230, p. 224.

15. The two independent discoveries were published as Ronald W. Gurney and Edward U. Condon, "Wave Mechanics and Radioactive Disintegration" in *Nature 122* (22 Sep 1928), 439, paper dated 30 Jul; and G. Gamow, "Zur Quantentheorie des Atomkernes" in *Zeitschrift für Physik 52* (17 Dec 1928), 204–212, paper received by journal 2 Aug 1928. Gamow's continued work on the same topic was published as G. Gamow and F. G. Houtermans, "Zur Quantenmechanik des radioaktiven Kerns" in *Zeitschrift für Physik 52* (17 Dec 1928), 496–509, paper received by journal from Göttingen 29 Oct 1928; G. Gamow, "Zur Quantentheorie der Atomzertrümmerung" in *ibid.*, 510–515, paper received by journal from Copenhagen 10 Nov 1928; idem., "The Quantum Theory of Nuclear Disintegration" in *Nature 122* (24 Nov 1928), 805–806, paper received by journal 29 Sep 1928; idem., "Bemerkung zur Quantentheorie des Radioaktiven Zerfalls" in *Zeitschrift für Physik 53* (25 Feb 1929), 601–604, paper received by journal 5 Jan 1929. See Roger H. Stuewer, "Gamow's Theory of Alpha-Decay" in Edna Ullmann-Margalit (ed.), *The Kaleidoscope of Science – The Israel Colloquium: Studies in History, Philosophy, and Sociology of Science, Volume 1* [Robert S. Cohen and Marx W. Wartofsky (eds.), *Boston Studies in the Philosophy of Science, Volume 94*] (Dordrecht, Boston, Lancaster, Tokyo: D. Reidel Publishing Company, 1986), pp. 147–186, which refers to Gamow's introduction of the liquid drop model on p. 179. Stuewer's forthcoming article, "The Origins of the Liquid-Drop Model of the Nucleus," in *Proceedings* of a Conference on "50 Years of Nuclear Fission" (West Berlin, 30–31 Mar 1989), describes the subject in more detail.

16. The perceived necessity to include the electron in the nucleus was not generally shared; see Bromberg, "Impact" (note 13), p. 308, which criticizes Edward M. Purcell, "Nuclear Physics without the Neutron; Clues and Contradictions" in *Dixième Congrès international d'histoire des sci-*

ences, 1962 [vol. I] (Paris: Hermann, 1964), pp. 121–133, on p. 128, for holding such a view. However, for our purposes this fine distinction is not significant.

17. A general account is Carsten Jensen, "A history of the beta spectrum and its interpretation, 1911–1934," PhD dissertation, Copenhagen University, forthcoming. See also Roger H. Stuewer, "The Nuclear Electron Hypothesis" in William R. Shea (ed.), *Otto Hahn and the Rise of Nuclear Physics* (Dordrecht, Boston, Lancaster: D. Reidel Publishing Company, 1983), pp. 19–67, on pp. 19–32; Pais, *Inward Bound* (note 3), p. 307; Purcell, "Nuclear Physics" (note 16), p. 128; Wolfgang Pauli, "Zur älteren und neueren Geschichte des Neutrinos," in *idem., Aufsätze und Vorträge über Physik und Erkenntnistheorie*, Victor F. Weisskopf (ed.) (Braunschweig: Friedr. Vieweg & Sohn, 1961), pp. 156–180, on pp. 156–158, reprinted in R. Kronig and V.F. Weisskopf (eds.), *Collected Scientific Papers by Wolfgang Pauli in Two Volumes, Vol. 2* (New York, London, Sydney: Interscience, 1964), pp. 1313–1337, on pp. 1313–1315.

18. Rudolf Peierls, "Introduction" in Erik Rüdinger (gen. ed.), *Niels Bohr Collected Works, Volume 9:* Rudolf Peierls (ed.), *Nuclear Physics (1929–1952)* (Amsterdam, etc.: North-Holland Physics Publishing, 1985), pp. 3–83, on pp. 4–5.

19. Pauli, "Zur älteren und neueren Geschichte" (note 17), pp. 156–157. Pais, *Inward Bound* (note 3), pp. 298–303, enumerates the increasing problems of incorporating an electron into the atomic nucleus.

20. G. Gamow, *Constitution of Atomic Nuclei and Radioactivity* (Oxford: Clarendon Press, 1931), quotation on p. 2; *idem., Der Bau des Atomkerns und die Radioaktivität* (Leipzig: S. Hirzel, 1932). Pais, *Inward Bound* (note 3), p. 297, claims that Gamow's was the first book on the nucleus by a theoretical physicist. For the skull and crossbones story, see the interview with Gamow by Charles Weiner 25 Apr 1968, deposited in the Niels Bohr Library of the American Institute of Physics in New York City, and Hendrik Casimir, *Haphazard Reality: Half a Century of Science* (New York, etc.: Harper & Row, 1983), pp. 117–118.

21. Heisenberg and Pauli, "Zur Quantendynamik" (note 12), p. 3. Heisenberg to Bohr 26 Feb and 10 Mar, Bohr to Heisenberg 18 Mar, Heisenberg to Bohr 18 Sep 1930, wherefrom quotation, BSC (20, 2) (note 11). See Bromberg, "Impact" (note 13), p. 323.

22. For the more general context of relativistic quantum physics during this period, see Cassidy, "Showers" (note 9); Peter Galison, "The Discovery of the Muon and the Failed Revolution in Quantum Electrodynamics" in *Centaurus 26* (1983), 262–316, further developed as "Chapter 3: Particles and Theories" in *idem., How Experiments End* (Chicago and London: University of Chicago Press, 1987), pp. 75–133.

23. The first proposal of nonconservation of energy was published as Niels Bohr, H. A. Kramers, and J. C. Slater, "The Quantum Theory of Radiation" in *Philosophical Magazine 47* (1924), 758–802; reprinted in Erik Rüdinger (gen. ed.), *Niels Bohr Collected Works, Volume 5:* Klaus Stolzenburg (ed.), *The Emergence of Quantum Mechanics (Mainly 1924–1926)* (Amsterdam, New York, Oxford, Tokyo: North-Holland, 1984), pp. 99–118. See John Hendry, "Bohr–Kramers–Slater: A Virtual Theory of Virtual Oscillators and Its Role in the History of Quantum Mechanics" in *Centaurus 25* (1981), 189–221. On other precursors of Bohr's proposal of nonconservation of energy, see Pais, *Inward Bound* (note 3), pp. 310–311. The manuscript written for *Nature* is Niels Bohr, "β-ray spectra and energy conservation," typed manuscript in MSS (12, 1) (note 6). It is reprinted in *Niels Bohr Collected Works, Volume 9* (note 18), pp. 85–89. See Peierls, "Introduction" (note 18), pp. 5–6.

24. Bohr to Pauli 1 Jul 1929, Pauli to Bohr 17 Jul 1929, wherefrom quotation; BSC (14, 3) (note 11). Both letters are reproduced in A. Hermann, K. v.

Meyenn, and V. F. Weisskopf (eds.), *Wolfgang Pauli, Scientific Correspondence with Bohr, Einstein, Heisenberg, a.o., Volume I: 1919–1929* (New York, Heidelberg, Berlin: Springer-Verlag, 1979), pp. 507–509, 512–514.

25. Dirac to Bohr 26 Nov 1929, Bohr to Dirac 5 Dec 1929; BSC (9,4) (note 11). The two letters are reproduced in full in Moyer, "Evaluations" (note 10), pp. 1057–1058.

26. Typewritten transcripts of Bohr's lectures at Cambridge University and the Chemical Society of London have been retained in MSS (12,2) (note 6); see *Niels Bohr Collected Works, Volume 6* (note 6), p. 317. The latter lecture was subsequently revised for publication: Niels Bohr, "Chemistry and the Quantum Theory of Atomic Constitution" in *Journal of the Chemical Society* (1932), 349–384; it is reprinted in *Niels Bohr Collected Works, Volume 6*, pp. 371–408.

27. Bohr's lecture at the Rome conference was published as Niels Bohr, "Atomic Stability and Conservation Laws" in *Atti del Convegno di Fisica Nucleare Ottobre 1931* (Rome: Reale Accademia d'Italia, 1932), pp. 119–130; reprinted in *Niels Bohr Collected Works, Volume 9* (note 18), pp. 99–114.

28. Heilbron, "Earliest missionaries" (note 14), p. 204.

29. The quotation from Eddington's brochure is taken from W. Bennett Lewis, "Early Detectors and Counters" in *Nuclear Instruments and Methods 162* (1979), 9–14, p. 12; I am grateful to the historian of science Tom Cornell at the Rochester Institute of Technology, New York, for supplying me with this reference. On subsequent uses of the term miraculous year, see the literature referred to in note 2.

30. For historical studies discussing the belated impact of the neutron, see Bromberg, "Impact" (note 13) and Stuewer, "Nuclear Electron Hypothesis" (note 17).

31. Rutherford's original speculations were presented as Ernest Rutherford, "Nuclear Constitution of Atoms – Bakerian Lecture" in *Proceedings of the Royal Society of London A97* (1920), 374–400, pp. 396–397; reprinted in *The Collected Papers of Lord Rutherford: Volume Three, Cambridge* (London: George Allen and Unwin Ltd., 1965), pp. 14–40, on p. 34. On the subsequent work of his students, see Purcell, "Nuclear Physics" (note 16), p. 124; see also Lawrence Badash, "Nuclear Physics in Rutherford's laboratory before the discovery of the neutron" in *American Journal of Physics 51* (1983), 884–889. The early papers seeking to establish the neutron experimentally were J. L. Glasson, "Attempts to Detect the Presence of Neutrons in a Discharge Tube" in *Philosophical Magazine 42* (Oct 1921), 596–600, paper dated July 1921; and J. Keith Roberts, "The Relation Between the Evolution of Heat and the Supply of Energy during the Passage of an Electric Discharge Through Hydrogen" in *Proceedings of the Royal Society of London A102* (2 Oct 1922), 72–88, paper received by journal 21 Jun 1922. The eventual announcement of the experimental discovery was made in James Chadwick, "Possible Existence of a Neutron" in *Nature 129* (27 Feb 1932), 312, paper dated 17 Feb 1932. Chadwick made his remark on the neutron's origins in "Some Personal Notes on the Search for the Neutron" in *Dixième Congrès* (note 16), pp. 159–162, on p. 162, which is reprinted in John Hendry (ed.), *Cambridge Physics in the Thirties* (Bristol: Adam Hilger, Ltd., 1984), pp. 42–45, on p. 45. The observation in Berlin was announced in W. Bothe and H. Becker, "Künstliche Erregung von Kern-γ-Strahlen" in *Zeitschrift für Physik 66* (3 Dec 1930), 289–306. The work in Paris was published as Irène Curie and Frédéric Joliot, "Emission de protons de grande vitesse par les substances hydrogénée sous l'influence des rayons très pénétrants" in *Comptes rendus des séances de l'academie des sciences 194* (18 Jan 1932), 273–275; *idem.*, "Effet d'absorption de rayons

de très haute fréquence par projection de noyaux légers" in *ibid. 194* (22 Feb 1932), 708–711. The two latter articles are reprinted in Frédéric and Irène Joliot-Curie, *OEvres Scientifiques Complètes* (Paris: Presses Universitaires de France, 1961), pp. 359–360, 361–363.

32. Chadwick informed Bohr of his discovery in a letter of 24 Feb 1932, BSC (18,1) (note 11). The quotation is from Fowler to Bohr 1 Mar 1932, BSC (19,2). Bohr to Heisenberg 21 Mar 1932, BSC (20,2); Bohr to Chadwick 25 Mar 1932, BSC (18,1).

33. Chadwick announced that he would be unable to come to the Copenhagen conference in a letter to Bohr 30 Mar 1932, BSC (18,1) (note 11). Bohr's lecture, wherefrom quotation, was called "Properties of the Neutron," a typewritten transcript of which has been retained in MSS (13,2) (note 6).

34. The quotation is from Heisenberg to Bohr 20 Jun 1932, BSC (20,2) (note 11). The trilogy comprises Werner Heisenberg, "Über den Bau der Atomkerne. I" in *Zeitschrift für Physik 77* (19 Jul 1932), 1–11; *idem.*, "Über den Bau der Atomkerne. II" in *ibid.* (21 Sep 1932), 156–164; *idem.*, "Über den Bau der Atomkerne. III" in *ibid. 80* (16 Feb 1933), 587–596, in which Heisenberg's comments on beta decay are on pp. 595–596. For a brief introduction to Heisenberg's work on his trilogy, see Laurie M. Brown, "Yukawa's Prediction of the Meson" in *Centaurus 25* (1981), 71–132, pp. 86–91. In Laurie M. Brown and Donald F. Moyer, "Lady or Tiger? – The Meitner-Hupfeld Effect and Heisenberg's Neutron Theory" in *American Journal of Physics 52* (Feb 1984), 130–136, pp. 133–134, the authors misdate Heisenberg's letter to 18 Sep 1930, mistakenly using it to argue why Heisenberg continued to maintain in his 1932 papers that the neutron was composite.

35. The discussions referred to were described in *British Association for the Advancement of Science: Report of the Annual Meeting 1932 (102 Year), York, August 31 – September 7* (London: Office of the British Association, 1932), pp. 306–308; the quotation is from p. 308.

36. The two letters quoted are Goudsmit to Bohr 4 Nov 1932 and Bohr to Goudsmit 28 Dec 1932, contained in the file "Amerikarejsen 1933" in the Bohr General Correspondence (BGC) in the Niels Bohr Archive, Copenhagen. In an important study restricted to the impact of the neutron, and describing Bohr's and Heisenberg's respective responses to its discovery until three months after it was made, Joan Bromberg concludes, in agreement with my own interpretation, that because of the continued belief in the nuclear electron, the present-day view of the neutron as a fundamental particle was established only slowly; Joan Bromberg, "Impact" (note 13). An overview of the continued problem of the nuclear electron after the discovery of the neutron is given in Stuewer, "Nuclear Electron Hypothesis" (note 17), pp. 46–56; see also the chronology in Pais, *Inward Bound* (note 3), pp. 316–319.

37. Rutherford to Bohr 21 Apr 1932, BSC (25,2) (note 11), wherefrom quotation. The results were published as J. D. Cockcroft and E. T. S. Walton, "Disintegration of Lithium by Swift Protons" in *Nature 129* (30 Apr 1932), 649, paper dated 16 Apr 1932.

38. Bohr to Rutherford 2 May 1932, BSC (25,2) (note 11).

39. *Ibid.*

40. A good account is Michelangelo De Maria and Arturo Russo, "The Discovery of the Positron" in *Rivista di storia della scienza 2* (1985), 237–286. For a history of the early ideas of particle creation, see Joan Bromberg, "The Concept of Particle Creation Before and After Quantum Mechanics" in *Historical Studies in the Physical Sciences 7* (1975), 161–191.

41. The announcement was first made in Carl D. Anderson, "The Apparent Existence of Easily Deflectable Positives" in *Science 76* (9 Sep 1932),

238–239, and more elaborately presented as *idem.*, "The Positive Electron" in *Physical Reveiw 43* (15 Mar 1933), 491–494. See the retrospective account, C. D. Anderson, "Early Work on the Positron and Muon" in *American Journal of Physics 29* (1961), 825–830. The observation in Paris was announced in Irène and Frédéric Joliot-Curie, *La projection de noyaux atomiques par un rayonnement très pénétrant: l'existence du neutron [Actualités scientifiques et industrielles, XXXII:* Louise de Broglie (gen. ed.), *Exposés de physique théorique]* (Paris: Hermann et Cie., 1932), 22 pp., on p. 21; reprinted in Joliot-Curie, *OEvres* (note 31), pp. 422–437, on p. 437.

42. The results were published as P. M. S. Blackett and G. P. S. Occhialini, "Some Photographs of the Tracks of Penetrating Radiation" in *Proceedings of the Royal Society of London A139* (3 Mar 1933), 699–726, p. 716. The endorsement appeared in P. A. M. Dirac, "Quantised Singularities in the Electromagnetic Field" in *ibid. A133* (1 Sep 1931), 60–72, p. 61. See Pais, *Inward Bound* (note 3), pp. 351–352.

43. On Klein's establishment in Stockholm in January 1931, see Bohr to the University Governing Board 8 Dec 1930, in the Bohr General Correspondence, Special File, film 8, section 5, retained in the Niels Bohr Archive, Copenhagen. Klein expressed his enthusiasm in a letter to Bohr 4 Apr 1933, BSC (22,1) (note 11). The quotation is from Bohr to Klein 7 Apr 1933, *ibid.*

44. The Caltech symposium is described in a printed announcement dated 20 Apr and signed in handwriting, "Mit herzlichem Gruss von R. Ladenburg" in "Amerikarejsen 1933," BGC (note 36). R. M. Langer, "Positive and Negative Electrons Apparently Produced in Pairs" – Wire Report, Science Service, Washington, D.C., 18 May 1933" in *ibid.* The other quotation is from Bohr to Heisenberg 19 May 1933, BSC (20,2) (note 11). The version of this letter contained in the BSC seems to be the original; it may therefore not have been sent.

45. The quotation is from Bohr to Heisenberg 17 Aug 1933, BSC (20,2) (note 11). On Bohr's prior changing his views on account of experimental evidence, see Hendry, "Bohr–Kramers–Slater" (note 23), p. 203. The following is a sample of the accumulated experimental evidence: J. Chadwick, P. M. S. Blackett, and G. Occhialini, "New Evidence for the Positive Electron" in *Nature 131* (1 Apr 1933), 473, paper dated 27 Mar 1933; L. Meitner and K. Philipp, "Die bei Neutronenanregung auftretenen Elektronbahnen" in *Die Naturwissenschaften 21* (14 Apr 1933), 286–287, paper dated 25 Mar 1933; Irène Curie and Frédéric Joliot, "Contribution à l'étude des électrons positifs" in *Comptes rendus des séances de l'académie des sciences 196* (10 Apr 1933), 1105–1107, reprinted in Joliot-Curie, *OEvres* (note 31), pp. 440–441; Carl D. Anderson, "Free Positive Electrons Resulting from the Impact Upon Atomic Nuclei of the Photons from ThC" " in *Science 77* (5 May 1933), 432; Carl D. Anderson and Seth H. Neddermeyer, "Positrons from Gamma-Rays" in *Physical Review 43* (15 June 1933), 1034; L. Meitner and K. Philipp, "Die Anregung positiver Elektronen durch γ-Strahlen von ThC" " in *Die Naturwissenschaften 21* (16 Jun 1933), 468; Irène Curie and Frédéric Joliot, "Sur l'origine des électrons positifs" in *Comptes rendus des séances de l'académie des sciences 196* (22 May 1933), 1581–1583, reprinted in Joliot-Curie, *OEvres*, pp. 442–443; James Chadwick, "The Neutron" in *Proceedings of the Royal Society of London A142* (1 Oct 1933), 1–25, pp. 24–25. The experimental work to induce positrons is summarized in J. Chadwick, P. M. S. Blackett, and G. P. S. Occhialini, "Some Experiments on the Production of Positive Electrons" in *ibid. A144* (1 Mar 1934), 235–249.

46. On the postponement of the annual conference, see, for example, Betty Schultz [Bohr's secretary] to Egil Hylleraas 7 Mar 1933, BSC

(21,2) (note 11). The quotations are from Bohr to Ellis 30 Aug 1933, BSC (19,1).

47. The lecture at the Solvay Conference was printed as Niels Bohr, "Sur la méthode de correspondence dans la théorie de l'électron" in *Structure et propriétés des noyaux atomiques: rapports et discussions du septième counseil de physique, tenu à Bruxelles du 22 au 29 Octobre 1933 sous les auspices de l'institut international de physique Solvay* (Paris: Gauthier-Villars, 1934), pp. 216–228, where the remarks on the nucleus are on pp. 226–228. These remarks are translated into English in *Niels Bohr Collected Works, Volume 9* (note 18), pp. 129–132.

48. Bohr's lecture at the Physics Society was transcribed as "Om de positive Elektroner," in MSS (13,4) (note 6). The quotation is from Bohr to Klein 10 Dec 1933, BSC (22,1) (note 11).

49. The paper provoking Bohr was L. Landau and R. Peierls, "Erweiterung des Unbestimmtheitsprinzips für die relativistische Quantentheorie" in *Zeitschrift für Physik 69* (23 Apr 1931), 56–69, paper received by journal 3 Mar 1931; the announcement of Bohr's impending comment is on p. 57. Landau was in Copenhagen 8 Apr to 3 May 1930, 20 Sep to 22 Nov 1930, 25 Feb to 19 Mar 1931; Peierls visited 8 to 24 Apr 1930, 24 to 28 Feb 1931; institute's "Guest Book" in the Archive for the History of Quantum Physics, film 35, section 2 (AHQP (35,2)), retained in the American Philosophical Society, Philadelphia, the Niels Bohr Archive, Copenhagen, the Niels Bohr Library of the American Institute of Physics in New York, and other places. Some of the dates are impossible to read on the microfilm, and can be checked against the original of the Guest Book contained in the Niels Bohr Archive. Somewhat dated information on the AHQP is contained in Thomas S. Kuhn, John L. Heilbron, Paul Forman, and Lini Allen, *Sources for the History of Quantum Physics: An Inventory and Report* (Philadelphia: The American Philosophical Society, 1967). Bohr's criticism was finally published as N. Bohr and L. Rosenfeld, "Zur Frage des Messbarkeit der elektromagnetischen Feldgrössen" in *Det Kongelige Danske Videnskabernes Selskab. Mathematisk-fysiske Meddelelser 12* (8, 1933), 65 pp. See also Léon Rosenfeld, "Niels Bohr in the Thirties: Consolidation and extension of the conception of complementarity" in Stefan Rozental (ed.), *Niels Bohr: His Life and Work as Seen by Friends and Colleagues* (Amsterdam: North-Holland Publishing Company, 1968), pp. 114–136, on pp. 125–127.

50. Landau and Peierls, "Erweiterung" (note 49), where quotation is on p. 56, and the application of their conclusion is noted on p. 69.

51. The quotation is from *ibid.*, p. 63. On the history of the correspondence principle, see J. Rud Nielsen, "Introduction" in L. Rosenfeld (gen. ed.), *Niels Bohr Collected Works, Volume 3*: J. Rud Nielsen (ed.), *The Correspondence Principle (1918–1923)* (Amsterdam, New York, London: North-Holland Publishing Company, 1976), pp. 3–45, and Klaus Michael Meyer-Abich, *Korrespondenz, Individualität und Komplementarität: Eine Studie zur Geistesgeschichte der Quantentheorie in den Beitragen Niels Bohrs* (Wiesbaden: F. Steiner, 1965). Bohr to Pauli 25 Jan 1933, BSC (24,2) (note 11); this letter is reproduced in Karl von Meyenn, Armin Hermann, and Victor F. Weisskopf (eds.), *Wolfgang Pauli. Scientific Correspondence with Bohr, Einstein, Heisenberg a.o., Volume II: 1930–1939* (Berlin, etc.: Springer-Verlag, 1985), pp. 152–156.

52. The quotation is from Bohr to Heisenberg 28 Oct 1932, BSC (20,2) (note 11); for similar remarks in connection with this work, see Bohr to Dirac 14 Nov 1932, BSC (18,4), Bohr to Jordan 27 Dec 1932, BSC (21,3), Bohr to Goudsmit 28 Dec 1932 (note 36), Bohr to Pauli 25 Jan 1933 (note 51). On Bohr's continued preoccupation with the article into 1934, see, in particular, Bohr to Heisenberg 28 Oct 1932, Heisenberg to Bohr 5 Nov 1932, BSC (20,2), Bohr to Dirac 14 Nov 1932, BSC (18,4), Bohr to Heisen-

berg 13 Mar 1933, 19 May 1933, BSC (20,2), Klein to Bohr 20 Jun 1933, BSC (22,1), Bohr to Heisenberg 17 Aug 1933, BSC (20,2), Klein to Bohr 12 Dec 1933, BSC (22,1).

53. The quotation is from Bohr to Pauli 15 Mar 1934, BSC (24,2) (note 11); this letter is reproduced in Meyenn *et al.* (eds.), *Pauli, Volume II* (note 51), pp. 307–311. Bohr reported work on his "small article" to Langmuir 17 Jan 1934, BSC (23,1), Breit 18 Jan 1934, BSC (17,4), Nishina 26 Jan 1934, BSC (24,1), Fermi 31 Jan 1934, BSC (19,2), and Fowler 14 Feb 1934, BSC (19,2). Bohr's "*Hjerteudgydelse*" to Pauli is dated 15 Feb 1934, BSC (24,2); reproduced in Meyenn *et al.* (eds.), *Pauli, Volume II*, pp. 285–289. The information on copies of this letter being sent to others is given in Bohr to Bloch 17 Feb 1934, BSC (17,3); this letter is reproduced in *Niels Bohr Collected Works, Volume 9* (note 18), pp. 540–541.

54. E. Fermi, "Tentativo di una teorie dell'emissione dei raggi 'beta'" in *La Ricerca Scientifica 2* (31 Dec 1933), 491–495; *idem.*, "Tentativo di una teorie dei raggi β" in *Nuovo Cimento 11* (1934), 1–19; *idem.*, "Versuch einer Theorie der β-Strahlen. I" in *Zeitschrift für Physik 88* (19 Mar 1934), 161–177, latter paper received by journal 16 Jan 1934. The latter two articles, which are practically identical in content, are reprinted in Enrico Fermi, *Collected Papers, Volume 1: Italy 1921–1938* (Chicago: University of Chicago Press, 1961), pp. 559–574, 575–590. The paper in German is translated into English in Charles Strachan, *The Theory of Beta Decay* (Oxford, etc.: Pergamon Press, 1961), pp. 107–128. On Fermi's theory in historical perspective, see Brown, "Yukawa's Prediction" (note 34), pp. 91–95.

55. On the general development of Pauli's idea, see Pauli's own account of the history of the neutrino, "Zur älteren und neueren Geschichte" (note 17), which reproduces Pauli's letter of 4 Dec 1930 on pp. 159–160; and Laurie M. Brown, "The Idea of the Neutrino" in *Physics Today 31* (Sep 1978), 23–28, where the letter is translated into English on p. 27. The letter is also reproduced in Meyen *et al.*, *Pauli, Volume II* (note 51), pp. 39–41.

56. Pauli, "Zur älteren und neueren Geschichte" (note 17), p. 161; Brown, "Idea" (note 55), p. 25, where the relevant part of Goudsmit's lecture is reprinted. The lecture was originally published as "Present Difficulties in the Theory of Hyperfine Structure" in *Atti del Convegno* (note 27), pp. 33–49, where the reference to Pauli's "neutron" is on p. 41.

57. Pauli, "Zur älteren und neueren Geschichte" (note 17), pp. 161–163, where Pauli's brief contribution at the Solvay Congress is reprinted in its original French on pp. 162–163. It has been translated into English both in Brown, "Idea" (note 55), p. 28, and, less completely, in Pais, *Inward Bound* (note 3), p. 315. The statements on the neutrino by Pauli and Bohr at the Solvay Congress were originally published in *Structure et propriétés* (note 47), pp. 324–325 and 327–328, respectively; Bohr described Beck's work on pp. 287–288. The theory of beta decay referred to by Bohr was published as Guido Beck and Kurt Sitte, "Theorie des β-Zerfalls" in *Zeitschrift für Physik 86* (1933), 105–119.

58. Fermi, "Tentativo di una teorie dell'emissione" (note 54). Bloch to Bohr 10 Feb 1934, BSC (17,3) (note 11), Bohr to Bloch 17 Feb 1934 (note 53). Bohr's public renunciation of nonconservation in beta decay is Niels Bohr, "Conservation Laws in Quantum Theory" in *Nature 138* (4 Jul 1936), 25–26, p. 26, reprinted in *Niels Bohr Collected Works, Volume 5* (note 23), pp. 215–216. Roger Stuewer, "Nuclear Electron Hypothesis" (note 17), p. 55, argues that the nuclear physics conference in London in October 1934 signifies the point when the nuclear electron ceased to be a viable hypothesis for physicists.

59. While not complete, the bound *Universitetets Institut for teoretisk Fysik: Afhandlinger,* deposited at the Niels Bohr Archive, Copenhagen, and

containing reprints of publications coming out of the institute between 1918 and 1959, is an invaluable source for evaluating the institute's contributions. Each volume contains a bibliography alphabetized by author listing the volume's contents as well as a few publications not included.

60. *Ibid.*
61. *Ibid.* The four articles on nuclear momentum from 1933 are Hans Kopfermann, "Über die Kernmomente der beiden Rubidiumisotope" in *Die Naturwissenschaften 21* (13 Jan 1933), 23; *idem.*, "Hyperfeinstruktur und Kernmomente des Rubidiums" in *Zeitschrift für Physik 83* (28 Jun 1933), 417–430; Hans Kopfermann and N. Wieth-Knudsen, "Die Kernmomente des Kryptons" in *Die Naturwissenschaften 21* (21 Jul 1933), 547–548; *idem.*, "Hyperfeinstruktur und Kernmomente des Kryptons" in *Zeitschrift für Physik 85* (14 Sep 1933), 353–359. The first of Kopfermann's articles is misplaced in the 1932 volume of the *Afhandlinger* (note 59), while the third is not mentioned there at all. For other work at the institute, see the *Afhandlinger*. According to the institute's Guest Book (note 49), Gamow visited from 12 Feb to 9 Jun 1934.
62. *Afhandlinger* (note 59).
63. Typescript in MSS (12,5) (note 6).
64. Ebbe Rasmussen's move to Bohr's previous home was confirmed in a letter from Erik Rüdinger of the Niels Bohr Archive to me 29 Feb 1988.
65. The autobiographical statement was published as Niels Bohr, "Physical Science and the Problem of Life" in *idem.*, *Atomic Physics and Human Knowledge* (New York: John Wiley & Sons, Inc., 1958), pp. 94–101; this collection has recently been reprinted as *The Philosophical Writings of Niels Bohr, Volume 2* (note 6). Bohr's article was based on a lecture he gave to the Danish Medical Society nine years earlier; a typewritten transcript in Danish of the lecture itself, which is considerably less complete on the origins of Bohr's biological interest, is deposited in MSS (18,2) (note 6). A personal account of the meetings of the four Danish academics is given in Harald Høffding, *Erindringer* (Copenhagen: Gyldendalske Boghandel, Nordisk Forlag, 1928), pp. 171–174.
66. The mandate and membership of the Biological Society, as well as Høffding's lecture, were printed in *Biologisk Selskabs Forhandlinger i Vinter-Halvaaret 1897–98* (Copenhagen: Biologisk Selskab, 1898), pp. III-IV, V-VI, 19–26; I am grateful to Knud Max Møller at the Biological Institute of the Carlsberg Foundation for this reference.
67. Niels and Harald Bohr's participation in the discussions is recounted in [David Jens Adler], "Childhood and Youth" in Stefan Rozental (ed.), *Niels Bohr* (note 49), p. 13.
68. For a debate of the extent of Høffding's influence on Bohr, see David Favrholdt, "Niels Bohr and Danish Philosophy" in *Danish Yearbook of Philosophy 13* (1976), 206–220; Jan Faye, "The Influence of Harald Høffding's Philosophy on Niels Bohr's Interpretation of Quantum Mechanics" in *ibid. 16* (1979), 37–72; David Favrholdt, "On Høffding and Bohr: A Reply to Jan Faye" in *ibid.*, 73–77; and, most recently, Jan Faye, "The Bohr-Høffding Relationship Reconsidered" in *Studies in History and Philosophy of Science 19* (1988), 321–346. For a discussion of the philosophical milieu around Bohr in his student days, see Johs. Witt-Hansen, "Leibniz, Høffding, and the 'Ekliptika Circle' " in *Danish Yearbook of Philosophy 17* (1980), 31–58, and David Favrholdt, "The Cultural Background of the Young Niels Bohr" in *Rivista di Storia della Scienza 2* (1985), 445–461. Bohr wrote an obituary for his friend Høffding: Niels Bohr, "Mindeord over Harald Høffding" in *Oversigt over Det Kongelige Danske Videnskabernes Selskabs Forhandlinger 1931– 1932*, pp. 131–136. Høffding introduced his idea of "psycho-physical parallelism" in a treatise published in several languages: Harald Høffding, *Psykologi i Omrids paa Grundlag af Erfaring* (Copenhagen: Gylden-

dal, 1882, etc.); *idem., Psychologie in Umrissen auf Grundlage der Erfahrung* (Leipzig: O.P Reisland, 1887, etc.); *idem., Outlines of Psychology* (London: Macmillan, 1891, etc.). In *Outlines*, the concept is introduced on p. 64, while the rejection of the parallelism concept is explained in the 1898 edition of *Psykologi*, p. 75, note 1, which has been added to explain Høffding's identity hypothesis. Rubin's article, "Psykologi," was printed in *Salmonsens Konversationsleksikon* [second edition], vol. 19 (Copenhagen: J. H. Schultz, 1925), pp. 681–683. I am grateful to David Favrholdt of Odense University, in letters of 6 Jan 1989, 10 Mar 1989, and 10 Apr 1989, for clarifying the distinction between the psycho-physical parallelism and Høffding's identity hypothesis, and for referring me to Rubin's published conflation of the two terms.

69. The quotations are from Høffding, *Outlines* (note 68), pp. 54, 67. Høffding argued strongly for determinism also on ethical grounds; see Harald Høffding, *Ethik: Eine Darstellung der ethischen Prinzipien und deren Anwendung auf besondere Lebensverhältnisse* (Leipzig: O. P. Reisland, 1901), especially Chapter Five, "Die Freiheit des Willens," pp. 96–114.

70. Bohr quoted his father in his "Physical Science" (note 65), p. 96. The statement was originally published in Christian Bohr, *Om den pathologiske Lungeudvidning (Lungeemphysem): Festskrift udgivet af Kjøbenhavns Universitet i Anledning af Universitetets Aarsfest November 1910* (Copenhagen: J. H. Schultz, 1910), pp. 3–48, on p. 5. The quotation is from Harald Høffding, "Mindetale over Christian Bohr" in *Tilskueren 1911*, pp. 209–212, on p. 209.

71. The term "founder of modern physiology" is taken from Heinz Schröer, *Carl Ludwig: Begründer der messenden Experimentalphysiologie 1816–1895* [Heinz Degen (ed.), *Grosse Naturforscher, Band 33*] (Stuttgart: Wissenschaftliche Verlagsgesellschaft m.b.H., 1967). On the influence of Ludwig on Christian Bohr's work in Copenhagen, see L. S. Fridericia, "Bohr, Christian Harald Lauritz Peter Emil" in *Dansk biografisk Leksikon III* (Copenhagen: J. H. Schultz, 1934), pp. 371–374. The dispute between Ludwig and Pflüger is described in J. S. Haldane, *Respiration* (Oxford: Oxford University Press; New Haven: Yale University Press, 1922), "Chapter I. Historical Introduction," pp. 1–14; see, however, Charles A. Culotta, "Tissue Oxidation and Theoretical Physiology: Bernard, Ludwig, and Pflüger" in *Bulletin of the History Medicine 44* (1970), 109–140, which argues, on p. 133, that Ludwig was not consistently maintaining an antidiffusion view, and that Pflüger was setting him up by provoking the dispute. Bohr made his claim in Christian Bohr, "Blutgase und respiratorischer Gaswechsel" in W. Nagel (ed.), *Handbuch der Physiologie des Menschen. Erster Band: Physiologie der Atmung, des Kreislaufs und des Stoffwechsels* (Braunschweig: Friedr. Vieweg, 1909), pp. 54–222; the quotation is from p. 155.

72. August Krogh, "On the Mechanism of the Gas-Exchange in the Lungs" in *Skandinavisches Archiv für Physiologie 23* (1910), 248–278, paper received by journal 5 Dec 1909. Krogh's explicit rejection of Bohr is on p. 249, the quotation on p. 278. Krogh's quoted conclusion, the last in a list of seven statements of "Summary," was the only statement in the article printed in bold italics. The article was the last of a series of seven papers – two of which were coauthored with his wife Marie – by Krogh on respiration experiments, which he published in the 1910 volume of the *Skandinavisches Archiv*. The claim of Christian Bohr's recantation is made in V. Henriques, "Chr. Bohrs videnskabelige Gerning" in *Oversigt over Det Kongelige Danske Videnskabernes Selskabs Forhandlinger 1911*, pp. 395–405, on p. 404.

73. The quotation is from J. S. Haldane, "Acclimatization to High Altitudes" in *Physiological Review 7* (Jul 1927), 363–384, p. 372, where Haldane also describes his relationship with Christian Bohr. As a particular-

ly illuminating example of Haldane's isolation in his view of the breathing mechanism, see Joseph Barcroft, *The Respiratory Function of the Blood* (Cambridge, England: Cambridge University Press, 1928), pp. 52–62.

74. For Haldane's general statements on respiration physiology as exemplifying the difference between physics and biology, see J. S. Haldane, *The Philosophy of a Biologist* (Oxford: Clarendon Press, 1935), pp. 150ff., and *idem.*, *The Sciences and Philosophy: Gifford Lectures, University of Glasgow 1927 and 1928* (London: Hodder and Stoughton, Limited, 1928). Haldane's most elaborate argument for the secretion theory as a basis for his philosophy is the collection of lectures, John Scott Haldane, *Organism and Environment as Illustrated by the Physiology of Breathing* (Oxford: Oxford University Press; New Haven: Yale University Press, 1917), which Haldane in 1922 developed into his textbook *Respiration* (note 71). The historian of biology Garland E. Allen has claimed that Haldane's "early work was markedly lacking in philosophical orientation"; see his "J. S. Haldane: The Development of the Idea of Control Mechanisms in Respiration" in *Journal of the History of Medicine and Allied Sciences 22* (1967), 392–412, p. 410. However, Haldane showed an early active philosophical interest in an article coauthored with his brother – R. B. Haldane and J. S. Haldane, "The Relation of Philosophy to Science" in Andrew Seth and R. B. Haldane (eds.), *Essays in Philosophical Criticism* (London: Longmans, Green and Co., 1883), pp. 41–66. Haldane's biological work is treated in a broader philosophical and political context in Steve Sturdy, "Biology as Social Theory: John Scott Haldane and Physiological Regulation" in *British Journal for the History of Science 21* (1988), 315–340.

75. Haldane denoted his "doctrine" as organicism in *Organism and Environment* (note 74), p. 3, note 1; he denounced the organicists twenty years later in *Philosophy* (note 74), p. 45.

76. The review article was published as Joseph Needham, "Recent Developments in the Philosophy of Biology" in *The Quarterly Review of Biology 3* (Mar 1928), 77–91.

77. Haldane stated explicitly his expectation that physics would be reduced to biology in J. S. Haldane, *The New Physiology* (London: Griffin, 1919), p. 19. The quotation is from Haldane, *Sciences and Philosophy* (note 74), p. 238.

78. Ralph S. Lillie, "Physical Indeterminism and Vital Action" in *Science 66* (12 Aug 1927), 139–144. Needham's approving comment is from "Recent Developments" (note 76), p. 84.

79. Bohr made his first published statements on biology in Niels Bohr, "Wirkungsquantum und Naturbeschreibung" in *Die Naturwissenschaften 17* (1929), 483–486, which was translated into English as "The Quantum of Action and the Description of Nature" in *idem.*, *Atomic Theory* (note 6), pp. 92–101.

80. Bohr first explained his turn to biology in Niels Bohr, "Indledende Oversigt" in *idem.*, *Atomteori og Naturbeskrivelse: Festskrift udgivet af Københavns Universitet i Anledning af Universitetets Aarsfest November 1929* (Copenhagen: Bianco Lunos Bogtrykkeri, 1929), pp. 5–19; the quoted translation is from *idem.*, "Introductory Survey (1929)" in *idem.*, *Atomic Theory* (note 6), pp. 1–24, on p. 15. Both versions are reprinted in *Niels Bohr Collected Works, Volume 6* (note 6), pp. 259–273, 279–302. For a rare account reporting Bohr's desire to spread his gospel of complementarity, see Léon Rosenfeld, "Niels Bohr's Contribution to Epistemology" in *Physics Today 16* (Oct 1963), 47–54, p. 54, which is reprinted in Robert S. Cohen and John J. Stachel (eds.), *Selected Papers of Léon Rosenfeld* [Robert S. Cohen and Marx W. Wartowsky (eds.), *Boston Studies in the Philosophy of Science, Volume XXI*] (Dor-

drecht, Boston, London: D. Reidel Publishing Company, 1979), pp. 522–535, on p. 535.

81. Bohr, "Wirkungsquantum" (note 79).
82. The quotations are from Bohr, "The Quantum of Action" (note 79), pp. 100, 101.
83. Niels Bohr, "The Atomic Theory and the Fundamental Principles underlying the Description of Nature" in *idem., Atomic Theory* (note 6), pp. 102–119, on p. 117. The lecture was originally published as *idem.*, "Atomteorien og Grundprincipperne for Naturbeskrivelsen" in *Fysisk Tidsskrift 27* (1929), 103–114.
84. Bohr, "Atomic Theory," (note 83), pp. 117–119, quotations on pp. 118–119.
85. The quotations are from Bohr, "Survey" (note 80), pp. 20–21.
86. The quotation is from Bohr to Heisenberg 10 Jan 1930, BSC (20,2) (note 11). A typewritten transcript of Bohr's Edinburgh lecture is in MSS (12,3) (note 6).
87. A brief vita of Jordan is contained in Kuhn *et al., Sources* (note 49), p. 51. Bohr's influence on Jordan's writings is particularly evident in J. Franck and P. Jordan, *Anregung von Quantensprüngen durch Stösse* (Berlin: Julius Springer, 1926), and M. Born and Pascual Jordan, *Elementare Quantenmechanik* (Berlin: Julius Springer, 1930), the latter of which was dedicated to Bohr; the two books constituted volumes III and IX, respectively, of Max Born and J. Franck (eds.), *Struktur der Materie in Einzeldarstellungen.* On Bohr's contribution to correcting Jordan's speech disorder, see James Franck to Bohr 9 Jul 1926, Bohr to Franck 21 Jul 1926, Franck to Bohr 29 Jul 1926, BSC (10,4) (note 11); Jordan to Bohr 29 Jul 1926, BSC (12,3). The appreciative letter to the IEB is Bohr to Tisdale 25 Feb 1927, BSC (12,3) [under Jordan]. On Jordan's scientific work and Bohr's interest in it, see Jammer, *Conceptual Development* (note 4), pp. 207–211, 295, 305–307, 362, 365, and Bohr to Jordan 14 May 1928, BSC (12,3).
88. P. Jordan, "Philosophical Foundations of Quantum Theory" in *Nature 119* (16 April 1927), 566–569, quotation on p. 566; J. Robert Oppenheimer had translated the lecture into English. Jordan to Einstein 11 Dec 1928, Einstein Papers, Tel Aviv, Israel, and Princeton, New Jersey; by permission of The Hebrew University of Jerusalem, Israel. I am grateful to Paolo Bernardini at the University of Urbino, Italy, for bringing this letter to my attention.
89. The two physicists' intention to publish on biology is stated in Jordan to Bohr 20 May 1931, BSC (21,3) (note 11), with which Jordan also enclosed his manuscript. The manuscript, "Statistik, Kausalität und Willensfreiheit," is deposited among "Fremmede Manuskripter" ("External Manuscripts") in the Niels Bohr Archive, Copenhagen.
90. Jordan to Bohr 20 May 1931 (note 89), Bohr to Jordan 5 Jun 1931, BSC (21,3) (note 11).
91. *Ibid.*
92. In spite of the similarity in their views, there are no letters exchanged between Bohr and Haldane in the BSC (note 11), BGC (note 36), or the Bohr Correspondence (BPC), which is also deposited in the Niels Bohr Archive.
93. Jordan to Bohr 22 Jun 1931, BSC (21,3) (note 11).
94. *Ibid.*
95. Bohr to Jordan 23 Jun 1931, BSC (21,3) (note 11).
96. Pascual Jordan, "Die Quantenmechanik und die Grundprobleme der Biologie und Psychologie" in *Die Naturwissenschaften 20* (4 Nov 1932), 814–821. The last section is titled "Das Wesen des Organischen." The classification of Jordan's theory as vitalism is made in Heilbron, "Earliest missionaries" (note 14), p. 214.

97. Jordan explains his membership in the Nazi party in Jordan to Bohr, May 1945, BSC (21,3) (note 11). The letters reporting Jordan's "literary slip" are Max Born to James Franck 14 Apr 1934 and Franck to Born 18 Apr 1934 in the James Franck Papers, Joseph Regenstein Library, University of Chicago and in Nachlass Born, 229 B1. 8 (carbon copies), Handschriftenabteilung, Staatsbibliothek, Preussische Kulturbesitz, West Berlin. Jordan's complaint to Bohr is undated, but probably written in June 1934, BSC (21,3). The reply is Bohr to Jordan 30 Jun 1934, *ibid.* I have been unable to find the relevant article by Jordan. In an interview with me on 25 Oct 1984 at Cornell University, Victor Weisskopf recalled that a newspaper article by Jordan sympathetic to the Nazi regime had caused considerable concern among physicists. However, in subsequent correspondence with me (my letter to Weisskopf 7 Mar 1989, Weisskopf to me 17 Mar 1989), Weisskopf has agreed that he was made aware of this article only in June 1936; it is therefore likely that it was a different one published later. The booklet published in 1935 was Pascual Jordan, *Physikalisches Denken in der neuen Zeit* (Hamburg: Hanseatische Verlagsgesellschaft, 1935); see, in particular, the fourth and last chapter, "Der Wert der Wissenschaft," pp. 41–59. I am grateful to Paul Forman for lending me his personal copy of the booklet. Bohr criticized Jordan's extension of complementarity to parapsychology in Bohr to Otto Meyerhof 5 Sep 1936, BSC (22,3). The deterioration of the Bohr–Jordan relationship is confirmed in Dieter Hoffmann, "Zur Teilnahme deutscher Physiker an den Kopenhagener Physikerkonferenzen nach 1933 sowie am 2. Kongress für Einheit der Wissenschaft, Kopenhagen 1936" in *NTM – Schriftenreihe für Geschichte der Naturwissenschaften, Technik und Medizin 25* (1988), 49–55, on p. 54.

98. Aside from Jordan to Bohr May 1945 (note 97), the correspondence between Bohr and Jordan after Jordan's response to Bohr's comment on the "literary slip" is as follows: Jordan to Bohr 21 Nov 1935 (birthday greeting), Bohr to Jordan 27 Nov 1935 (thank-you note), Jordan to Bohr 16 Aug 1936 (explanation of parapsychology article), 24 Oct 1945, all BSC (21,3) (note 11).

99. Bohr explained the background for the forthcoming German edition in his letter to Jordan 23 Jun 1931 (note 95), the last letter in their correspondence regarding Jordan's manuscript. The new addition was to become Niels Bohr, "Addendum" in *idem., Atomtheorie und Naturbeschreibung* (Berlin: Julius Springer, 1931), pp. 14–15. It was subsequently translated into English as *idem.*, "Addendum (1931)" in *idem.*, "Survey" (note 80), pp. 21–24. The German and English booklets contain the same articles.

100. The quotations are from *ibid.*, pp. 22–24. The text was also italicized in the original German version (note 99). A typewritten transcript of Bohr's Helsingør lecture is deposited in MSS (12,5) (note 6). In his letter of 23 Jun 1931 (note 95), Bohr told Jordan that he had written the "Addendum," without, however, noting any influence from Jordan. In view of the clarification of Bohr's ideas on biology, I disagree with J. L. Heilbron's assertion that Bohr's views on complementarity and biology remained unchanged from 1929 to 1932; Heilbron, "Earliest missionaries" (note 14), p. 212.

101. *Deuxième Congrès International de la Lumière: Biologie, Biophysique, Thérapeutique – Copenhague 15–18 Août 1932* (Copenhagen: Engelsen & Schrøder, no date).

102. The quotations are from Hansen to Bohr 17 Jul 1932, BSC (20,1) (note 11). The "philosophers' congress" referred to by Kissmeyer was probably a meeting of the Copenhagen "Society for Philosophy and Psychology" in late 1929, where Bohr gave the lecture (in English translation) "Remarks about the Relation of the More Recent Physics to the Causali-

ty Principle." See Bohr to Kramers 7 Dec 1929, BSC (13,2), and *Niels Bohr Collected Works, Volume 6* (note 6), p. 196, and pp. 428–430, where the letter from Kramers is reproduced both in the original Danish and in English translation.

103. Bohr's answer to Hansen's request of 22 Jul 1932 is in BSC (20,1) (note 11). Bohr's manuscripts are in MSS (note 6). In chronological order, Bohr's lecture was published as follows: Niels Bohr, "Light and Life" in *Nature 131* (25 Mar 1933), 421–423 and (1 Apr 1933), 457–459; *idem.*, "Licht und Leben" in *Die Naturwissenschaften 21* (31 Mar 1933); *idem.*, "Lys og Liv" in *Naturens Verden 17* (1933), 49–59; *idem.*, "Light and Life" in *Congrès* (note 101), pp. XXXVII–XLVI; *idem.*, "Light and Life" in *idem.*, *Atomic Physics* (note 6), pp. 3–12. The reference to the Danish version as the "original" was made in Bohr to Klein 19 Jan 1933, BSC (22,1).

104. The quotations are from the most developed version of the "Light and Life" lecture in *Atomic Physics* (note 103), p. 9.

105. Pascual Jordan, "Quantenphysikalische Bemerkungen zur Biologie und Psychologie" in *Erkenntnis 4* (1934), 215–252 – Mach and Reichenbach on pp. 217–218, psychoanalysis on pp. 247–248. On the Unity of Science group, see Victor Kraft, *The Vienna Circle: The Origin of Neo-Positivism, A Chapter in the History of Recent Philosophy* (New York: Philosophical Library, 1953), and Herbert Feigl, "The Wiener Kreis in America" in Donald Fleming and Bernard Bailyn (eds.), *The Intellectual Migration: Europe and America, 1930–1960 [Perspectives in American History Volume 2, 1968]* (Cambridge, Massachusetts: Harvard University Press, 1969), pp. 630–673.

106. In the proceedings from the conference Edgar Zilsel, "P. Jordans Versuch, den Vitalismus quantenmechanisch zu retten" in *Erkenntnis 5* (1935), 56–64, quotation on p. 61, was followed by Hans Reichenbach, "Metaphysik bei Jordan?"; Otto Neurath, "Jordan, Quantentheorie und Willensfreiheit"; Moritz Schlick, "Ergänzende Bemerkungen über P. Jordans Versuch einer quantentheoretischen Deutung der Lebenserscheinungen"; Philipp Frank, "Jordan und der radikale Positivismus" in *ibid.*, pp. 178–179, 179–181, 181–183, 184. The defense was published as Pascual Jordan, "Ergänzende Bemerkungen über Biologie und Quantenmechanik" in *ibid.*, pp. 348–352. See also Heilbron, "Earliest Missionaries" (note 14), pp. 217–218. On Zilsel, see Feigl, "Wiener Kreis" (note 105), pp. 641–642.

107. Jordan continued to promote the amplifier theory in "Die Verstärkertheorie der Organismen in ihrem gegenwärtigen Stand" in *Die Naturwissenschaften 26* (19 Aug 1938), 537–545. The two other works referred to are *idem.*, *Die Physik und das Geheimnis des organischen Lebens* [Wilhelm Westphal (ed.), *Die Wissenschaft: Einzeldarstellungen aus der Naturwissenschaft und der Technik, Bd. 95]* (Braunschweig: Friedr. Vieweg, 1945), and *idem.*, *Verdrängung und Komplementarität* (Hamburg-Bergedorff: Stromverlag, 1947).

108. The quotation is from Bohr, "Light and Life" (*Atomic Physics* version) (note 103), p. 11. Bohr to Meyerhof 5 Sep 1936 (note 97). The only candidate for a publication by Eddington that Bohr might have referred to in his "Light and Life" address is Sir Arthur Eddington, "The Decline of Determinism – Presidential Address to the Mathematical Association, 1932" in *The Mathematical Gazette 16* (May 1932), 65–80, where Eddington provides some brief remarks on "Mind and Indeterminism" on pp. 79–80; these remarks, however, were hardly sufficient as a point of attack for Bohr.

109. Delbrück stated the importance of the "Light and Life" address for his career in Carolyn Kopp, "Max Delbrück – How It Was" in *Engineering & Science* (California Institute of Technology), first installment *43* (Mar–

Apr 1980), 21–26, second installment *43* (May-Jun 1980), 21–27, see first installment, p. 26. On Delbrück's "intellectual impetus" see Lily E. Kay, "Conceptual Models and Analytical Tools: The Biology of the Physicist Max Delbrück, 1931–1946" in *Journal of the History of Biology 18* (1985), 207–246, p. 215. See also *idem.*, "The Secret of Life: Niels Bohr's Influence on the Biology Program of Max Delbrück" in *Rivista di storia della scienza 2* (1985), 487–510.

110. Delbrück to Bohr 30 Nov 1934, BSC (18,3) (note 11).
111. The resumé is filmed with Delbrück's letter, *ibid.* The positive reply is Bohr to Delbrück 8 Dec 1934, *ibid.*
112. The quotation is from Delbrück to Bohr 28 Jun 1932, BSC (18,3) (note 11).
113. The quotation is from Delbrück to Bohr 5 Apr 1935, BSC (18,3) (note 11). The *Dreimännerarbeit* was published as N. W. Timofeéff-Ressovsky, K. G. Zimmer, and M. Delbrück, "Über die Natur der Genmutation und der Genstruktur" in *Nachrichten von der Gesellschaft der Wissenschaften zu Göttingen: Mathematisch-physikalische Klasse, Fachgruppe VI: Biologie 1* (1935), 189–245; paper read by A. Kuhn 12 Apr, published 29 Jun 1935. For a general introduction to the circumstances of the *Dreimännerarbeit*, see Robert Olby, *The Path to the Double Helix* (London: Macmillan, 1974), pp. 232–233.
114. Frank's letter to Bohr of 9 Jan 1936 complaining about Jordan's brochure is quoted without reference in Heilbron, "Earliest Missionaries" (note 14), p. 221; it is in BSC (19,3) (note 11). Bohr's lecture was first published as Niels Bohr, "Kausalität und Komplementarität" in *Erkenntnis 6* (27 Apr 1937), 293–303; it was translated as *idem*, "Causality and Complementarity" in *Philosophy of Science 4* (Jul 1937), 289–298, wherefrom quotations on pp. 289 and 295; the disclaimer of spiritualism is on p. 297.
115. Frank's lecture was printed as Philipp Frank, "Philosophische Deutungen und Missdeutungen der Quantentheorie" in *Erkenntnis 7* (27 Apr 1937), 303–317, quotations on pp. 314–315. The comment on the discussion is in the Warren Weaver Diary 24–26 Jun 1936, filed in Record Group 12, Rockefeller Archive Center, Pocantico Hills, New York.
116. One single exception is J. Gray, "The Mechanical Way of Life" in *Nature 132* (28 Oct 1933), 661–664, which contains a brief reference to Bohr's "Light and Life" address. During the same period, *Science*, the American counterpart of the British and German journals, also contained little or no reference to Bohr's biological views. Furthermore, *The Quarterly Review of Biology*, the general approach journal started in 1926 that published Needham's review article (note 76), did not refer to Bohr's views. In 1934, another American journal, *The Philosophy of Science*, was started. Like *Erkenntnis* in Germany, it was dedicated to promoting the philosophy of science as a new academic discipline. Even though devoting several articles during its first year to the philosophy of biology, a brief reference to Bohr was only given in Ralph S. Lillie, "The Problem of Vital Organization" in *Philosophy of Science 1* (1934), 296–312, p. 306. In subsequent years, the philosophy of biology received less attention by the journal. Even the journal's publication in English of Bohr's lecture at the Copenhagen Unity of Science conference (note 114) did not provoke discussion of Bohr's views.
117. Otto Meyerhof, "Betrachtungen über die naturphilosophischen Grundlagen der Physiologie" in *Abhandlungen der Friesschen Schule 6* (1933), 33–65, abstracted as *idem.*, "Betrachtungen" in *Die Naturwissenschaften 22* (18 May 1934), 311–314. Three years later Bohr sought, unsuccessfully, to straighten out his differences with Meyerhof; see Bohr to Meyerhof 5 Sep 1936 (note 97) and Meyerhof to Bohr 14 Sep 1936, BSC (23,3) (note 11). Joseph Needham, *Order and Life* (Oxford: Oxford University Press; New Haven: Yale University Press, 1936), quotations on

pp. 32, 15. The quotations on p. 32 are from Bohr, "Light and Life" (*Nature* version) (note 103), p. 458.

118. The quotation is from Muller to Mohr 13 Apr 1933, Institutt for medisinsk genetikk, University of Oslo; I am grateful to Nils Roll-Hansen for bringing this letter to my attention. A copy of the letter is retained in the Hermann Muller Papers at the Lilly Library of Indiana University.

119. On the Copenhagen spirit as "imperialism," see Heilbron, "Earliest missionaries" (note 14), pp. 196, 213.

120. For discussions of the influence of Bohr's intellectual environment on his ideas, see the section "4.2. The Philosophical Background of Non-classical Interpretations" in Jammer, *Conceptual Development* (note 4), pp. 166–180; Gerald Holton, "The Roots of Complementarity" in *idem., Thematic Origins of Scientific Thought: Kepler to Einstein* (Cambridge, Massachusetts: Harvard University Press, 1973), pp. 115–161; the chapter, "Niels Bohr: The *Ekliptika* Circle and the Kierkegaardian Spirit" in Lewis S. Feuer, *Einstein and the Generations of Science* (New York: Basic Books, 1974), pp. 109–157; and the articles by Favrholdt, Faye, and Witt-Hansen referred to in note 68. There is little evidence of any interest of Bohr's in biology before 1929 in the BSC (note 11) and the BGC (note 36). The BPC (note 92), which is currently being processed, also contains hardly any such evidence.

121. Bohr to Jordan 5 Jun 1931 (note 90); Bohr, "Atomic Stability" (note 27).

122. The quotation is from Charles Weiner's interview with Christian Møller 25 Aug and 21 Oct 1971, p. 73 of original transcript; the interview is deposited in the Niels Bohr Library of the American Institute of Physics in New York City.

CHAPTER 3

1. Alan D. Beyerchen, *Scientists under Hitler: Politics and the Physics Community in the Third Reich* (New Haven and London: Yale University Press, 1977), pp. 2–14.

2. *Ibid.*, pp. 15–50.

3. Charles Weiner, "A New Site for the Seminar: The Refugees and American Physics in the Thirties" in Donald Fleming and Bernard Bailyn (eds.), *The Intellectual Migration: Europe and America, 1930–1960 [Perspectives in American History Volume 2, 1968]* (Cambridge, Massachusetts: Harvard University Press, 1969), pp. 190–234, on pp. 204, 213; the newspaper page is reproduced in facsimile on p. 234. A more recent collection of articles on the intellectual refugees, which considers also emigration to other countries than the United States, is Jarrell C. Jackman and Carla M. Borden (eds.), *The Muses Flee Hitler: Cultural Transfer and Adaptation 1930–1945* (Washington, D.C.: Smithsonian Institution Press, 1983). See also Roger H. Stuewer, "Nuclear Physicists in the New World: The Emigrés of the 1930s in America" in *Berichte zur Wissenschaftsgeschichte 7* (1984), 23–40, and Richard Rhodes, *The Making of the Atomic Bomb* (New York, etc.: Simon & Schuster, Inc., 1986), pp. 184–197.

4. Beyerchen, *Scientists under Hitler* (note 1), p. 14, wherefrom quotation. Weiner, "New Site" (note 3), pp. 192ff., introduces the term "the travelling seminar."

5. For the history of the PTR, see David Cahan, "Werner Siemens and the origins of the Physikalisch-Technische Reichsanstalt, 1872–1887" in *Historical Studies in the Physical Sciences 12* (1982), 253–285, and *idem., An Institute for an Empire: The Physikalisch-Technische Reichsanstalt 1871–1918* (Cambridge University Press: Cambridge, New York, Melbourne, Sydney, 1989). On the arrangements for Rasmussen's

stay in Berlin, see Bohr to Paschen 28 Jun 1932, Paschen to Bohr 1 Jul 1932, Bohr to Tisdale 31 Aug 1932, Tisdale to Bohr 28 Oct 1932, Bohr to Rasmussen 5 Jan 1933. The correspondence with Paschen and Rasmussen is retained in the Bohr Scientific Correspondence, microfilm 24, respectively sections 2 and 4 (BSC (24,2), BSC (24,4)). The correspondence with Tisdale is filed under "Rockefeller Foundation" in the Bohr General Correspondence, Special File, microfilm 5, section 6 (BGC-S (5,6)). The BGC is retained only in the Niels Bohr Archive, Copenhagen. The microfilmed BSC is also retained in the Niels Bohr Library of the American Institute of Physics in New York and several other places.

6. Rasmussen to Bohr 23 Feb 1933, BSC (24,4) (note 5).
7. Rasmussen to Bohr 6 Apr 1933, BSC (24,4) (note 5).
8. *Ibid.* The Kaiser Wilhelm Society and its Institutes are described in Paul Forman, *The Environment and Practice of Atomic Physics in Weimar Germany: A Study in the History of Science,* PhD dissertation at the University of California at Berkeley, 1967 (microfilms international 6810322), pp. 257–261. The comment on the *Notgemeinschaft* is from *idem.,* "The Financial Support and Political Alignment of Physicists in Weimar Germany" in *Minerva 12* (1974), 39–66, p. 39. For information on Wolfson I am grateful to H.B.G. Casimir and Etti Atagem, director of the Central Archives at the Hebrew University of Jerusalem.
9. Rasmussen to Bohr 6 Apr 1933 (note 7). On the Einstein Laboratory, see Ronald W. Clark, *Einstein: The Life and Times* (New York: World Publishing Company, 1971), pp. 390–392.
10. Rasmussen to Bohr 31 Jul 1933, BSC (24,4) (note 5).
11. A file on Bohr's visit, "Amerikarejsen 1933" is deposited in the Bohr General Correspondence (BGC), Niels Bohr Archive, Copenhagen. Bohr described the visit in a letter to Yoshio Nishina 26 Jan 1934, BSC (24,1) (note 5).
12. The request for a meeting was made in Bohr to Weaver 13 Apr 1933, "713D University of Copenhagen. Biophysics. 1933–34," Rockefeller Foundation archives, Record Group 1.1, Series 713, Box 4, Folder 46 (RF RG 1.1, 713, 4, 46), Rockefeller Archive Center (RAC), Pocantico Hills, New York. Mason arranged for Bohr to meet with him in his letter to Bohr 26 Apr 1933, *ibid.* The report of the meeting is from an excerpt from the Max Mason Diary 1 May 1933, *ibid.*
13. Bohr made his optimistic assessment in his letters to Heisenberg 19 May 1933, BSC (20,2) (note 5), and to Miller 24 May 1933, BSC (26,4) [filed under Williams], wherefrom the quotation; the letter to Tisdale is referred to in both of these letters. On Teller, see Donnan to Bohr 10 Nov 1933, BSC (18,4) and the Lauder Jones Diary 20 Oct 1933; the diaries of the Rockefeller Foundation officers are filed under RF RG 12, RAC (note 12). On Frisch, see Bohr to Blackett 2 Sep 1933, BSC (17,3).
14. On the background for Kopfermann's stay in Copenhagen, see Bohr to Haber 16 Jan 1932 [under Kopfermann], Kopfermann to Bohr 24 August 1932, BSC (22,2) (note 5). The quotations are from Kopfermann to Bohr 23 May 1933, BSC (22,2). On Haber's decision to resign, see, for example, J. L. Heilbron, *The Dilemmas of an Upright Man: Max Planck as Spokesman for German Science* (Berkeley, etc.: University of California Press, 1986), p. 161.
15. Historical assessments of Nazi Germany also tend to find that, at least in the early months of Hitler's rule, the Nazi party organization, if not the German masses more generally, constituted the most radical element. See, for example, Martin Broszat, *The Hitler State: The Foundation and Development of the Internal Structure of the Third Reich* (London and New York: Longman, 1981), in particular, Chapter 6, "Party and State in the Early Stages of the Third Reich," pp. 193–240.

16. Laue was soon to change his optimistic attitude, however. See Beyerchen, *Scientists under Hitler* (note 1), pp. 64–65.
17. Kopfermann to Bohr 23 May 1933 (note 14). Beyerchen, *Scientists under Hitler* (note 1), note 46 on p. 288, makes the observation that Kopfermann probably meant to refer to Jordan. As Beyerchen notes, just after the war, Jordan made a similar interpretation of his Nazi connection in an explanatory letter addressed to Bohr and others; Jordan to Bohr, May 1945, BSC (21,3) (note 5).
18. Heisenberg to Bohr 30 Jun 1933, BSC (20,2) (note 5).
19. The official announcement of the Committee's establishment was made as a letter on behalf of the Danish Committee for the Support of Refugee Intellectual Workers, from Aage Friis to the executive of the Carlsberg Foundation 26 Oct 1933, which enclosed a printed "Opfordring" ("Appeal") expressing the Committee's purpose and providing a full list of the forty-nine members; both items are in the Archive of the Carlsberg Breweries, Copenhagen. On Friis, see Erik Stiig Jørgensen, "Friis, Aage" in *Dansk Biografisk Leksikon* [third edition] *IV* (Copenhagen: Gyldendal, 1980), pp. 649–653. I am grateful to Henning Friis, the son of the Committee's chairman, for permission to see Aage Friis's extensive papers in Rigsarkivet, Copenhagen, which, however, contains little information on the Committee's work. Most of my information on the Committee is taken from my interview in Copenhagen on 18 Jan 1983 with Gerhard Breitscheid, who became the Committee's secretary in 1935.
20. *Ibid.* For a broader perspective on the achievements and problems of the Danish efforts for the refugees, see Aage Friis, "De tyske politiske Emigranter i Danmark 1933–46" in *Politiken* 8 and 10 May 1946, pp. 9–10, 8–9.
21. The destruction of material on the German occupation is reported in Stefan Rozental, "The Forties and the Fifties" in *idem.* (ed.), *Niels Bohr: His Life and Work as Seen by His Friends and Colleagues* (Amsterdam: North-Holland Publishing Company, 1968), pp. 149–190, on p. 155. See also Charles Weiner's interview with Bohr's secretary Betty Schultz 25 Mar 1972, original transcript pp. 57–59, deposited in the Niels Bohr Library of the American Institute of Physics, New York City. On the destruction of the Committee's papers, see my Breitscheid interview 18 Jan 1983 (note 19). On the Bohr Scientific Correspondence (BSC), see note 5.
22. For general information on Beck's situation, see the application signed by Reinhold Fürth and Philipp Frank to "das holländische Hilfskommite für deutsche Gelehrte zu Handen von Herrn Prof. Dr. Fryda, Amsterdam" 20 Oct 1933. A copy of the application is retained as enclosure with Fürth to Bohr 21 Oct 1933, BSC (19,3) [under Frank] (note 5).
23. Heisenberg's request is mentioned in Fürth to Bohr 5 Oct 1933, BSC (19,3) [under Frank] (note 5), which also contains the information on Beck's situation. Frank made his remark in a letter to Bohr 5 Oct 1933, *ibid.* The agreement to seek foreign sources was expressed in Bohr to Frank and Fürth 18 Oct 1933, *ibid.*
24. Frank and Fürth's application and the accompanying letter to Bohr 21 Oct 1933 is referred to in note 22. Bohr reported his success in interesting the Danes and Swedes, as well as Beck's early decision to leave for Prague, in his letter to Frank and Fürth 25 Nov 1933, BSC (19,3) [under Frank] (note 5). The final decision of the Swedes was reported in Klein to Bohr 12 Dec 1933, BSC (22, 1). On the positive decision of the Amsterdam Committee, see Fürth to Bohr 23 Dec 1933, BSC (22,1) [under Frank] and Kramers to Bohr 9 Jan 1934, BSC (22,3). Beck stayed in Copenhagen from 31 Jul to 23 Sep and from 14 Oct to 21 Nov 1933; institute's "Guest Book," Archive for the History of Quantum Physics,

microfilm 35, section 2 [AHQP (35,2)]. The microfilmed AHQP is deposited in the American Philosophical Society, Philadelphia, the Niels Bohr Library of the American Institute of Physics, New York, the Niels Bohr Archive, Copenhagen, and other places. For a description, which is not up to date, see Thomas S. Kuhn, John L. Heilbron, Paul Forman and Lini Allen, *Sources for the History of Quantum Physics: An Inventory and Report* (Philadelphia: American Philosophical Society, 1967).

25. A short vita of Gordon is contained in *ibid.*, p. 42. Klein requested Bohr's help in a letter of 12 Dec 1933, BSC (22,1) (note 5), where he also described the Rockefeller Foundation's lukewarm response to his application and the conditional promise from the Swedish Committee. Bohr suggested that Klein accept support in his letter 18 Feb 1934, BSC (22,1). Klein reported back on the positive Swedish response 6 Mar 1934, BSC (22,1). On the Klein-Gordon equation, see, for example, Helge Kragh, "The Genesis of Dirac's Relativistic Theory of Electrons" in *Archive for History of Exact Sciences 24* (1981), 31–67, pp. 34–36.

26. For the pertinent biographical details, see Otto Robert Frisch, *What Little I Remember* (Cambridge, etc.: Cambridge University Press, 1979), pp. 1–56. On Stern's exemption and his search for jobs for his dismissed staff, see Beyerchen, *Scientists under Hitler* (note 1), p. 49. Stern soon resigned from the University of Hamburg, and emigrated to the United States. He was one of the minority of first-class German physicists unable to transform their successful careers to the new environment. See *ibid.*, pp. 49–50.

27. Blackett made his request to Bohr in his letter 13 Aug 1933, BSC (17,3) (note 5). On the Rockefeller Foundation's negative reply, see Bohr to Miller 23 Aug 1933, Miller to Bohr 29 Aug 1933, "Rockefeller Foundation," BGC-S (5,7) (note 5), and Bohr to Blackett 2 Sep 1933 (note 13). Blackett reported the successful application to the Council in his letter to Bohr 16 Sep 1933, BSC (17,3); see also Frisch, *What Little* (note 26), pp. 52–53, which also notes Frisch's arrival in England on p. 56. Bohr's first application to the Rask–Ørsted Foundation 28 Sep 1934 for support to Frisch is deposited in "Rask–Ørsted Foundation," BGC-S (4,3). The classic account of the *Privatdozent* is Alexander Busch, *Die Geschichte des Privatdozenten: Eine soziologische Studie zur Grossbetrieblichen Entwicklung der deutschen Universitäten* (Stuttgart: Ferdinand Enke Verlag, 1959). For a particularly lucid description in English of the position, see Forman, *Environment and Practice* (note 8), pp. 70–79.

28. Beck visited the institute from 5 Apr to 18 Jun 1932 and 31 Jul to 23 Sep 1933; see the institute's Guest Book (note 24).

29. The quotation is from the letter of refusal from the Rask–Ørsted Foundation to the Intellectual Workers Committee 20 Dec 1933, "Rask–Ørsted Foundation," BGC-S (4,3) (note 5).

30. On Rabinowitch's situation in Germany, see Beyerchen, *Scientists under Hitler* (note 1), p. 28. Bohr submitted his application 27 Feb 1933, and received the approval 31 May 1933; both documents are in "Rask–Ørsted Foundation," BGC-S (4, 3) (note 5). On Rabinowitch's subsequent career, see "300 Notable Émigrés" in Fleming and Bailyn (eds.), *Intellectual Migration* (note 3), pp. 675–718, on p. 706.

31. On the relevant biographical details, see Victor Weisskopf, "My Life as a Physicist" in *idem.*, *Physics in the Twentieth Century: Selected Essays* (Cambridge, Massachusetts and London, England: Massachusetts Institute of Technology Press, 1972), pp. 1–21, on pp. 2–9, which implies, on p. 13, that Weisskopf received support from the Danish Committee for the Support of Refugee Intellectual Workers for his second stay in Copenhagen. The dates of Weisskopf's first visits to Copenhagen are taken from the institute's Guest Book (note 24), which records Weisskopf's final stay before leaving for America as extending from 1 Apr 1936 to 18

Sep 1937. Bohr commented on Weisskopf's leaving Copenhagen in Bohr to Tisdale 14 Jan 1933, "Rockefeller Foundation," BGC-S (5,7) (note 5).

32. On the relevant biographical details, see Stanley A. Blumberg and Gwinn Owens, *Energy and Conflict: The Life and Times of Edward Teller* (New York: G. P. Putnam's Sons, 1976), pp. 43–63; see also Beyerchen, *Scientists under Hitler* (note 1), pp. 29–30. On Donnan's successful application to the Rockefeller Foundation, see Donnan to Bohr 10 Nov 1933 and the Lauder Jones Diary 20 Oct 1933, both of which items are referred to in note 13.

33. The relevant biographical details are given in Thomas S. Kuhn's interview with Bloch 14 May 1964, AHQP (note 24); see also Bloch to me 21 May 1980. From Bloch's departure from Copenhagen until New Year 1933, as many as fourteen letters of his correspondence with Bohr are retained; BSC (17,3) (note 5).

34. Bloch's letter to Bohr was dated 6 Apr 1933, BSC (17,3) (note 5). Rasmussen to Bohr 6 Apr 1933 and Heisenberg to Bohr 30 Jun 1933 are referred to in notes 7 and 18, respectively.

35. Miller to Bohr 29 Aug 1933, "Rockefeller Foundation," BGC-S (5,7) (note 5).

36. Bloch to Bohr 10 Feb 1934, BSC (17,3) (note 5).

37. Personal communication from Levi to me 7 Dec 1987; my interview with Levi at the Niels Bohr Institute, Copenhagen 28 Oct 1980. See also Charles Weiner's interview with Hilde Levi 28 Oct 1971, pp. 16–18 of original transcript; the interview is deposited in the Niels Bohr Library of the American Institute of Physics in New York City. The information on the International Federation of University Women is taken from *The World of Learning 1988* (Thirty-Eighth Edition) (London: Europa Publications Limited, 1987), p. 22.

38. See references in note 37; institute's Guest Book (note 24).

39. The quotation, as well as the information cited, are from "The problem of the refugee scholars," p. 2, typewritten, undated, but from circa 1940, in "717 German Exiles, Special Research Aid: Apr. 1933–1940," RF RG 1.1, 717, 1, 6 (note 12). Another typewritten manuscript with some corrections by hand, "Special Research Aid Fund for Deposed Scholars, 1933–1939," from about the same time and in the same file, contains more detailed information, including an alphabetical list of 121 scholars helped in the United States. See also [Raymond Fosdick,] "President's Review" in *The Rockefeller Foundation: Annual Report 1936* (New York: Rockefeller Foundation, undated), pp. 3–60, on pp. 57–59. The temporary nature of the support was emphasized in Max Mason, "Foreword" in *The Rockefeller Foundation: Annual Report 1933* (New York: Rockefeller Foundation, undated), pp. xvii–xix, on p. xix.

40. For the employing institutions' power to make their own decision, see, for example, *Report of the Emergency Committee in Aid of Displaced German Scholars, January 1, 1934*, p. 7, "717 German Exiles," RF RG 1.1, 717, 1, 1 (note 12) and "Problem" (note 39), p. 3.

41. The first quotation is from "Problem" (note 39), p. 3. The name change of the Emergency Committee is described in Weiner, "New Site" (note 3), p. 213. The collaboration with the Emergency Committee and the last quotation are taken from *Report* (note 40), p. 11.

42. The meeting is described in the Warren Weaver Diary 1 Jul 1933, RF RG 12 (note 13).

43. Both of Weaver's tours of Europe are described in detail in *ibid.*; the second visit took place from 3 May to 13 Jun 1933.

44. The timing of Hevesy's decision is concluded from Rutherford to Hevesy 12 May 1933, Hevesy Scientific Correspondence (HSC), deposited in the Niels Bohr Archive, Copenhagen. I am grateful to Hilde Levi at the Archive, who has arranged the collection, for access to this rich re-

source, which includes file cards with important excerpts from each let-
ter. Jones's visit in the spring is reported in the Lauder Jones Diary 5
Apr 1933, RF RG 12 (note 13). The other quotations are from the Warren
Weaver Diary 13 Jun 1933, *ibid.*

45. That Bohr's visit to England took place on the way back from the United
States is indicated in Bohr to Heisenberg 17 Aug 1933, BSC (20,2) (note
5). The letter containing information on Bohr and Jones's meeting in
Cambridge is ambiguous as to whether Bohr at this time proposed spe-
cific names of refugees that might be funded by the Special Research
Aid Fund to visit the institute: Bohr to Jones 11 Oct 1933, "Rockefeller
Foundation," BGC-S (5,7) (note 5).

46. Bohr and Brønsted to Jones 11 Oct 1933, RF RG 1.1, 713, 4, 46 (note 12).
In their letter, Bohr and Brønsted referred to the Intellectual Workers
Committee as a "private Danish committee"; for the proper identification, see Harald Bohr to Aage Friis 19 Dec 1933, Aage Friis Papers (note
19).

47. Quotation from Bohr and Brønsted to Jones 11 Oct 1933 (note 46).

48. Bohr to Jones 11 Oct 1933 (note 45). Bohr's application to the Rask–
Ørsted Foundation is dated 16 Oct 1933, "Rask–Ørsted Foundation,"
BGC-S (4,3) (note 5).

49. Bohr's meeting with Jones in connection with the Solvay Congress is
noted in the Lauder Jones Diary 30 Oct 1933 (note 13). On Meitner's
problems in Nazi Germany, see Beyerchen, *Scientists under Hitler* (note
1), p. 47. The formal application was submitted as Bohr to Jones 18 Nov
1933, "713D, University of Copenhagen: Meitner (Physics) 1933–1934,"
RF RG 1.1, 173, 5, 57 (note 12). The foundation announced the acceptance of the application in the official letter, as well as Jones to Bohr, and
Jones to the president of Copenhagen University, all dated 23 Nov 1933,
ibid. Bohr explained Meitner's decision not to come in his letter to Jones
15 Jan 1934, which was prompted by the letter from Jones 4 Jan 1933,
ibid. The action to support Meitner was subsequently rescinded; see
[president] Nørlund to Jones 10 Feb, Jones to Nørlund 2 Mar 1934, *ibid.*

50. The actions of the Paris office have been retained in RF RG 1.1, 713, 4,
46 (note 12); the approval from New York was announced in a telegram
from Weaver to the Rockefeller Foundation, Paris 11 Jan 1934, "713D,
University of Copenhagen: Franck (Physics) 1933–1935," RF RG 1.1,
713, 5, 55. The announcements of funding for Hevesy, Franck, and
equipment were made in Jones to Brønsted 18 Jan 1934, RF RG 1.1, 713,
4, 46, and in two letters from Jones to Bohr 18 Jan 1934, "Rockefeller
Foundation," BGC-S (5,8) (note 5). That funding was granted as part of
the Special Research Aid Fund and as a Research Aid Grant is reported
in, respectively, *The Rockefeller Foundation: Annual Report 1933* (New
York: Rockefeller Foundation, no date), pp. 223, 369, and *ibid. 1934*,
pp. 151–152, 305. Regarding support from Danish sources, see Nørlund
to Bohr and Brønsted 20 Dec 1933, "Rask–Ørsted Foundation," BGC-
S (4,3) (note 5).

51. The quotation is from the action documents (note 50). The information
on the grant to Franck at Göttingen is taken from "717D, University of
Göttingen Physics (Franck, J., equipment) 1931–1932," RF RG 1.1, 717,
13, 122 (note 12); see, in particular, Tisdale to Jones 1 May 1931, the
recommendation for approval Oct 1931, and Tisdale to Franck 22 Oct
1931 announcing that support had been granted. I am grateful to historian
Alan D. Beyerchen at the Ohio State University for guiding me to this
information. On the connection of the new equipment with the construction of the mathematics institute, see Bohr to Jones 11 Oct 1933 (note
45); Bohr refers explicitly to working on an "expansion" of the institute
in, for example, Bohr to Nishina 26 Jan 1934 (note 11).

52. The biographical information is taken from H. G. Kuhn, "James Franck

1882–1964'' in *Biographical Memoirs of Fellows of the Royal Society 11* (1965), 53–74, pp. 53–56. The biographical material on Franck is scarce. In addition to the article by Kuhn, which contains a bibliography of Franck's work (pp. 67–74), see E[ugene] R[abinowitch], "James Franck, 1882–1964, Leo Szilard, 1898–1964" in *Bulletin of the Atomic Scientists 20* (Oct 1964), 16–20. See also, Thomas S. Kuhn *et al.*, six sessions of interviews with Franck, 9–14 Jul 1962, AHQP (note 24), which discusses Franck and Hertz's early unawareness of Bohr's work in session II, p. 12, and the first meeting between the two in Berlin in session IV, p. 22; this and other interviews in the AHQP have not been microfilmed. The historian Alan D. Beyerchen of Ohio State University is working on a biography of Franck, to be titled *James Franck and the Social Responsibility of the Scientist*.

53. The first extant letter between the two is Bohr to Franck 18 Oct 1920, BSC (2,4) (note 5). On the newspaper coverage, see Peter Robertson, *The Early Years: The Niels Bohr Institute 1921–1930* (Copenhagen: Akademisk Forlag, 1979), p. 39. The relationship with the Copenhagen physicists is described in the Franck interview (note 52), session IV, p. 11.

54. On the development of Born and Franck's scientific relationship, see Kuhn, "Franck" (note 52), p. 58. On Franck's successor in Berlin, see for example, Forman, *Environment and Practice* (note 8), p. 258. The quotations from Born Are from Max Born, *My Life: Recollections of a Nobel Laureate* (London: Taylor & Francis Ltd., 1978), p. 211. The last quotation is from the Franck interview (note 52), session II, p. 9; see, however, Franck's admission of Bohr's fallibility in *ibid.* session IV, pp. 11–12. Franck expressed his surprise at Born's reaction in *ibid.* session IV, p. 14.

55. Beyerchen, *Scientists under Hitler* (note 1), p. 18.

56. The classic account of the development of quantum mechanics is Max Jammer, *The Conceptual Development of Quantum Mechanics* (New York: McGraw-Hill, 1966). On Bohr and Heisenberg's first meeting, see Robertson, *Early Years* (note 53), pp. 60–63. On Franck's role in acquiring Heisenberg for the institute, see Bohr to Franck 18 Nov 1925, Franck to Bohr 20 Nov 1925, Bohr to Franck 26 Nov 1925, 22 Dec 1925, BSC (10,4) (note 5).

57. The correspondence between Bohr and Franck is retained in the BSC (note 5); there are no additional letters during the period between Bohr and Franck in the James Franck Papers, Joseph Regenstein Library, University of Chicago. The quotation is from the Franck interview (note 52), session II, p. 4.

58. *Ibid.*, session IV, p. 10; Kuhn, "Franck" (note 52), p. 59.

59. Franck and Hertz received their Nobel Prize "for their discovery of the laws governing the impact of an electron upon an atom"; *Nobel Lectures, Including Presentation Speeches and Laureates' Biographies: Physics, 1922–1941* (Amsterdam, London, New York: Elsevier Publishing Company, 1965), p. 93. See the Nobel Prize speeches, James Franck, "Transformation of Kinetic Energy of Free Electrons into Excitation Energy of Atoms by Impacts," Gustav Hertz, "The Results of Electron-Impacts in the Light of Bohr's Theory of the Atom," in *ibid.*, pp. 98–111, 112–129. The quotation is from the Franck interview (note 52), session V, p. 13.

60. On the opening for Franck in Berlin, see the Lauder Jones Diary 30 Mar 1931 (note 13), and Beyerchen, *Scientists under Hitler* (note 1), pp. 18–19. On Nernst's succession of Rubens, see K. Mendelsohn, *The World of Walther Nernst: The Rise and Fall of German Science, 1864–1941* (Pittsburgh: University of Pittsburgh Press, 1973), p. 138.

61. On the possibility of succeeding Haber, see Beyerchen, *Scientists under Hitler* (note 1), p. 37, and Jost Lemmerich (ed.), *Max Born, James Franck, Physiker im ihrer Zeit: Der Luxus des Gewissens* (Berlin: Staatsbibliothek Preussischer Kulturbesitz, 1982), p. 116. On the acceptance of the Baltimore offer, see Beyerchen, *Scientists under Hitler*, p. 37, and Daniel Willard to Edmund E. Day 22 Aug 1933, RF RG 1.1, 713, 5, 55 (note 50).

62. On Franck's meetings with Harald Bohr and Lauder Jones in Göttingen, see Franck to Niels Bohr 9 Aug 1933, Box 10, Folder 5, Franck Papers (note 57), which also contains Franck's expressions of enthusiasm and reservation. Franck requested a personal meeting in Germany in his letter 26 Aug 1933, BSC (19,2) (note 5), to which Bohr replied positively 2 Sep 1933, *ibid.* Beyerchen, *Scientists under Hitler* (note 1), p. 37, implies that the meeting actually took place.

63. Bohr to Franck 23 Oct 1933, Box 10, Folder 5, Franck Papers (note 57).

64. For the offer from MIT and Harvard, see K. T. Compton to Stephen Duggan 14 Jun 1933, RF RG 1.1, 713, 5, 55 (note 50); for Stanford, see R. L. Wilbur to Mason 10 Jul [1933], telegrams Weaver to Wilbur, Wilbur to Weaver 28 Jul 1933, *ibid.;* for Princeton, see Joseph S. Ames to Weaver 28 Sep 1933, *ibid.* The possibility that Franck would continue at Hopkins was discussed in an interview with Bohr, Lauder Jones Diary 30 Oct 1933, RF RG 12 (note 13).

65. The meeting with Harald Bohr is reported in Niels Bohr to Lauder Jones 18 Nov 1933, RF RG 1.1, 713, 5, 55 (note 50). Franck's expression of uncertainty about his wish to stay in Copenhagen is from the H. M. Miller Diary, excerpt in *ibid.* Jones telegraphed the confirmation of Franck's intention to go to Copenhagen to the New York office 20 Nov 1933, *ibid.* Weaver refused support for Franck's Baltimore stay in his letter to Ames 28 Nov 1933, *ibid.* Ames expressed his confidence about Franck's preference in his letter to Weaver 1 Dec 1933, *ibid.* The date of Franck's departure is taken from Beyerchen, *Scientists under Hitler* (note 1), p. 37.

66. Wilbur Tisdale Diary 8 Apr 1934, RF RG 12 (note 13).

67. Kuhn, "Franck" (note 52), p. 72; for Franck's publications, see the bibliography in Kuhn's obituary.

68. See William O. McCagg, Jr., *Jewish Nobles and Geniuses in Modern Hungary* (Boulder: East European Quarterly, 1972), which discusses Hevesy, in particular, on pp. 158–160.

69. Hevesy and Bohr were born on 1 Aug and 7 Oct 1885, respectively. The biographical information on Hevesy is taken from John D. Cockcroft, "George de Hevesy 1885–1966" in *Biographical Memoirs of Fellows of the Royal Society 13* (1967), 125–166, pp. 125–127. For biographical material on Hevesy, see G. Hevesy, "A Scientific Career" in *Perspectives in Biology and Medicine 1* (1958), 345–365, reprinted in *idem., Adventures in Radioisotope Research: Collected Papers* [2 volumes, consecutively paginated, 2nd volume beginning on p. 517] (New York, etc.: Pergamon Press, 1972), pp. 11–30; R. Spence, "George Charles de Hevesy" in *Chemistry in Britain 3* (1967), 527–532; Hilde Levi, "George de Hevesy, 1 August 1885 – 5 July 1966" in *Nuclear Physics A98* (1967), 1–24; F. Szabadváry, "George Hevesy" in *Journal of Radioanalytical Chemistry 1* (1968), 97–102. Hilde Levi, *George de Hevesy: Life and Work* (Bristol: Adam Hilger; Copenhagen: Rhodos, 1985) is the most insightful treatment of the man Hevesy. Levi's two contributions contain bibliographies of Hevesy's work that are somewhat more comprehensive than the one included in Cockcroft's account. For Bohr's early publications, see Léon Rosenfeld (gen. ed.), *Niels Bohr Collected Works, Volume 1:* J. Rud Nielsen (ed.), *Early Work (1905–1911)* and *Volume 2:*

Ulrich Hoyer (ed.), *Work on Atomic Physics (1912–1917)* (Amsterdam, etc.: North-Holland Publishing Company, 1972 and 1981).

70. Hevesy's tour of Europe can be followed in his letters: Hevesy to Johannes Stark 2 Feb 1912, Nachlass Stark, Handschriftenabteilung, Staatsbibliothek, Preussischer Kulturbesitz, West Berlin; Hevesy to Rutherford 14 Feb 1912, Rutherford Correspondence Collection, Cambridge University Library, Cambridge, England; copies of both items are deposited in the HSC (note 44); the first letter was written from Manchester, the second from Graz, Austria. The Bunsen lecture was published in the Society's journal as G. Hevesy, "Radioaktive Methoden in der Elektrochemie" in *Zeitschrift für Elektrochemie 18* (1912), 546–549, as well as in a slightly modified version with the same title in *Physikalische Zeitschrift 13* (1912), 715–719.

71. For Bohr's and Hevesy's involvement in the discussion, see, in particular, Hevesy to Rutherford 14 Oct 1913, Rutherford Collection (note 70) and HSC (note 44); this letter will be discussed later. The most central papers resolving the problem of the chemical identity of radioactive substances have been reprinted in English in Alfred Romer (ed.), *Radiochemistry and the Discovery of Isotopes [Classics of Science, Volume VI]* (New York: Dover, 1970) with an illuminating introduction by the editor – "The Science of Radioactivity, 1896–1913: Rays, Particles, Transmutations, Nuclei and Isotopes," pp. 3–60.

72. The quotation is from Bohr to Oseen 5 Feb 1913, BSC (5,4) (note 5); the letter is reproduced in *Bohr Collected Works, Volume 2* (note 69), pp. 551–552. Bohr was referring to G. Hevesy, "Die Valenz der Radioelemente" in *Physikalische Zeitschrift 14* (1913), 49–62. The authoritative account of Bohr's achievement is John L. Heilbron and Thomas S. Kuhn, "The Genesis of the Bohr Atom" in *Historical Studies in the Physical Sciences 1* (1969), 211–290, pp. 266–283. Bohr received his Nobel Prize "for his services in the investigation of the structure of atoms and of the radiation emanating from them"; *Nobel Lectures, Including Presentation Speeches and Laureates' Biographies: Physics, 1922–1941* (Amsterdam, etc.: Elsevier Publishing Company, 1965), p. 1; see his Nobel Prize Lecture, "The structure of the atom" in *ibid.*, pp. 7–43.

73. On Hevesy's failure in Manchester, see Hevesy, "Scientific Career," p. 14, Levi, *Hevesy*, pp. 23–27; Levi, "Hevesy," p. 2, Cockcroft, "Hevesy," p. 130, Spence, "Hevesy," p. 527, Szabadváry, "Hevesy," p. 98 (all note 69); see also Hevesy to Rutherford 14 Feb 1912 (note 70). Hevesy launched the collaboration with Paneth with the quoted letter from Hevesy to Paneth 3 Jan 1913, Nachlass Paneth, Archiv & Bibliothek der Geschichte der Max Planck Gesellschaft, Berlin, with copy in HSC (note 44). On Paneth, see Herbert Dingle and G. R. Martin, "Introduction" in *idem.* (eds.), *Chemistry and Beyond: A Selection from the Writings of the Late Professor F. A. Paneth* (New York, London, Sydney: Interscience Publishers, 1964), pp ix–xxi, which also gives references to other biographical publications on Paneth.

74. Soddy made his claim in Frederick Soddy, *The Chemistry of the Radio-Elements* [Alexander Findlay (ed.), *Monographs in Physical Chemistry*] (London: Longmans, Green & Co., 1911), pp. 58–59. Hevesy reported Meyer's suggestion in Hevesy to Rutherford 1 Dec 1912, HSC (note 44).

75. The introduction of the indicator technique was only the second of four lectures given by Hevesy and Paneth, presented in print as "Mitteilungen aus dem Institut für Radiumforschung" nos. 42–45, and published by the Austrian Academy in two separate ways: "Über Radioelemente als Indikatoren in der analytischen Chemie" in *Sitzungsberichte der Akademie der Wissenschaften* [Vienna], *mathematisch-naturwissenschaftliche Klasse, Abteilung IIa, 122* (1913), 1001–1007, and in *Monats-*

hefte für Chemie und verwandte Teile anderer Wissenschaften: Gesammelte Abhandlungen aus den Sitzungsberichten der kaiserlichen Akademie der Wissenschaften 34 (1913), 1401–1407. The article was published in a slightly revised form as G. Hevesy and F. Paneth, "Die Löslichkeit des Bleisulfids und Bleichromats" in *Zeitschrift für anorganischer Chemie 82* (1913), 323–328. The latter version is reprinted in English translation in Hevesy, *Adventures* (note 69), pp. 31–35, and in G. Hevesy, *Selected Papers of George Hevesy* (London, etc.: Pergamon Press, 1967), pp. 1–5. Hevesy received the Nobel Prize "for his work on the use of isotopes as tracers in the study of chemical processes"; see *Nobel Lectures, Including Presentation Speeches and Laureates' Biographies: Chemistry, 1942–1962* (Amsterdam, etc.: Elsevier Publishing Company, 1964), p. 1. Hevesy's Nobel Prize lecture, "Some Applications of Isotopic Indicators," is on pp. 9–41, and is reprinted in Hevesy, *Adventures*, pp. 929–969.

76. G. Hevesy, "Radio-Elements as Indicators in Chemistry and Physics" in *Report of the Eighty-Third Meeting of the British Association for the Advancement of Science, Birmingham: 1913, September 10–17* (London: John Murray, 1914), pp. 448–449.

77. Hevesy to Rutherford 14 Oct 1913 (note 71). In the BAAS *Report* (note 76), p. 403, only the title of Thomson's presentation, "X₃ and the Evolution of Helium," is provided.

78. The self-promotion was contained in G. Hevesy, "Bericht über die Verhandlungen der British Association in Birmingham" in *Zeitschrift für Elektrochemie 20* (1914), 88–93, p. 92. On the war's impact on Hevesy's work, see Cockcroft, "Hevesy" (note 69), pp. 132–133. New applications of the radioactive indicator technique were published as: G. Hevesy and E. Rona, "Die Lösungsgeschwindigkeit molekülarer Schichten" in *Zeitschrift für physikalische Chemie 89* (1915), 294–305, reprinted in Hevesy, *Adventures* (note 69), pp. 89–96; G. Gróh and G. Hevesy, "Die Selbstdiffusionsgeschwindigkeit des geschmolzenen Bleis" in *Annalen der Physik 63* (1920), 85–92; idem., "Die Selbstdiffusion in festem Blei" in *ibid.* 65 (1921), 216–222, reprinted in *Adventures*, pp. 110–113, and abstracted in a lecture to the Bunsen Gesellschaft in April 1920, G. Hevesy, "Die Selbstdiffusion in geschmolzenem Blei" in *Zeitschrift für Elektrochemie 26* (1920), 363–364; G. Hevesy and L. Zechmeister, "Über den intermolekülaren Platzwechsel gleichartiger Atome" in *Berichte der Deutschen Chemischen Gesellschaft* [Berlin] *53* (1920), 410–415, reprinted in *Adventures*, pp. 103–108 and in Hevesy, *Selected Papers* (note 75), pp. 24–29; G. Hevesy and L. Zechmeister, "Über den Verlauf des Umwandlungsvorgangs isomerer Ionen" in *Zeitschrift für Elektrochemie 26* (1920), 151–153. On Hevesy's situation in Budapest, see Cockcroft, "Hevesy" (note 69), p. 133, and the more detailed account of Hevesy's experiences during World War I, in Levi, *Hevesy* (note 69), pp. 34–46.

79. On the situation in Hungary, see Peter Pastor, *Hungary Between Wilson and Lenin: The Hungarian Revolution of 1918–1919 and the Big Three* (Boulder: East European Quarterly, 1976), pp. 144, 151. Other information is from the first letters exchanged between Bohr and Hevesy after the war: Hevesy to Bohr 2 Mar 1919, Bohr to Hevesy 14 Mar 1919, Hevesy to Bohr 19 Apr 1919, Bohr to Hevesy 9 Aug 1919, Hevesy to Bohr 1 Sep 1919, BSC (3,3) (note 5). Bohr announced the arrangement of a fellowship in a letter to Hevesy 10 Apr 1920, *ibid.* Bohr's renewals of the fellowship is reported in Robertson, *Early Years* (note 53), p. 49.

80. For contemporary evidence on the limited applicability of Hevesy's indicator technique, see Fritz Paneth and Walther Bothe, "Radioelemente als Indikatoren" in *Handbuch der Arbeitsmethoden in der anorganisch-*

en Chemie, Zweiter Band, Zweite Hälfte (Berlin and Leipzig: Walter de Gruyter & Co., 1925), pp. 1027–1047, on pp. 1027–1028. On the timing of Hevesy's arrival, see Bohr to Hevesy 1 Jun 1920, BSC (3,3) (note 5). The separation work was published as J. N. Brønsted and George Hevesy, "The separation of the isotopes of mercury" in Nature 106 (30 Sep 1920), 144 (wherefrom quotation), and idem., "The separation of the isotopes of chlorine" in ibid. 107 (14 Jul 1921), 619. This contribution is described in Cockcroft, "Hevesy," pp. 134–135, Szabadvary, "Hevesy," pp. 99–100, Spence, "Hevesy," p. 528, Hevesy, "Scientific Career," p. 19 (all note 69), and G. Hevesy, "Gamle Dage" in Niels Bohr: Et Mindeskrift [Fysisk Tidsskrift 60 (1962)] (Copenhagen: Fysisk Tidsskrift, 1963), pp. 26–30, on pp. 27–29. None of Hevesy and Brønsted's resulting publications is contained in the bound volumes, retained in the Niels Bohr Institute, which contains the institute's published work: Universitetets Institut for teoretisk Fysik: Afhandlinger [1918–1959].

81. The relationship of x-ray spectroscopy to Bohr's theory, as well as the collaboration of Coster and Hevesy and its consequences, are described in Helge Kragh, "Niels Bohr's Second Atomic Theory" in Historical Studies in the Physical Sciences 10 (1979), 123–186, pp. 167ff., 184–186.

82. Bohr's first coauthored paper was Niels Bohr and Dirk Coster, "Röntgenspektren und periodisches System der Elemente" in Zeitschrift für Physik 12 (1923), 342–374, paper dated 22 Oct 1922; reprinted, with English translation, in J. Rud Nielsen (ed.), Niels Bohr Collected Works, Volume 4: The Periodic System (1920–1923) (Amsterdam, New York, Oxford: North-Holland Publishing Company, 1977), pp. 485–518, 519–548.

83. The discovery of hafnium is dealt with in Helge Kragh, "Anatomy of a Priority Conflict: The Case of Element 72" in Centaurus 23 (1980), 275–301. The news of the discovery was broken in Niels Bohr, "The Structure of the Atom" in Nobel Lectures, Physics: 1922–1941 (note 72), pp. 7–43, on p. 42; the lecture is printed in its original Danish and in English translation in Niels Bohr Collected Works, Volume 4 (note 82), pp. 427–465, 467–482. Of the thirty-five papers with Hevesy as author listed in the Afhandlinger (note 80), twenty were directly concerned with hafnium. The bibliographies in Levi, Hevesy, and Cockcroft, "Hevesy" (both note 69) list thirteen more hafnium papers, and only add to their number relative to Hevesy's work on other subjects.

84. When Hevesy first commented in writing on his motivation, he emphasized his interest in Bohr's theory: [G. Hevesy,] "Aufzeichnungen über die Entdeckung des Hafnium. Geschrieben in Tapio-Sap im Juli 1923," p. 2; typed manuscript deposited in the Niels Bohr Archive, Copenhagen. The possibility of nuclear disintegration as a motivation for turning to x-ray spectroscopy is referred to in Cockcroft, "Hevesy" (note 69), p. 136, which was actually composed by Hevesy; see Cockcroft to Léon Rosenfeld 5 May 1967, HSC (note 44). This motivation was connected to Hevesy's wish to expand on his earlier paper, G. Hevesy, "An Attempt to Influence the Rate of Radioactive Disintegration by Use of Penetrating Radiation" in Nature 110 (12 Aug 1922), 216, paper dated 11 Jul 1922. The interest in mineral analysis as a motivation is mentioned in Hevesy, "Career" (note 80), p. 21. This interest is substantiated by the content of Hevesy's lecture "Jordens Alder" ("The Age of the Earth") to the Danish Geological Association on 13 Nov 1922, a report of which is contained in Meddelelser fra Dansk Geologisk Forening 6 (2, 1923), 13–19. The subsequent extent of Hevesy's x-ray work is recounted in Cockcroft, "Hevesy," pp. 140–141.

85. George Hevesy, "The Absorption and Translocation of Lead by Plants: A Contribution to the Application of the Method of Radioactive Indicators in the Investigation of the Change of Substance in Plants" in The

Biochemical Journal 17 (1923), 439–445. An abstract of this paper was published in *Nature 112* (1923), 772.

86. J. A. Christiansen, G. Hevesy, and S. Lomholt, "Recherches, par une méthode radiochimique, sur la circulation du bismuth dans l'organisme" in *Comptes rendus des séances de l'académie des sciences* [Paris] *179* (7 Apr 1924), 1324–1326; idem., "Recherches, par une méthode radiochimique, sur la circulation du plomb dans l'organisme" in *ibid. 179* (28 Jul 1924), 291–292. These papers have been published in English translation in Hevesy, *Adventures* (note 69), pp. 143–145, 146–147; and in idem., *Selected Papers* (note 75), pp. 53–55, 57–58.

87. Hevesy presented the review of his results in George Hevesy, "Über die Anwendung von radioaktiven Indikatoren in der Biologie" in *Biochemische Zeitschrift 173* (1926), 175–180. Hevesy's correspondence has been retained in HSC (note 44). Although Hevesy pursued his work on biology outside the context of the institute, the collection of publications of work there prepared in the 1950s does contain all but one – Christiansen, Hevesy, and Lomholt, "Recherches sur la circulation du bismuth" (note 86) – of Hevesy's such publications in the 1920s; see *Afhandlinger* (note 80).

88. On Hevesy's move to Freiburg, see Levi, *Hevesy* (note 69), pp. 59–62, and Cockcroft, "Hevesy" (note 69), pp. 140–141. The published article was G. Hevesy, "The Use of X-Rays for the Discovery of New Elements" in *Chemical Reviews 3* (1927), 321–329. The Johannesburg lecture was reprinted as G. Hevesy, "Quantitative Chemical Analysis by X-Rays and Its Application" in *Nature 124* (1929), 841–843. The lecture series at Cornell was published as G. Hevesy, *Chemical Analysis by X-Rays and Its Applications* [*The George Fisher Baker Non-Resident Lectureship in Chemistry at Cornell University, volume 10*] (New York: McGraw-Hill, 1932).

89. The paper on samarium was G. Hevesy and M. Pahl, "Radioactivity of Samarium" in *Nature 130* (1932), 846–847. Auer's providing rare earths to Hevesy while in Copenhagen is reported in Cockcroft, "Hevesy," p. 141, Spence, "Hevesy," p. 529, and Levi, *Hevesy*, p. 57 (all note 69). The relationship with Auer, including obtaining rare earths, dated from before Hevesy's Copenhagen period. See G. Hevesy, "Freiherr Auer von Welsbach" in *Akademische Mitteilungen aus Freiburg 4* (1929), 17–18; Fritz Paneth, "Zum 70. Geburtstag Auer von Welsbach" in *Die Naturwissenschaften 16* (1928), 1037–1038, translated into English as "Auer von Welsbach" in Dingle and Martin (eds.), *Chemistry and Beyond* (note 73), pp. 73–76.

90. The non-biological applications of the radioactive indicator technique were published as: G. Hevesy and W. Seith, "Der radioaktive Rückstoss im Dienste von Diffusionsmessungen" in *Zeitschrift für Physik 56* (1929), 790–801, with an editors' "Berichtigung" in *ibid., 57* (1929), 869; G. Hevesy and R. Hobbie, "Lead Content of Rocks" in *Nature 128* (1931), 1038–1039. English translations, excepting the "Berichtigung," are printed in Hevesy, *Adventures* (note 69), pp. 127–137, 43–44, and in idem., *Selected Papers* (note 75), pp. 37–47, 6–7. The biological applications was printed as G. Hevesy and O. H. Wagner, "Die Verteilung des Thoriums im tierischen Organismus" in *Archiv für experimentelle Pathologie und Pharmakologie 149* (1930), 336–342.

91. On the discovery of deuterium, see, for example, F. G. Brickwedde, "Harold Urey and the Discovery of Deuterium" in *Physics Today 35* (Sep 1982), 34–39 and Daniel J. Kevles, *The Physicists: The History of a Scientific Community in Modern America* (New York: Alfred A. Knopf, 1978), pp. 225–226. See also Roger H. Stuewer, "The naming of the deuteron" in *American Journal of Physics 54* (1986), 206–218, p. 206.

The help from Urey is recounted in Hevesy, "Career" (note 80), p. 25, wherefrom the first quotation. The second quotation is from Hevesy to Rutherford 1 Apr 1934, Rutherford Collection (note 70); copy in HSC (note 44). See Hevesy's correspondence with Bohr in BSC (20, 3) (note 5).

92. Bohr and Brønsted reported their expectations of Hevesy's place of work in their letter to Jones 11 Oct 1933 (note 46). Hevesy noted his anticipation of working with Jacobsen in his letter to Bohr 6 Nov 1933, BSC (20, 3) (note 5). His subsequent letters to Bohr not containing specific research plans were dated 27 and 31 Dec 1933, *ibid.* Hevesy confided his plan to move to India in a letter to Franck 20 Jul 1934, Franck Papers (note 57), Box 4, Folder 2; copy in HSC (note 44). He expressed relief for not having to tell Bohr about his planned trip to India in a second letter to Franck 18 Aug 1934, *ibid.* The last two quotations are from Hevesy to Bohr 11 Aug 1934, BSC (20, 3).

93. On Hevesy's marriage, see Levi, *Hevesy* (note 69), p. 59.

94. Bohr to Rutherford 1 Feb 1934, BSC (25, 2). Bohr to Bloch 17 Feb 1934, BSC (17, 3) (note 5); this letter is reproduced in Erik Rüdinger (gen. ed.), *Niels Bohr Collected Works, Volume 9:* Rudolf Peierls (ed.), *Nuclear Physics (1929–1952)* (Amsterdam, etc.: North-Holland Physics Publishing 1986), pp. 540–541.

95. The "new kind of radioactivity" was announced as Irène Curie and Frédéric Joliot, "Un nouveau type de radioactivité" in *Comptes rendus des séances de l'académie des sciences 19* (15 Jan 1934), 254–256; see also *idem.,* "Artificial Production of a New Kind of Radioelement" in *Nature 133* (10 Feb 1934), 201–202. The two articles are reprinted in Frédéric and Iréne Joliot-Curie, *Oevres Scientifiques Complètes* (Paris: Presses Universitaires de France, 1961), pp. 515–516, 520–521.

96. Bohr to Curie 10 Mar 1934, BSC (18, 2) (note 5).

97. The quotation is taken from *Nobel Lectures, Including Presentation Speeches and Laureates' Biographies: Chemistry, 1922–1941* (Amsterdam, etc.: Elsevier Publishing Company, 1966). Their Nobel Prize Lectures, given 12 Dec 1935, were, Irène Joliot-Curie, "Artificial Production of Radioactive Elements" and Frédéric Joliot, "Chemical evidence of the transmutation of elements"; reprinted in *ibid.*, pp. 366–368, 369–373. The original French versions can be read in Joliot-Curie, *Oevres* (note 95), pp. 516–548, 549–552.

98. For two complementary accounts of Fermi's research program, see Gerald Holton, "Striking Gold in Science: Fermi's Group and the Recapture of Italy's Place in Physics" in *Minerva 12* (1974), 158–198 (reprinted as "Fermi's Group and the Recapture of Italy's Place in Physics" in *idem., The Scientific Imagination: Case Studies* (Cambridge, England: Cambridge University Press, 1978), pp. 155–198, and Emilio Segrè, *Enrico Fermi: Physicist* (Chicago and London: University of Chicago Press, 1970), pp. 64–93. Rutherford's early suggestion was published as Ernest Rutherford, "Nuclear Constitution of Atoms – Bakerian Lecture" in *Proceedings of the Royal Society A97* (1920), pp. 374–400, 396–397; reprinted in *The Collected Papers of Lord Rutherford: Volume Three, Cambridge* (London: George Allen and Unwin Ltd, 1965), pp. 14–40, on p. 34. The first in the Rome group's series of publications was Enrico Fermi, "Radioattività indotta da bombardemento di neutroni" in *Ricerca Scientifica 5* (15 Mar 1934), 283, paper dated 25 Mar; reprinted and translated into English in Enrico Fermi, *Collected Papers, volume 1: Italy 1921–1938* (Chicago: University of Chicago Press, 1961), pp. 645–646, 674–675.

99. On Franck's vacillation, see Franck to Max Born 18 May 1934, Nachlass Born 229 B1. 8, Staatsbibliothek, Berlin (note 70); copy in Franck Papers (note 57), Box 1, Folder 7. On Franck and Levi's collaboration, see my

interview with Hilde Levi at the Niels Bohr Institute, Copenhagen, 16 Sep 1981.
100. Bohr reported on Franck's work in his letter to Heisenberg 20 Apr 1934, BSC (20, 2) (note 5). On the Radium Station as a provider of radioactive sources, see Levi, *Hevesy* (note 69), p. 87. Franck asked Frisch for confirmation of his arrival in a letter 26 April 1934, Document F41¹, Otto Robert Frisch Papers, Trinity College, Cambridge University, England. Frisch answered Franck in a letter 1 May 1934, Franck Papers (note 57), Box 1, Folder 7. Franck reported back to Bohr 19 Jul 1934, BSC (19, 2) (note 5).
101. On Metropolitan-Vickers's outfitting of Rutherford's laboratory, see T. E. Allibone, "Metropolitan-Vickers Electrical Company and the Cavendish Laboratory" in John Hendry (ed.), *Cambridge Physics in the Thirties* (Bristol: Adam Hilger Ltd, 1984), pp. 150–173, on pp. 161ff. Hevesy reported to Franck on his trip to London in a letter 30 Apr 1934, Franck Papers (note 57), Box 4, Folder 2. On the plans for Bohr's trip to the Soviet Union, see Bohr to Abraham Joffe 10 Mar 1934, Joffe to Bohr 23 Mar 1934, Bohr to Joffe 26 Mar 1934 and 13 April 1934, BSC (21, 3) (note 5).
102. The quotation is from Bohr to Bjerge 21 Jun 1934, BSC (17, 2) (note 5). On Bjerge's visit to Cambridge, see Bjerge to Bohr 20 Nov 1933, *ibid.*, Bohr to Rutherford 19 Dec 1933, BSC (25, 2).
103. The article was published as J. Ambrosen, "Über den aktiven Phosphor und das Energiespektrum seiner 'β-Strahlen' " in *Zeitschrift für Physik 91* (1934), 43–48; paper received by journal 19 Jul 1934 – the acknowledgements are on p. 48. In regard to the related work in Rome, Ambrosen referred to E. Fermi, "Possible Production of Elements of Atomic Number Higher than 92" in *Nature 133* (16 Jun 1934), 898–899, p. 898. While presenting the same empirical data, the previous article from the Rome group dated 10 May 1934 did not make the same point explicitly: E. Amaldi, O. D'Agostino, E. Fermi, F. Rasetti, E. Segrè, "Radioattività 'beta' provocata da bombardementa di neutroni. – III" in *Ricerca Scientifica 5* (30 Apr 1934), 452–453. The articles from Rome have been reprinted, and the latter translated into English, in Fermi, *Collected Papers* (note 98), pp. 748–750 (on p. 749), 649–650, 677–678. Franck claimed responsibility for submitting Ambrosen's paper in his letter to Bohr 19 Jul 1934 (note 100).
104. Bohr noted his disagreement with Fermi in a letter to Rutherford 30 Jun 1934, to which Rutherford replied on 9 Jul 1934; both letters are in BSC (25, 2) (note 5). The former is also reproduced in *Niels Bohr Collected Works, Volume 9* (note 94), pp. 651–652. In a letter to Bohr 14 Sep 1934, BSC (19, 4), Gamow noted the receipt of Bohr's statement. The work of the Rome experimentalists, including the collaboration with Cambridge, is reported in Segrè, *Fermi* (note 98), pp. 77–78. On Bohr's view, see R. Peierls, "Introduction" in *Niels Bohr Collected Works, Volume 9* (note 94), pp. 3–83, on pp. 14–15.
105. T. Russell Wilkins, "A Visit to the Institute of Theoretical Physics – Copenhagen, June, 1934"; three-page typed report in the Rush Rhees Papers, Department of Rare Books and Special Collections, The University of Rochester Library. I am indebted to Tom Cornell for directing me to this source.
106. The accident is first mentioned in the Bohr Scientific Correspondence in Schultz [Bohr's secretary] to Trumpy 13 Jul 1934, BSC (26, 1) (note 5). It is described in several places; see, for example, Niels Blædel, *Harmoni og Enhed: Niels Bohr – En Biografi* (Copenhagen: Rhodos, 1985), pp. 184–186, published in English as *idem., Harmony and Unity: The Life of Niels Bohr* (Madison, Wisconsin: Science Tech Publishers; Berlin, etc.: Springer Verlag, 1988), pp. 152–155; Ulrich Röseberg, *Niels*

Bohr: Leben und Werk eines Atomphysikers 1885–1962 (Berlin: Akademie-Verlag, 1985), p. 208. The first extant letter from Bohr's hand after the accident is Bohr to Heisenberg 8 Sep 1934, BSC (20, 2). On Bohr's decision not to go to the London conference, see Bohr to Irène Joliot-Curie 22 Sep 1934, BSC (21, 3). The proceedings of the London conference were published as *International Conference on Physics London 1934: A Joint Conference Organized by the International Union of Pure and Applied Physics and the Physical Society. Papers and Discussions. In two volumes. Vol. 1. Nuclear Physics* (London: The Physical Society, 1935). Volume 2 was devoted to "The Solid State of Matter." On the importance of this conference, see Roger H. Stuewer, "The Nuclear Electron Hypothesis" in William R. Shea (ed.), *Otto Hahn and the Rise of Nuclear Physics* (Dordrecht, Boston, Lancaster: D. Reidel Publishing Company, 1983), pp. 19–67, on p. 51.

107. Fermi to Hevesy 15 Oct 1934, BSC (19, 2) (note 5) [under Fermi]; wrongly filed and microfilmed as being addressed to Bohr. The count of rare earths irradiated by Fermi is based on the published reports from the induced radioactivity project in Rome appearing before Fermi's letter to Hevesy: Fermi, "Radioattività" (note 98), *idem.*, "Radioattività provocata da bombardemento di neutroni" in *Ricerca Scientifica 5* (31 Mar 1934), 330–331, Amaldi *et al.*, "Radioattività – III" (note 103), *idem.*, "Radioattività provocata da bombardemento di neutroni. – IV" in *Ricerca Scientifica 5* (15 Jun 1934), 652–653; *idem.*, "Radioattività provocata da bombardemento di neutroni. – V" in *ibid.* (15–31 Jul 1934), 21–22, reprinted in Fermi, *Collected Papers, volume 1* (note 98), pp. 645–646, 647–648, 649–650, 651–652, 653–654, with English translation, pp. 674–675, 676, 677–678, 679–680, 681–682. The papers were dated 25 Mar, undated, 10 May, 23 Jun, 12 Jul 1934. In the course of these papers the Rome group published detailed, positive results for thirty-six elements.

108. Hevesy to Fermi 26 Oct 1934, BSC [under Fermi] (19, 2) (note 5).

109. The relevant publications were G. Hevesy, "Artificial Radioactivity of Scandium" in *Det Kongelige Danske Videnskabernes Selskab: Mathematisk-fysiske Meddelelser 13* (3, 1935), 17 pp. [dated Jan 1935]; G. Hevesy and H. Levi, "Radiopotassium and Other Artificial Radio-elements" in *Nature 135* (13 Apr 1935), 580; *idem.*, "Artificial Radioactivity of Dysprosium and Other Rare Earth Elements" in *Nature 136* (20 Jul 1935), 103; *idem.*, "The Action of Neutrons on the Rare Earth Elements" in *Det Kongelige Danske Videnskabernes Selskab: Mathematisk-fysiske Meddelelser 14* (4, 1936), 33 pp. On Bohr's inducement, see Levi, *Hevesy* (note 69), p. 77, and particularly Weiner's Levi interview 28 Oct 1971 (note 37), pp. 21–22.

110. Franck expressed his frustrations in a letter to Born 29 Oct 1934, Nachlass Born (note 99) 229 Bl. 9–12; copy in Franck Papers (note 57), Box 1, Folder 7. The papers referred to by Franck are most probably J. Franck and H. Levi, "Beitrag zur Untersuchung der Fluoreszenz in Flüssigkeiten" in *Zeitschrift für physikalische Chemie 27* (1935), 409–420 and Erich Schneider, "Der Prozess der Auslöschung der Fluoreszenz von Flüssigkeiten durch Halogenionen" in *ibid.* 28 (1935), 311–322, papers received by journal 5 Oct 1934 and 14 Feb 1935, respectively; although not formally an author of the latter paper, Franck was strongly involved in working on it. Frisch expressed his views on the Copenhagen effort in a letter to Segrè 29 Oct 1934, Document B75[9], Frisch Papers (note 100).

111. Bohr to Fowler 12 Dec 1934, BSC (19, 2) (note 5).

112. Franck to Born 8 Jan 1935, Nachlass Born (note 99) 229 Bl. 13 (carbon copy); copy in Franck Papers (note 57) Box 1, Folder 7. On the discovery of the effect of "slow neutrons," see Segrè, *Fermi* (note 98), pp. 79–83, which also notes that the disagreement between Bohr and Fermi was

resolved in the course of these experiments. The published letter first publicizing the result is translated in Segrè's book on pp. 81–82, and was originally published as E. Fermi, E. Amaldi, B. Pontecorvo, F. Rasetti, E. Segrè, "Azione di sostanze idrogenate sulle radioattivitá provocata da neutroni. I" in *Ricerca Scientifica 5* (2, 1934), 282–283; paper dated 22 Oct 1934. It is reprinted in the original Italian and English translation in Fermi, *Collected Papers, volume I* (note 98), pp. 757–758, 761–762.

113. Franck to Born 8 Jan 1935 (note 112).
114. Franck interview (note 52), session IV, p. 12.
115. The anecdote is recalled in Charles Weiner's interview with Christian Møller 25 Aug and 21 Oct 1971, pp. 86–87 of original transcript; the interview is deposited in the Niels Bohr Library of the American Institute of Physics in New York City. See also Levi, *Hevesy* (note 69), p. 81. For a particularly astute and forthright comparison of Franck's and Hevesy's research styles, see Weiner's Levi interview 28 Oct 1971 (note 37), pp. 28–29.
116. The quotation is from the Franck interview (note 52), session IV, p. 11. Hevesy described his first visit to Göttingen in a letter to Bohr 21 Aug 1932, BSC (20, 3) (note 5). For another interpretation of Franck's and Hevesy's different reactions to the Copenhagen experience, see Levi, *Hevesy* (note 69), p. 82.
117. My interview with Victor Weisskopf 5 Jun 1981 at the Massachusetts Institute of Technology.
118. Weiner's Møller interview 25 Aug and 21 Oct 1971 (note 115), p. 72.

CHAPTER 4

1. Robert E. Kohler, "A Policy for the Advancement of Science: The Rockefeller Foundation, 1924–29" in *Minerva 16* (1978), 480–515. See also Raymond B. Fosdick, *The Story of the Rockefeller Foundation* (New York: Harper and Brothers, 1952), pp. 135–137.
2. Kohler, "Policy" (note 1), pp. 502–510; Abraham Flexner, *Funds and Foundations: Their Policies Past and Present* (New York: Harper & Brothers Publications, 1952), p. 83; Fosdick, *Story* (note 1), p. 154; George W. Gray, *Education on an International Scale: A History of The International Education Board 1923–1938* (New York: Harcourt Brace and Company, 1941), p. 3.
3. Robert E. Kohler, "The Management of Science: The Experience of Warren Weaver and the Rockefeller Foundation Programme in Molecular Biology" in *Minerva 14* (1976), 279–306, pp. 283–284. A slightly revised version of this article is *idem.*, "Warren Weaver and the Rockefeller Foundation Program in Molecular Biology: A Case Study in the Management of Science" in Nathan Reingold (ed.), *The Sciences in the American Context: New Perspectives* (Washington, D.C.: Smithsonian Institution Press, 1979), pp. 249–293.
4. Warren Weaver, *Scene of Change: A Lifetime in American Science* (New York: Charles Scribner's Sons, 1970), pp. 28–32; Fosdick, *Story* (note 1), p. 156.
5. Kohler, "Management" (note 3), pp. 284–286.
6. Mason's plea is reproduced in "A Brief Summary of the Conferences of Trustees and Officers at Princeton" in "900, Programs and Policy: General," Rockefeller Foundation archive, Record Group 3, Series 900, Box 22, Folder 166 (RF RG 3, 900, 22, 166) in the Rockefeller Archive Center (RAC), Pocantico Hills, New York. The information on Spoehr is taken from Kohler, "Management" (note 3), p. 286.
7. The funding information is taken from "Agenda for Special Meeting April 11, 1933" in RF RG 3, 900, 22, 168 (note 6).

8. A copy of Hevesy's application is deposited in "717D, University of Freiburg: Physical Chemistry," RF RG 1.1, 717, 13, 119 (note 6).

9. The timing of Hevesy's arrival in New York is taken from his letter to Spoehr 8 Sep 1930, RF RG 1.1, 717, 13, 119 (note 8). The quotation is from Jones to Hevesy 3 Sep 1930, *ibid.* Hevesy's lectures at Cornell were published as George Hevesy, *Chemical Analysis by X-Rays and Its Applications* [*The George Fisher Baker Non-Resident Lectureship in Chemistry at Cornell University*, volume 10] (New York: McGraw-Hill, 1932).

10. Spoehr arranged the meeting in a letter to Hevesy 24 Sep 1930, RF RG 1.1, 171, 13, 119 (note 8). For the approval of Hevesy's grant, see the Rockefeller Foundation officers' recommendation to the trustees 10 Dec 1930, and Norma S. Thompson to Hevesy s.d., *ibid.*

11. Weaver, *Scene of Change* (note 4), pp. 2, 28–55. The full reference to their joint book is Max Mason and Warren Weaver, *The Electromagnetic Field* (Chicago: University of Chicago Press, 1929).

12. On the introduction of quantum mechanics in the United States, see Stanley Coben, "The Scientific Establishment and the Transmission of Quantum Mechanics to the United States, 1919–32" in *American Historical Review* 76 (1971), 442–466; and Katherine R. Sopka, *Quantum Physics in America 1920–1935* (New York: Arno Press, 1980), republished as *idem., Quantum Physics in America: The years through 1935 [The History of Modern Physics 1800–1950, Volume 10]* (New York: Tomash Publishers and American Institute of Physics, 1988). Despite the publication date, the latter book is identical to its predecessor, and hence does not employ or refer to literature published after the mid-1970s. Mason and Weaver's opposition to quantum mechanics is described in Weaver, *Scene of Change* (note 4), pp. 56–57, which also deals with Weaver's appointment on pp. 58–63.

13. *Ibid.*, pp. 65–66, quotation on p. 65. Additional information on Jones is taken from J. McKeen Cattell and Jaques Cattell (eds.), *American Men of Science: A Biographical Directory* [sixth edition] (New York: Science Press, 1938).

14. The quotation is from the Lauder Jones Diary 19 and 20 Jul 1932; diaries of Rockefeller Foundation officers are filed in RF RG 12, RAC (note 6). That Bohr had given prior notice of his absence is seen from Jones to Bohr 30 Jun 1932, "Rockefeller Foundation," Bohr General Correspondence, Special File, film 1, section 6 (BGC-S (1,6)), Niels Bohr Archive, Copenhagen.

15. The pertinent letter, which also contains the quotation, is Bohr to Weaver 13 Apr 1933, "713D, University of Copenhagen: Biophysics 1933–34," RF RG 1.1, 713, 4, 46 (note 6).

16. Kohler, "Warren Weaver" (note 3), p. 259.

17. Weaver's report is in *The Rockefeller Foundation: Annual Report 1932* (New York: Rockefeller Foundation, undated), pp. 235–256, quotation on p. 236. The size of the appropriations can be seen by matching the information on institutions in Weaver's report with the list of all Rockefeller Foundation appropriations on pp. 324–327.

18. The quotation is from "Staff Conference, Tuesday, March 14, 1933" in RF RG 3, 900, 21, 160 (note 6). For a commentary on Rose's view of social science, see Fosdick, *Story* (note 1), p. 141.

19. This report is the "Agenda" referred to earlier (note 7).

20. The quotation is from *ibid.*, p. 62. The authorship of the "Proposed Future Program" was not provided in the 1933 account, but was later revealed in "Report of the Committee of Appraisal and Plan," p. 25, in RF RG 3, 900, 22, 170 (note 6).

21. *Ibid.*, pp. 62–63.

22. *Ibid.*, pp. 63–65.

23. Weaver's contribution was called "Natural Sciences – Proposed Program," "Agenda" (note 7), pp. 76–87, quotation on pp. 76–77.
24. On "colonization," see Pnina Abir-Am, "The Discourse of Physical Power and Biological Knowledge in the 1930s: A Reappraisal of the Rockefeller Foundation's 'Policy' in Molecular Biology" in *Social Studies of Science 12* (1982), 341–382, p. 368. The forthcoming Robert E. Kohler, *Managers of Science: Foundations and the Natural Sciences 1900–1950* treats Weaver and Mason as true interdisciplinarians.
25. The quotation is from Weaver, "Natural Sciences" (note 23), p. 77. The "biological institute" is identified as the Rockefeller Institute in Abir-Am, "Discourse" (note 24), p. 349.
26. *The Rockefeller Foundation: Annual Report 1933* (New York: Rockefeller Foundation, undated). Mason's "Foreword" is on pp. xvii–xix, where the quotation is on p. xvii, and the Special Research Aid for European Scholars is announced on p. xix. Weaver's section is on pp. 193–230, where the quotation is on p. 200.
27. *Ibid.;* useful "Summary of Appropriations for 1933" and "Payments on Former Appropriations" are on pp. 226 and 227–229, respectively.
28. The quotation is from the Warren Weaver Diary 10 Jul 1933, RF RG 12 (note 14).
29. Weaver prepared a memorandum for his meeting with Fosdick: "Modern Biology" in "915, Programs and Policy: Natural Science and Agriculture," RF RG 3, 915, 4, 38 (note 6). On the appointment of Fosdick's committee, see Kohler, "Management" (note 3), p. 292.
30. Weaver quoted Osterhout's letter in his "Modern Biology" (note 29).
31. On Fricke, see O. A. Allen, "Hugo Fricke and the Development of Radiation Chemistry: A Perspective View" in *Radiation Research 17* (1962), 255–261. Bohr's letter of recommendation for Fricke of 9 Jun 1919 as well as his letter of 5 Jun 1919 expressing his expectation that Fricke would return to the institute are in the Bohr Scientific Correspondence, microfilm 2, section 4 (BSC (2,4)); the BSC is retained in the Niels Bohr Archive, Copenhagen, with microfilm copies also in the Niels Bohr Library of the American Institute of Physics in New York and other places.
32. The first quotation is from Weaver to Fosdick 22 Mar 1934, RF RG 3, 915, 4, 38 (note 29), which was the covering letter for Weaver's "Modern Biology" (note 29) memorandum. The two quotations from Hill's letter are taken from this memorandum.
33. Krogh's words as quoted in *ibid.*
34. The quotation is from Weaver to Fosdick 22 Mar 1934 (note 32).
35. Quotation from *ibid.*
36. The published writings quoted in Weaver's "Modern Biology" (note 29) memorandum were R. G. Hoskins, *The Tides of Life: The Endocrine Glands in Bodily Adjustment* (New York: W. W. Norton & Company, Inc., 1933), pp. 347–348; H. I. Brock, "Conant States His Creed for Harvard – To Inspire the Undergraduate with an Enthusiasm for Creative Scholarship Is the President's Ambition" in *The New York Times Magazine* 18 Mar 1934, pp. 2, 22, on p. 22; H. G. Wells, Julian Huxley, and G. P. Wells, *The Science of Life* (London, etc.: Cassell and Company, Limited, 1931), pp. 879, 880, 879; Frederick Gowland Hopkins, "Some Chemical Aspects of Life [Presidential Address Delivered at Leicester on September 6, 1933]" in *Nature* (Supplement) *132* (9 Sep 1933), 381–394, pp. 382, 391.
37. The quotation is from Weaver to Fosdick 22 Mar 1934 (note 32).
38. The information on Tisdale and O'Brien is taken from Jaques Cattell (ed.), *American Men of Science: A Biographical Directory* [seventh edition] (New York: Science Press, 1944).
39. Tisdale and O'Brien Diary 8 to 11 Apr 1934, RF RG 12 (note 14).
40. *Ibid.*

41. The quotations are from *ibid*. The meeting with Morgan in the United States is noted in Bohr to Heisenberg 22 Nov 1933, BSC (20,2) (note 31). On Morgan, see Garland E. Allen, *Thomas Hunt Morgan: The Man and His Science* (Princeton: Princeton University Press, 1978).

42. The first quotation is from Tisdale to Bohr 28 Apr 1934, "Rockefeller Foundation," BGC-S (5,8) (note 14). Morgan did visit Bohr in Copenhagen; see Warren Weaver Diary 20 Jun 1934, RF RG 12 (note 14). The second quotation is from Tisdale to Weaver 30 Apr 1934, RF RG 1.1, 713, 4, 46 (note 15).

43. Tisdale made his comments in his letter to Weaver 30 Apr 1934 (note 42). The other quotation is from Weaver to Tisdale 5 Jun 1934, RF RG 1.1, 713, 4, 46 (note 15).

44. The quotation is from Bohr to Tisdale 26 Apr 1934, "Rockefeller Foundation," BGC-S (5,8) (note 14).

45. Tisdale and O'Brien Diary 8 Apr 1934, RF RG 12 (note 14).

46. *Ibid*.

47. The first quotation is from the Tisdale and O'Brien Diary 10 Apr 1934, RF RG 12 (note 14). On a visit to Freiburg in April 1933 – Lauder W. Jones Diary 5 Apr, *ibid*. – Hevesy explained his slow expenditure of the grant as a means to save money. Weaver expressed his satisfaction with Hevesy's use of the grant when he and Jones made Freiburg the last stop on their European trip in June 1933 – Weaver Diary 13 Jun 1933, *ibid*. The second quotation is from the Jones Diary 14 Oct 1933, *ibid*. Hevesy complained of his treatment in his letter to Bohr 15 Oct 1933, BSC (20,3) (note 31).

48. Tisdale and O'Brien Diary 10 Apr 1934, RF RG 12 (note 14).

49. Tisdale and O'Brien's diary notes are in *ibid*. Tisdale's other observations were written in two separate memos dated 28 Apr 1934, both in RF RG 1.1, 713, 4, 46 (note 15).

50. Kohler, "Management" (note 3), pp. 289, 292.

51. Fosdick's evaluation of the subcommittee is described in *ibid*., p. 294. The term "experimental biology" was first used in the committee's "Report" (note 20), p. 57.

52. Kohler, "Management" (note 3), pp. 270, 293.

53. "Report" (note 20), p. 58, wherefrom quotation, and p. 61.

54. *The Rockefeller Foundation: Annual Report 1934* (New York: Rockefeller Foundation, undated). Mason's foreword, pp. xi–xiv, discusses the continued narrowing of effort on p. xii. In Weaver's section, "The Natural Sciences," pp. 121–165, the need to specialize and the change in the fellowship program is discussed on pp. 125–126. "Summary of Appropriations Made in 1934" and "1934 Payments" are on pp. 158–161 and 161–164, respectively. Pauling's project is treated in Abir-Am, "Discourse" (note 24), pp. 357–361.

55. See reference to *Rockefeller Foundation: Annual Report 1934* in note 54.

56. Kohler, "Management" (note 3), pp. 292, 297–298.

57. *The Rockefeller Foundation: Annual Report 1935* (New York: Rockefeller Foundation, undated). Weaver's report, "The Natural Sciences," pp. 119–187, includes "Summary of Appropriations Made in 1935," pp. 181–183, and "1935 Payments," pp. 183–186.

58. The quotation is from "Report" (note 20), p. 24. The assessment of Weaver and his program as pioneering is taken from Kohler, "Management" (note 3), pp. 303–306. One project described in the 1935 *Annual Report* (note 57), that of William T. Astbury at the University of Leeds, is described in Abir-Am, "Discourse" (note 24), pp. 353–357, which also describes, on pp. 361–367, a research proposal from 1935 by Joseph Needham and others at Cambridge University, which was never accept

ed; details on rejected research proposals were not printed in the *Annual Reports.*

59. *Annual Report* 1935 (note 57); the project supported at Bohr's institute is described on pp. 129–130.
60. Wilbur Tisdale Diary 29 Oct 1934, RF RG 12 (note 14). On the Radium Station, see C. A. Clemmensen, *Radiumfondet, Oprettet til Minde om Kong Frederik VIII, 1912–1929: Et Afsnit af Kampen mod Kræften i Danmark* (Copenhagen: Radiumfondet, 1931).
61. The first quotation is from August Krogh, "The Use of Deuterium in Biological Work" in *Enzymologia 5* (1938), 185–189, p. 185. Hevesy's assessment is from his autobiographical remarks, G. Hevesy, "A Scientific Career" in *Perspectives in Biology and Medicine 1* (1958), 345–365, reprinted in *idem., Adventures in Radioisotope Research: Collected Papers* [2 volumes, consecutively paginated, 2nd volume beginning on p. 517], pp. 11–30, on p. 25. Krogh made his second remark in an address at Harvard Tercentenary Conference of Art and Sciences, which was published as August Krogh, "The Use of Isotopes as Indicators in Biological Research" in *Science 85* (19 Feb 1937), 187–191, p. 191. Incidentally, the argumentation in Krogh's lecture was used by the Rockefeller Foundation officers when they recommended to the trustees in 1937 that an additional sum be provided for the cyclotron at Bohr's institute; see resolution dated 19 Mar 1937 in RF RG 1.1, 713, 4, 46 (note 15).
62. Tisdale Diary 29 Oct 1934, RF RG 12 (note 14).
63. The quotation is from *ibid.* The earlier interview with Hevesy in Stockholm is referred to in the Tisdale Diary 30 Oct 1934, RF RG 12 (note 14). The exchange rate is taken from *Banking and Monetary Statistics* (Washington, D.C.: Board of Governors of the Federal Reserve System, November 1943), p. 681.
64. The quotation is from the Tisdale Diary 30 Oct 1934, RF RG 12 (note 14). Tisdale informed Bohr of the approval in his letter 17 Dec 1934, "Rockefeller Foundation," BGC-S (5,8) (note 14).
65. The quotation is from Tisdale to Weaver 16 Nov 1934, RF RG 1.1, 713, 4, 46 (note 15). On the terminology and history of these particle accelerator devices, see, for example, M. Stanley Livingston, *Particle Accelerators: A Brief History* (Cambridge, Massachusetts: Harvard University Press, 1969), pp. 1–21.
66. The first quotations are from Tisdale to Weaver 16 Nov 1934 (note 65), whereas Weaver's comments are from his letter to Tisdale 21 Jan 1935; excerpt in "713D, University of Copenhagen: Biophysics 1935–1937," RF RG 1.1, 713, 4, 47 (note 6).
67. The biographical information on Miller is taken from Cattell (ed.), *American Men of Science* [seventh edition] (note 38). For Miller's expectation, see his letter to Bohr 19 Jan 1935, "Rockefeller Foundation," BGC-S (6,1) (note 14). The quotations and the description of Bohr and Hevesy's plans are taken from the H. M. Miller Diary 25 Jan 1935, RF RG 12 (note 14).
68. The quotation is from *ibid.* Bohr's correspondence with Delbrück during the relevant years (1932–1945) is retained in BSC (18,3) (note 31). Delbrück and Weisskopf's nonparticipation in the experimental biology project was confirmed by Delbrück in a letter to me 13 May 1980, and by Weisskopf in an interview with me at the Massachusetts Institute of Technology 5 Jun 1981.
69. H. M. Miller Diary 25 Jan 1935, RF RG 12 (note 14).
70. Although in his diary Miller referred instead to a grant of $5,000 per year for "general research needs" from the Carlsberg Foundation, I contend that Bohr instead meant to refer to the prospective high-voltage grant. As we saw in Chapter 1, the "general research needs" grant had been

received by 1934, so Bohr could hardly call it "unexpected" at the time of Miller's interview. I therefore assume that Miller's identification of the grant mentioned by Bohr was due to a misunderstanding. The quotation in this paragraph is from Bohr to Tisdale 24 Nov 1934, "Rockefeller Foundation," BGC-S (5,8) (note 14). Bohr's statement a week later is from his letter to Tisdale 30 Nov 1934, *ibid*.

71. Both the draft, dated 26 Jan 1935, and the final application, dated 28 Jan 1935, of Bohr's application are in "Carlsberg Foundation," BGC-S (1,3) (note 14). The submitted application, identical to the copy retained by Bohr, can be seen in the Carlsberg Foundation Archive, Copenhagen.

72. The Carlsberg Foundation's board of directors are listed in Kristof Glamann, *Carlsbergfondet* (Copenhagen: Rhodos, 1976), p. 199. On Bohr's friendships with Bjerrum and Pedersen, see, for example, Niels Blædel, *Harmoni og Enhed: Niels Bohr, En biografi* (Copenhagen: Rhodos, 1985), pp. 176–178, 237, published in English as *idem.*, *Harmony and Unity: The Life of Niels Bohr* (Madison, Wisconsin: Science Tech Publishers; Berlin, etc.: Springer-Verlag, 1988), pp. 145–148; the comment on Bohr and Pedersen is provided in a figure caption that is not reproduced in the English translation of Blædel's book. On Bohr's relationship with the Royal Danish Academy, see Johannes Pedersen, "Niels Bohr and the Royal Danish Academy of Sciences and Letters" in Stefan Rozental (ed.), *Niels Bohr: His Life and Work as Seen by Friends and Colleagues* (Amsterdam: North-Holland Publishing Company, 1968), pp. 266–280, and Bengt Strömgren, "Niels Bohr and the Royal Danish Academy of Sciences and Letters" in Jorrit de Boer, Erik Dal, and Ole Ulfbeck (eds.), *The Lesson of Quantum Theory: Niels Bohr Centenary Symposium October 3–7, 1985* (Amsterdam: North-Holland Physics Publishing, 1986), pp. 3–12.

73. The Carlsberg Foundation reported its conditional approval of the application in a letter to Bohr 7 Feb 1935, "Carlsberg Foundation," BGC-S (1,3) (note 14).

74. Bohr to Tisdale 22 Feb 1935. A version of this letter transcribed by the Rockefeller Foundation is deposited in RF RG 1.1, 713, 4, 47 (note 66); Bohr's carbon copy is in "Rockefeller Foundation," BGC-S (6,1) (note 14). The two versions of the letter are identical except for a few language corrections in the former.

75. Bohr's application is *ibid.*, wherefrom quotation. On the application's deadline, see Tisdale to Bohr 6 Feb 1935, Bohr to Miller 11 Feb 1935, Bohr to Tisdale 14 Feb 1935, "Rockefeller Foundation," BGC-S (6,1) (note 14). The acceptance of Bohr's argument for Hevesy's biological tradition at the institute is expressed in the Rockefeller Foundation's eventual resolution of 17 Apr 1935 to provide the institute with a grant from the experimental biology program, RF RG 1.1, 713, 4, 46 (note 15).

76. Bohr to Tisdale 22 Feb 1935 (note 74).

77. *Ibid.*

78. On Bohr's continued vacation, see his letters to Heisenberg 28 Jan 1935 and 9 Mar 1935, BSC (20,2) (note 31). For the arrangement of Hevesy's personally delivering the application, see Bohr to Tisdale 14 Feb 1935, Tisdale to Bohr 28 Feb 1935, "Rockefeller Foundation," BGC-S (6,1) (note 14), as well as the application itself, Bohr to Tisdale 22 Feb 1935 (note 74). The quotation is from Tisdale to Weaver 27 Feb 1935, RF RG 1.1, 713, 4, 47 (note 66).

79. *Ibid.*

80. Hevesy's description of other candidate institutions is in *ibid*. Chievitz wrote about his ideas of applying atomic physics to biology in a letter to Bohr 20 Apr 1924, Bohr General Correspondence, deposited in the Niels Bohr Archive, Copenhagen. Bohr wrote an obituary for his friend, "Ole

Chievitz" in *Ord och Bild 55* (1947), 49–53. On the Finsen Institute, see the comprehensive Vilhelm Møller-Christensen (ed.), *Finsen Instituttet 1896–23. Oktober–1946* (Copenhagen: Det Berlingske Bogtrykkeri, 1946); I am grateful to Knud Max Møller for this reference.

81. Tisdale to Weaver 27 Feb 1935 (note 78).
82. Bohr to Tisdale 22 Feb 1935 (note 74).
83. Tisdale put forward his confidential request in his letter to Bohr 6 Feb 1935, and Bohr expressed his favorable attitude toward support in his reply 14 Feb 1935, "Rockefeller Foundation," BGC-S (6,1) (note 14). The information brought by Hevesy was reported in Tisdale to Weaver 27 Feb 1935 (note 78). The quotations are from Tisdale to Hevesy 7 Mar 1935, "Rockefeller Foundation," BGC-S (6,1) (note 14), where Hevesy's telegram and Krogh's letter are, respectively, quoted and cited.
84. Tisdale wrote approvingly of support for Krogh in a letter to Weaver 6 Mar 1935, RF RG 1.1, 713, 4, 47 (note 66).
85. The quotations are from Tisdale to Hevesy 7 Mar 1935 (note 83). Tisdale used the term "Krogh – v. Hevesy – Bohr proposal" as the heading of his letter to Weaver 27 Feb 1935 (note 78).
86. For Bohr's correspondence with the Rockefeller Foundation on other matters, see Bohr to Tisdale 8 Mar 1935, Bohr to Miller 8 Mar 1935, Tisdale to Bohr 11 Mar 1935, Miller to Bohr 11 Mar 1935, "Rockefeller Foundation," BGC-S (6,1) (note 14). The substance of Hevesy's response to Tisdale was reported in Krogh to Tisdale 12 Mar 1935, RF RG 1.1, 713, 4, 47 (note 66). On Rehberg's permission to teach, see also "Oprettelse af et Lektorat i Zoofysiologi ved Universitetet for Amanuensis, Dr. phil. P.B. Rehberg" in *Aarbog for Københavns Universitet, Kommunitetet og Den Polytekniske Læreanstalt, Danmarks Tekniske Højskole, indeholdende Meddelelser for det akademiske Aar 1935–1936 [Copenhagen University Yearbook 1935–36]* (Copenhagen: A/S J. H. Schultz Bogtrykkeri, 1939), pp. 35–36. The quotation is from Tisdale to Hevesy 11 Mar 1935, "Rockefeller Foundation," BGC-S (6,1).
87. Hevesy to Tisdale 13 Mar 1935, *ibid.*
88. Chievitz statement 12 Mar 1935, *ibid.*
89. Krogh to Tisdale 12 Mar 1935 (note 86).
90. *Ibid.*
91. Tisdale to Hevesy 16 Mar 1935, "Rockefeller Foundation," BGC-S (6,1) (note 14).
92. Tisdale to Bohr 18 Apr 1935, Norma S. Thompson (Tisdale's secretary) to President Østrup 25 Apr 1935, Weaver to Østrup 25 Apr 1935, *ibid.* See also the Rockefeller Foundation's *Annual Report 1935* (note 57), pp. 129–130. The exchange rates are taken from *Banking* (note 63), p. 669. On the support from the Carlsberg Foundation, see Chapter 1.
93. Tisdale Diary 24 May 1935, RF RG 12 (note 14).

CHAPTER 5

1. The first grant mentioned was announced in a letter from the Nordic Insulin Foundation to Bohr 22 Mar 1935, "Nordic Insulin Foundation," Bohr General Correspondence, Special File, film 3, section 4 (BGC-S (3,4)), deposited in the Niels Bohr Archive, Copenhagen. Bohr applied for the second grant in a letter of 10 Nov 1938, receiving a positive response on 7 Jul 1939, *ibid.* For general information on the Nordic Insulin Foundation as well as other Danish organizations, I am indebted to the erudition of Knud Max Møller at the Biological Institute of the Carlsberg Foundation. The quotation is from the Warren Weaver Diary, 14 Feb 1937; diaries of the Rockefeller Foundation officers are filed in the

Rockefeller Foundation archive, Record Group 12 (RF RG 12), Rockefeller Archive Center (RAC), Pocantico Hills, New York.

2. The quotation is from the Warren Weaver Diary 14 Feb 1937, RF RG 12 (note 1). A photocopy of the relevant internal records of the Thrige Foundation was kindly supplied to me by Erik Rüdinger at the Niels Bohr Archive, Copenhagen. The records have been provided to the Archive by the Thrige Foundation in connection with research for a book on the history of the cyclotron at the institute, written by Niels Ove Roy-Poulsen, Niels Ove Lassen, and Mikael Jensen, to be published in early 1990. I have found no correspondence regarding the gift from the Thrige Foundation. The amounts in dollars here and subsequently are calculated on the basis of *Banking and Monetary Statistics* (Washington, D.C.: Board of Governors of the Federal Reserve System, November 1943), p. 669.

3. The first quotation is from a copy of the official letter announcing the gift, dated 7 Oct 1934, which is retained in the file "Radiumgaven," Bohr General Correspondence (BGC), the Niels Bohr Archive, Copenhagen; the file also contains detailed information on the purchases made with the gift. For a calculation of the exact amount of the gift, see Albert V. Jørgensen to Bohr 2 Sep 1936, *ibid*. I have found no correspondence regarding the gift before it was made. Hevesy's remark to Rutherford was made in his letter of 8 Oct 1935, Rutherford Correspondence Collection, Cambridge University Library, with a copy in the Hevesy Scientific Correspondence (HSC), Niels Bohr Archive, Copenhagen. Rutherford's quoted reply, in the HSC, is dated 14 Oct 1935.

4. Bohr's application to the Carlsberg Foundation of 30 Sep 1936 and the positive reply of 25 Nov 1936 are in "Carlsberg Foundation," BGC-S (1,3) (note 1). A carbon copy of Bohr's application to the Zeuthen Fund of 30 Sep 1936, as well as the reply of 14 Nov 1934, wherefrom quotation, are in "Zuethen Memorial Fund," BGC-S (6,8). For the subsequent support from the Zeuthen Fund, see the Fund's approval of 29 Nov 1939, *ibid*. I am indebted to Knud Max Møller for information about the Fund's establishment.

5. Robert E. Kohler, "The Management of Science: The Experience of Warren Weaver and the Rockefeller Foundation Programme in Molecular Biology" in *Minerva 14* (1976), 279–306, pp. 296–301. The first quotation is from the Warren Weaver Diary 14 Feb 1937, RF RG 12 (note 1). The last two statements are from Weaver to Wilbur Tisdale 23 Feb 1937, "713D, University of Copenhagen: Biophysics 1935–1937," Rockefeller Foundation archives, Record Group 1.1, Series 713, Box 4, Folder 47 (RF RG 1.1, 713, 4, 47), RAC (note 1). The amounts appropriated by the Rockefeller Foundation in 1936 and 1937 can be seen in *The Rockefeller Foundation: Annual Report 1936* (New York: Rockefeller Foundation, undated), and ditto 1937, pp. 348–349 and 392–393, respectively.

6. The resolution to approve Bohr's grant, dated 19 Mar 1937, is in "713D, University of Copenhagen: Biophysics 1933–1934," RF RG 1.1, 713, 4, 46 (note 5). The Copenhagen cyclotron is presented together with the Minnesota Van de Graaff and the Paris cyclotron as "Apparatus for Tagging the Atoms" in *Rockefeller Foundation: Annual Report 1937* (note 5), pp. 194–198, quotation on p. 194. The support to Franck is described in *Rockefeller Foundation: Annual Report 1936* (note 5), pp. 209–211, quotation on p. 210.

7. The quotation is from Bohr to Weaver 4 Apr 1937, RF RG 1.1, 713, 4, 47 (note 5). On Lawrence's Berkeley Radiation Laboratory, see J. L. Heilbron, Robert W. Seidel, and Bruce R. Wheaton, *Lawrence and His Laboratory: Nuclear Science at Berkeley 1931–1961* (Berkeley: Office of the History of Science and Technology, University of California, 1981), and the much more substantial John L. Heilbron and Robert W. Seidel,

A History of the Lawrence Berkeley Laboratory, vol. 1: *Lawrence and His Laboratory* (Berkeley: University of California Press, in press).

8. On the prehistory and establishment of the Cancer Society, see the "Indledning" [Introduction] of its annual report, *Landsforeningen til Kræftens Bekæmpelse: Aarsberetning 1929,* pp. 5–6; on Bohr's involvement, see "Niels Bohr 7/10 1885–18/11 1962" in *ibid. 1962,* pp. 11–12.

9. For the cooperation between the institute and the Cancer Society, see the Cancer Society to Bohr 5 Oct 1938, Bohr to the Cancer Society 7 Oct 1938, wherefrom first quotation, and Cancer Society to Bohr 22 Jun 1939, "National Committee for the Eradication of Cancer," BGC-S (3,3) (note 1). Bohr made his recommendation to install a Van de Graaff generator in a letter to the Society of 18 May 1940, *ibid.;* the final decision not to heed Bohr's advice can be seen from Bohr's letter to the Society of 4 May 1942, *ibid.* In a letter, wherefrom quotation, of 26 Feb 1941 *(ibid.),* Bohr informed the Cancer Society of his application to the Carlsberg Foundation of 30 Sep 1940; this application is in "Carlsberg Foundation," BGC-S (3,3). The failure of the high-voltage installation at the Radium Station was kindly reported to me by Knud Max Møller, who solicited his information from Morten Christensen, a physicist at the Finsen Institute's Department for Radiation Physics, who was involved in the installation of the new equipment.

10. Bohr's May 1935 application to the Danish government for operating expenses is reproduced in "Institutet for teoretisk Fysik," *Aarbog for Københavns Universitet, Kommunitetet og Den polytekniske Læreanstalt, Danmarks tekniske Højskole indeholdende Meddelelser for det akademiske Aar 1937–38 [Copenhagen University Yearbook 1937–38* (1944)] (Copenhagen: J. H. Schultz A/S, 1944), pp. 204–207. The new permanent assistantship is described in the *Copenhagen University Yearbook 1935–36* (1939), p. 17. See other relevant volumes of this *Yearbook* for Bohr's applications to the Danish government and the government's response to them.

11. Bohr to the Carlsberg Foundation 30 Sep 1938, "Carlsberg Foundation," BGC-S (1,3) (note 1). The fate of this and other applications of Bohr's can be seen in the handwritten "Budget Books" of the institute, retained in the Niels Bohr Archive, Copenhagen.

12. Bohr presented his request to the government for an increased annual amount to complement the support from the Rockefeller Foundation for experimental biology in his letter to the University Governing Board 7 May 1935; for its approval, see the University Governing Board to Bohr 13 Jun 1935, "University Board of Directors/University," BGC-S (8,7) (note 1). In my calculations, the arbitrarily fluctuating support from the Rask-Ørsted Foundation for foreign visitors has been set to a constant Dkr 15,000 per year, which is the average taken from the institute's Budget Books (note 11). B. R. Mitchell, *European Historical Statistics 1705–1975* (Second Revised Edition) (New York: Facts on File, 1981) provides Danish "Wholesale Indices" and "Cost of Living Indices," respectively, on pp. 774 and 781.

13. G. Hevesy and E. Hofer, "The Elimination of Water from the Human Body" in *Nature 134* (8 Dec 1934), 879; G. Hevesy, E. Hofer, and A. Krogh, "The Permeability of the Skin of Frogs to Water as Determined by D_2O and H_2O" in *Skandinavische Archiv für Physiologie 72* (1935), 199–214, paper received by journal 19 Jul 1934. On Schönheimer's relationship with the Rockefeller Foundation, see "200D, Columbia University: Biological Chemistry 1935–1951," RF RG 1.1, 200, 130, 1604–1608 and 1.1, 200, 131, 1609–1611 (note 5). A historical account of Schönheimer's work is Robert E. Kohler, "Rudolf Schoenheimer, Isotopic Tracers, and Biochemistry in the 1930's" in *Historical Studies in the Physical Sciences 8* (1977), 257–298.

14. Charles Weiner's interview with Hilde Levi 28 Oct 1971, p. 22 of original transcript; this interview is deposited in the Niels Bohr Library of the American Institute of Physics, New York City.
15. *Ibid.* See also my interviews with Levi at the Niels Bohr Institute, Copenhagen 28 Oct 1980 and 10 Sep 1981; Hilde Levi, *George de Hevesy: Life and Work* (Bristol: Adam Hilger; Copenhagen: Rhodos, 1985), pp. 77–79. Levi's Hevesy biography also contains a list of his publications on pp. 135–147.
16. The quotation is from O. Chievitz and G. Hevesy, "Radioactive Indicators in the Study of Phosphorus Metabolism in Rats" in *Nature 136* (9 Nov 1935), 754–755; paper dated 13 Sep 1935. The editor's comments were presented as "Points from Foregoing Letters" in *ibid.*, p. 761. An expanded version of the results was published as O. Chievitz and G. Hevesy, "Studies on the Metabolism of Phosphorus in Animals" in *Det Kongelige Danske Videnskabernes Selskab. Biologiske Meddelelser 13* (9, 1937), 22 pp. The remark on Chievitz is taken from John D. Cockcroft, "George de Hevesy 1885–1966" in *Biographical Memoirs of Fellows of the Royal Society 13* (1967), 125–166, p. 146; this obituary was actually written by Hevesy himself before he died – see Cockcroft to Léon Rosenfeld 5 May 1967, HSC (note 3).
17. G. Hevesy, K. Linderstrøm-Lang and C. Olsen, "Atomic Dynamics of Plant Growth" in *Nature 137* (11 Jan 1936), 66–67; paper dated 9 Dec 1935. For general information on the Carlsberg Laboratory, see H. Holter and K. Max Møller (eds.), *The Carlsberg Laboratory 1876–1976* (Copenhagen: Rhodos, 1976), which also contains biographical articles on Linderstrøm-Lang and Olsen: H. Holter, "K. U. Linderstrøm-Lang," pp. 88–117, Poul Larsen, "Carsten Olsen," pp. 130–138. On Linderstrøm-Lang's earlier negative attitude to Hevesy's work, see Cockcroft, "Hevesy," (note 16), pp. 145–146.
18. G. Hevesy, "Isotopernes Anvendelse som Indikatorer i Kemi og Biologi" in *Naturens Verden 20* (1936), 289–301; *idem.*, "Isotopernes Anvendelse i den kemiske Analyse" in *ibid.*, pp. 357–366; *idem.*, "Isotopernes Anvendelse til Undersøgelse af Stofskiftet i levende Organismer" in *ibid.*, pp. 401–414.
19. Regarding the planning of the conferences, see the Wilbur Tisdale Diary 24–26 Jun 1936, RF RG 12 (note 1), and Bohr to Tisdale 17 Sep 1936, RF RG 1.1, 713, 4, 47 (note 5), as well as Lily E. Kay, "Conceptual Models and Analytical Tools: The Biology of Physicist Max Delbrück, 1931–1946" in *Journal of the History of Biology 18* (1985), 207–246, p. 223. On the absence of Hevesy and Krogh at the first conference, see Bohr to H. M. Miller 2 Oct 1936, copy in RF RG 1.1, 713, 4, 47. Krogh's lecture at Harvard University on 10 Sep 1936 was reprinted as "The Use of Isotopes as Indicators in Biological Research" in *Science 85* (19 Feb 1937), 187–191. The first conference is described in M. Delbrück and N. W. Timofeéff-Ressovsky, "Summary of discussions on mutations (held by: Muller, Timofeéff-Ressovsky, Bohr, Delbrück and others) at Copenhagen 28–29/IX 1936," a copy of which was kindly supplied to me by the late Max Delbrück. Dirac's comment was noted in the H. M. Miller Diary 18–19 Nov 1936, RF RG 12 (note 1).
20. The chronology leading to the Rockefeller Foundation's support of the two conferences is as follows. Bohr to Tisdale 17 Sep 1936: a second conference on Hevesy's indicator technique is suggested for the first time; Miller (in Tisdale's absence) to Bohr 21 Sep 1936: more information asked for; Bohr to Miller 23 Sep 1936: two separate conferences proposed; Miller to Bohr 25 Sep 1936: request approved. The quotations are taken from Bohr to Tisdale 17 Sep 1936, RF RG 1.1, 713, 4, 47 (note 5). A report of the conference, including a list of participants, is retained in "713D, University of Copenhagen: Biophysics 1938–1939," RF RG

1.1, 713, 4, 48 (note 5). On Meyerhof's situation, see Laura Fermi, *Illustrious Immigrants: The Intellectual Migration from Europe 1930–41* (Chicago and London: University of Chicago Press, 1968), p. 313.

21. See Levi's Hevesy bibliography (note 15).

22. G. Hevesy and E. Lundsgaard, "Lecithinaemia following the Administration of Fat" in *Nature 140* (14 Aug 1937), 275–276. On Lundsgaard, see Egill Snorrason (L. S. Fridericia), "Lundsgaard, Einar" in *Dansk Biografisk Leksikon* [third edition] *IX* (Copenhagen: Gyldendal, 1981), pp. 196–197. G. Hevesy, J. Holst, and A. Krogh, "Investigations on the Exchange of Phosphorus in Teeth Using Radioactive Phosphorus as Indicator" in *Det Kongelige Danske Videnskabernes Selskab. Biologiske Meddelelser 13* (13, 1937), 34 pp., printing completed 23 Nov 1937. The work for the Danish State Farm was reported as G. C. Hevesy, H. B. Levi, and O. H. Rebbe, "The Origin of the Phosphorus Compounds in the Embryo of the Chicken" in *The Biochemical Journal 32* (12, 1938), 2147–2155, paper received by journal 31 Aug 1938. The cooperation with the hospitals is referred to in several places. See Chievitz and Hevesy, "Studies" (note 16), pp. 7–10; G. Hevesy, "Radioactive Phosphorous as Indicator in Biology" [lecture given in October 1938] in *Nuovo Cimento 15* (May 1938), 279–312, pp. 293–295; *idem.*, "The Application of Isotopic Indicators in Biological Research" in *Enzymologia 5* (10 Oct 1938), 138–157, pp. 142–144; and [*idem.*] "Brief summary of the physicobiological researches which have been carried out in the last three years at Copenhagen with the support of the Rockefeller Foundation" [Sep 1938], RF RG 1.1, 713, 4, 48 (note 20).

23. G. Hevesy, T. Baranowski, A. J. Guthke, P. Ostern, and J. K. Parnas, "Untersuchungen über die Phosphorübertragungen in der Glykolyse und Glykogenolyse" in *Acta Biologiae Experimentalis 12* (4, 1938), 34–39, paper presented by J. K. Parnas for the Polish Physiological Society on 1 Dec 1937. G. C. Hevesy and Ida Smedley-MacLean, "The Synthesis of Phospholipin in Rats Fed on the Fat-Deficient Diet" in *The Biochemical Journal 34* (1940), 903–905, paper received by journal 9 Mar 1940. The information on the Lister Institute is taken from *The World of Learning 1988* (thirty-eighth edition) (London: Europa Publications Limited, 1987), p. 1350.

24. [Hevesy,] "Brief Summary" (note 22). On Hevesy's enthusiastic and successful efforts at interesting other individuals and institutions, see, in particular, the sprightly discussion in Weiner's Levi interview 28 Oct 1971 (note 14), pp. 54–56.

25. My interview with Hilde Levi at the Niels Bohr Institute 28 Oct 1980. For publications from the institute, see the relevant volumes of *Universitetets Institut for teoretisk Fysik, Afhandlinger*, retained in the Niels Bohr Archive and containing the publications resulting from work there from 1918 to 1959. Levi's first coauthorship in biology was Hevesy, Levi, and Rebbe, "Origin" (note 22). On Hevesy's continued interest in biological questions, as well as Levi's move to Krogh's institute see Levi, *Hevesy* (note 15), pp. 76ff, 103–105. On the move of zoophysiology to the August Krogh Institute, as well as the history of Copenhagen zoophysiology generally, see C. Barker Jørgensen, "Dyrefysiologi og gymnastikteori" in *Københavns Universitet 1479–1979* [14 volumes], *bind XIII: Det matematisk-naturvidenskabelige Fakultet, 2. del* (Copenhagen: G.E.C. Gads Forlag, 1979), pp. 447–488, on p. 477.

26. A record of visitors to the institute was kept in the institute's "Guest Book," microfilmed for the Archive for the History of Quantum Physics, film 35, section 2 (AHQP (35,2)); the AHQP is retained in the American Philosophical Society, Philadelphia; the Niels Bohr Archive, Copenhagen; the Niels Bohr Library of the American Institute of Physics in New York; and other places. On the AHQP, see Thomas S. Kuhn, John L.

Heilbron, Paul Forman, and Lini Allen, *Sources for the History of Quantum Physics: An Inventory and Report* (Philadelphia: The American Philosophical Society, 1967). The original of the Guest Book, some dates in which are impossible to read on microfilm, is deposited in the Niels Bohr Archive, Copenhagen. On the sources of support for Rebbe, Levi, and Aten, see Bohr to the Secretary of the University of Copenhagen 13 Nov 1936 (on Levi only) and 27 Sep 1937 [account of expenditures of the Rockefeller Foundation's annual support 1 Jul 1936 to 30 Jun 1937], RF RG 1.1, 713, 4, 47 (note 5). In the institute's Budget Books (note 11), information is supplied on Hahn's situation as well.

27. A useful source to the papers reported to the Rockefeller Foundation as experimental biology work is "Papers on the Application of Isotopic Indicators, Copenhagen 1935 to 1940," "713D, University of Copenhagen: Biophysics 1940," RF RG 1.1, 713, 4, 49 (note 5). Most of the publications originating from the institute can also be gleaned from *Afhandlinger* (note 25). The development of Hevesy's program is discussed in general terms in my interview with Hilde Levi 10 Sep 1981 (note 15); see also Levi, *Hevesy* (note 15), pp. 76–101. The stays of Hevesy's visiting collaborators can be determined from the institute's Guest Book (note 26).

28. For Armstrong's work, see G. Hevesy and W. D. Armstrong, "Exchange of Radiophosphate by Dental Enamel" in *Proceedings of the American Society of Biological Chemists, Thirty-Fourth Annual Meeting, New Orleans, Louisiana, March 13–16, 1940* [received for publication 25 Jan 1940], published as addendum to *Journal for Biological Chemistry 133* (1940), xliv. Arnold described his Copenhagen work in a letter to me 24 Aug 1980. For Kamen and Ruben's contribution, see Martin D. Kamen, *Radiant Science, Dark Politics: A Memoir of the Nuclear Age* (Berkeley, Los Angeles, London: University of California Press, 1985), pp. 122–160.

29. The information on publications from Krogh's laboratory is taken from "Papers" (note 27).

30. The quotation is from [Hevesy,] "Brief Summary" (note 22), which is also a useful description of the experimental biology project at Copenhagen University. For Hevesy's relationship with Lawrence and his laboratory, see the extensive Hevesy-Lawrence correspondence from 1938 to 1940, beginning with Lawrence to Hevesy 5 Jan 1938, HSC (note 3). On the use of the Copenhagen cyclotron, see J. C. Jacobsen, "Om Cyklotronen" in *Fysisk Tidsskrift 39* (1941), 33–50, pp. 49–50. The first publication from Copenhagen on the use of potassium as a radioactive indicator was L. Hahn, G. Hevesy, and O. Rebbe, "Permeability of Corpuscles and Muscle Cells to Potassium Ions" in *Nature 143* (17 Jun 1939), 1021–1022, paper dated 9 May 1939.

31. Relevant contemporary review articles of the emerging field are: David M. Greenberg, "Mineral Metabolism, Calcium, Magnesium, and Phosphorus"in *Annual Review of Biochemistry 8* (1939), 269–300; John R. Loofbourow, "Borderland Problems in Biology and Physics" in *Reviews of Modern Physics 12* (Oct 1940), 267–358. Arnold's recollection was made in his letter to me 24 Aug 1980 (note 28). The quotation about Hevesy is taken from Urey to Weaver 9 May 1939, RF RG 1.1, 713, 4, 48 (note 20). Hevesy's review article was his "Application of Radioactive Indicators in Biology" in *Annual Review of Biochemistry 9* (1940), 641–662, which reviewed papers published prior to 1 Nov 1939.

32. The scientific workers paid by the experimental biology program are reported in Bohr to the Secretary of the University of Copenhagen 13 Nov 1936, 27 Sep 1937, 30 Jun 1938, Sep 1939, RF RG 1.1, 713, 4, 48 (note 20). The contamination problem is discussed in my interview with Hilde Levi 10 Sep 1981 (note 15).

33. The Niels Bohr Scientific Correspondence (BSC) is retained in the Niels Bohr Archive, Copenhagen, with microfilm copies also at the Niels Bohr Library of the American Institute of Physics in New York and other places. Whereas the material on the films themselves is arranged alphabetically by correspondents, the Bohr Archive has also prepared a useful chronological listing of the letters. The plans for a sequel to the article with Rosenfeld were reported in Bohr to Heisenberg 16 Mar 1935, BSC, *ibid.*, film 20, section 2, (BSC (20,2)).

34. The quotation is from Bohr to Heisenberg 9 May 1935 [not sent], *ibid.* The criticism of quantum mechanics was published as A. Einstein, B. Podolsky, and N. Rosen, "Can Quantum-Mechanical Description of Physical Reality Be Considered Complete?" in *Physical Review 47* (15 May 1935), 777–780. It was rebutted in Niels Bohr, "Can Quantum-Mechanical Description of Physical Reality Be Considered Complete?" in *Physical Review 48* (15 Oct 1935), 697–702. Bohr and Rosenfeld's sequel was finally published as "Field and Charge Measurements in Quantum Electrodynamics" in *Physical Review 78* (15 Jun 1950), 794–798; paper received by journal 19 Oct 1949.

35. The publications coming out of the institute are listed in *Afhandlinger* (note 25). On Gamow, see George Gamow, *My World Line: An Informal Autobiography* (New York: Viking Press, 1970). The time of Gamow's stay is inferred from the institute's Guest Book (note 26).

36. *Afhandlinger* (note 25).

37. Institute's Guest Book (note 26). Otto Robert Frisch, *What Little I Remember* (Cambridge, etc.: Cambridge University Press, 1979), pp. 81–119; Levi interview 10 Sep 1981 (note 15). For Frisch's basis of support, see the institute's Budget Books (note 11) and Bohr to the Secretary of the University of Copenhagen 13 Nov 1936, 27 Sep 1937, 30 Jun 1938, Sep 1939 (all note 32). Frisch tells of his subsequent career in *What Little*, pp. 120ff.

38. *Afhandlinger* (note 25). The number of visitors is taken from "Visitors from abroad who for longer periods have worked at the Institute for Theoretical Physics, University of Copenhagen 1918–1948 (arranged according to their native countries)," prepared for application purposes circa 1950 and retained in the Niels Bohr Archive, Copenhagen. A copy of the list is deposited in the J. Robert Oppenheimer Papers, container 21, Library of Congress, Washington, D.C. More detailed information on the visitors is in the institute's Guest Book (note 26).

39. Niels Bohr, "Neutron Capture and Nuclear Constitution" in *Nature 137* (29 Feb 1936), 344–348; reprinted in Erik Rüdinger (gen. ed.), *Niels Bohr Collected Works, Volume 9:* Rudolf Peierls (ed.), *Nuclear Physics (1929–1952)* (Amsterdam, etc.: North-Holland Physics Publishing, 1986), pp. 152–156.

40. R. Peierls, "Introduction" in *Niels Bohr Collected Works, Volume 9* (note 39), pp. 2–83, on pp. 15–16.

41. Bohr's renunciation of his idea of energy nonconservation is briefly described in *ibid.*, pp. 13–14, where the letter itself, Niels Bohr, "Conservation Laws in Quantum Theory" is reprinted on pp. 215–216. It was originally published in *Nature 138* (4 Jul 1936), 25–26; paper dated 6 Jun 1936. Shankland's claim was published as Robert S. Shankland, "An Apparent Failure of the Photon Theory of Scattering" in *Physical Review 49* (1 Jan 1936), 8–13. Dirac's turnabout was published as P. A. M. Dirac, "Does Conservation of Energy Hold in Atomic Processes?" in *Nature 137* (22 Feb 1936), 298–299.

42. Bohr to Heisenberg 8 Feb 1936 (reproduced in *Niels Bohr Collected Works, Volume 9* (note 39), pp. 579–581), with reply 12 Feb 1936, BSC (20,2) (note 33); Bohr to Klein 8 Feb 1936, with reply 2 Mar 1936, BSC (22, 1); Bohr to Delbrück 11 Feb 1936, with reply 18 Feb 1936, BSC (18,

3); Bohr to Houston 25 Feb 1936, with reply 18 Mar 1936, BSC (21, 1); Bohr to Gamow 26 Feb 1936 (reproduced in *Niels Bohr Collected Works, Volume 9*, pp. 572–573), with reply 25 Mar 1936, BSC (19, 4). Dirac's letter, BSC (18, 4), is dated 9 Jun 1936, and was written in reply to a letter not included in the BSC. The joint article was published as N. Bohr and F. Kalckar, "On the Transmutation of Atomic Nuclei by Impact of Material Particles" in *Det Kongelige Danske Videnskabernes Selskab. Mathematisk-fysiske Meddelelser 14* (10, 1937), 40 pp.; printing completed 27 Nov 1937. The article is reprinted in *Niels Bohr Collected Works: Volume 9* (note 39), pp. 225–264. The close collaboration between Bohr and Kalckar is evidenced in the several manuscripts for their article as well as for related work; see Peierls, "Introduction" (note 40), pp. 30–32.

43. This incident is recounted in Peierls, "Introduction" (note 40), pp. 22–24; the relevant letters are Bohr to Delbrück 18 Mar 1936, Delbrück to Bohr 29 Mar 1936, Rosenfeld to Delbrück 22 [?] Mar 1936, and Delbrück to Rosenfeld 25 Mar 1936, all BSC (18, 3) (note 33), and all reproduced in *Niels Bohr Collected Works, Volume 9* (note 39), pp. 544–547.

44. Delbrück to Bohr 6 Nov [1935], Bohr to Delbrück 22 Nov 1935, BSC (18, 3).

45. Bohr mentioned the postponement of the conference to be held in 1935 in his letter to Klein 9 Aug 1935, BSC (22, 1) (note 33). For Bohr's plans for the conference, which was eventually held in 1936, see his letters to Delbrück 10 Aug 1935, BSC (18, 3), Heisenberg 10 Aug 1935, BSC (20, 2), and Kramers 14 Mar 1936, BSC (22, 3).

46. The number of visitors at each conference can be deduced from the institute's Guest Book (note 26).

47. The account of the domination of Bohr's model was given in Hans A. Bethe, "The Happy Thirties" in Roger H. Stuewer (ed.), *Nuclear Physics in Retrospect: Proceedings of a Symposium on the 1930s* (Minneapolis: University of Minnesota Press, 1979), pp. 11–26, on pp. 23–24; Bethe repeated the account when interviewed by me at the Niels Bohr Institute 1 Apr 1981. For a similar reaction to Bohr and his nuclear model, see Peierls, "Introduction" (note 40), pp. 76–77.

48. *Nordiska (19. skandinaviska) naturforskarmötet i Helsingfors den 11–15 augusti 1936* (Helsinki, 1936). These proceedings contained abstracts of the talks: Niels Arley, "Om spredningen af neutroner med termiske hastingheder ved bundne protoner," pp. 248–249; Torkild Bjerge, "Induceret Radioaktivitet med kort Halveringstid," pp. 251–253; Fritz Kalckar, "Teoretiske Undersøgelser over Forløbet af Atomkerneprocesser," p. 263; Christian Møller, "Om Positronudsendelsen fra β-radioaktive Stoffer," pp. 266–267; Ebbe Rasmussen, "Hyperfinstruktur og Kernespin," pp. 269–270.

49. Peierls, "Introduction" (note 40), pp. 39–42. An account of Bohr's lectures in the United States, for example, was published as "Transmutations of Atomic Nuclei" in *Science 87* (1937), 161–165; reprinted in *Niels Bohr Collected Works, Volume 9* (note 39), pp. 207–211.

50. *Afhandlinger* (note 25).

51. Niels Bohr, "The Rutherford Memorial Lecture 1958: Reminiscences of the Founder of Nuclear Science and of Some Developments Based on His Work" in *Proceedings of the Physical Society 78* (1961), 1083–1115; reprinted in J. B. Birks (ed.), *Rutherford at Manchester* (London: Heywood & Company Ltd., 1962), pp. 114–167, and in Niels Bohr, *Niels Bohr: Essays 1958–1962 on Atomic Physics and Human Knowledge* (New York and London: Interscience, 1963), pp. 30–73; this collection has been republished as *The Philosophical Writings of Niels Bohr, Volume 3: Essays 1958–1962 on Atomic Physics and Human Knowledge* (Woodbridge, Connecticut: Ox Bow Press, 1987). Kalckar died 6 Jan

1938; see, for example, Bohr to Oppenheimer 7 Jan 1938, BSC (24, 1) (note 33).

52. Bohr's first published note on the nuclear photoeffect was Niels Bohr, "Nuclear Photoeffects" in *Nature 141* (1938), 326; reprinted in *Niels Bohr Collected Works: Volume 9* (note 39), pp. 297–298. There is extensive correspondence on this publication: Bohr to Mott 31 Jan 1938, with reply 10 Feb 1938, BSC (23, 4) (note 33); Bohr to Oppenheimer 31 Jan 1938, BSC (24, 1); Bohr to Bloch 1 Feb 1938, with reply 15 Feb 1938, BSC (17, 3); Bohr to Heisenberg 1 Feb 1938, with reply 9 Feb 1938, BSC (20, 2); Bohr to Peierls 1 Feb 1938, with reply 8 Feb 1938, BSC (24, 2); Bohr to Perrin 1 Feb 1938, BSC (24, 2); Bohr to Bothe 2 Feb 1938, with reply 10 Feb 1938, Bohr to Bothe 14 Feb 1938, BSC (17, 4); Bohr to Klein 2 Feb 1938 – Klein's reply seems to be missing, see Bohr to Klein 9 Feb 1938 – BSC (22, 1); Bohr to Kramers 8 Feb 1938, BSC (22, 3); Gamow to Bohr 11 Feb 1938 – Bohr's solicitation of Gamow's advice seems to have been lost – BSC (19, 4); Pauli to Bohr 11 Feb 1938, BSC (24, 2) (reproduced in Karl von Meyenn, Armin Hermann, and Victor F. Weisskopf (eds.), *Wolfgang Pauli, Scientific Correspondence with Bohr, Einstein, Heisenberg o.a., Volume II: 1930–1939* (Berlin, etc.: Springer-Verlag, 1985), pp. 550–551 – this letter was prompted by Bohr's letter to Bloch, which Bloch had shown to Pauli; Weizsäcker to Bohr 21 Feb 1938, BSC (26, 2). Bohr's second published note on this topic was Niels Bohr, "Resonance in Nuclear Photo-Effects" in *Nature 141* (1938), 1096; reprinted in *Niels Bohr Collected Works: Volume 9* (note 39), pp. 331–332. Bohr's work on the nuclear photoeffect is described in Peierls, "Introduction" (note 40), pp. 43–52, which also mentions Bohr's unpublished manuscripts on pp. 42 and 49ff. On the original proposal and discovery of the effect, see Roger H. Stuewer, "The Nuclear Electron Hypothesis" in William R. Shea (ed.), *Otto Hahn and the Rise of Nuclear Physics* (Dordrecht, Boston, Lancaster: D. Reidel Publishing Company, 1983), pp. 19–67, on pp. 54–55.

53. Hippel has published his autobiography, Arthur R. von Hippel, *Life in Times of Turbulent Transitions* (Anchorage, Alaska: Stone Age Press, 1988), which recounts his Copenhagen experience on pp. 94–102; Hippel notes on p. 101 that the Dresden firm was his idea, and remembers the journey with Franck on pp. 103–104. For Hippel's and subsequent correspondence with the German firm regarding the purchase of the high-voltage equipment, see the substantial file "Koch & Sterzel" in BGC-S (note 1); this file has not yet been microfilmed. On Hippel's marriage to Franck's daughter, see Thomas S. Kuhn *et al.*, six sessions of interview with James Franck 9–14 Jul 1962, Session III, 11 Jul 1962, p. 12, AHQP (note 26). On Hippel's move to the United States, as well as the role of Ebbe Rasmussen, see Hippel to Koch & Sterzel 27 Jun, 14 Aug 1936 in the "Koch & Sterzel" file. On the development and first use of the equipment, see T. Bjerge, K. J. Brostrøm, J. Koch, and T. Lauritsen, "A High Tension Apparatus for Nuclear Research" in *Det Kongelige Danske Videnskabernes Selskab. Mathematisk-fysiske Meddelelser 18* (1, 1940), 37 pp., p. 4. On American physicists' experience of the institute, see Shannon Davies, *American Physicists Abroad: Copenhagen, 1920–1940*, PhD dissertation at the University of Texas at Austin, 1985 (microfilms international 8609491).

54. Bohr to Weaver 4 Apr 1937, copy of excerpt in RF RG 1.1, 713, 4, 47 (note 5). Lawrence's objection and his assessment of Laslett were recorded in the Warren Weaver Diary 5 and 9 May 1937, respectively, RF RG 12 (note 1). Tisdale to Joliot 24 May 1937, RF RG 1.1, 713, 4, 47 reported that Joliot would not have to share an assistant with Bohr.

55. Laslett's Copenhagen experience is described in Laslett to me 29 Apr and 15 Jul 1980. His arrival is noted in the institute's Guest Book (note

26). On the problems involved in putting the cyclotron to work, see John L. Heilbron, "The First European Cyclotrons" in *Rivista di storia della scienza 3* (1986), 1–44, p. 25. The circumstances for Laslett's departure was noted in the Wilbur Tisdale Diary 29 Oct to 3 Nov 1938, RF RG 12 (note 1). Laslett reported neutron production by the Copenhagen cyclotron in a letter to Floyd Lyle 3 Dec 1938, RF RG 1.1, 713, 4, 48 (note 20). The quotation is from Bohr to Tisdale 19 Nov 1938, *ibid*. In a conversation with me on 12 May 1981, the physicist N. O. Lassen at the Niels Bohr Institute provided the exact starting date of the Copenhagen cyclotron from a "Special Protocol" from the cyclotron's construction. On the first cyclotrons to operate outside the United States, see Charles Weiner, "Cyclotrons and Internationalism: Japan, Denmark and the United States, 1935–1945" in *XIVth International Congress of the History of Science* [Tokyo and Kyoto 19–27 Aug 1974]: *Proceedings No. 2* (Tokyo: Science Council of Japan, 1975), pp. 353–365; Heilbron, "European Cyclotrons," pp. 27, 28; I. Kh. Lemberg, V. O. Najdenov, and V. Ya. Frenkel, "The Cyclotron of the A. F. Ioffe Physico-Technical Institute of the Academy of the Sciences of the USSR (on the fortieth anniversary of its startup)" in *Soviet Physics Uspekhi 30* (1987), 993–1006, p. 994; letter from Frenkel to me 18 Apr 1989. On the installation of the Van de Graaff generator at the institute, see Barry Richman and Charles Weiner's oral history interview with Thomas Lauritsen conducted 16 Feb 1967, pp. 14–17, deposited in the Niels Bohr Library of the American Institute of physics, New York City.

56. The development from Fermi's 1934 investigations until the discovery of fission in 1938 is described and analyzed in Spencer Weart, "The Discovery of Fission and a Nuclear Physics Paradigm" in Shea (ed.), *Otto Hahn* (note 52), pp. 91–133. Fermi's June 1934 report of having produced heavier elements than any previously known is his "Possible Production of Elements of Atomic Number Higher Than 92" in *Nature 133* (16 Jun 1934), 898–899; reprinted in Enrico Fermi, *Collected Papers, Volume 1: Italy 1921–1938* (Chicago: University of Chicago Press, 1961), pp. 748–750. The Nobel Prize lecture is Enrico Fermi, "Artificial Radioactivity Produced by Neutron Bombardment" in *Nobel Lectures Including Presentation Speeches and Laureates' Biographies: Physics 1922–1941* (Amsterdam, London, New York: Elsevier Publishing Company, 1965), pp. 414–421, where the new elements were reported on pp. 416–417; reprinted in *Collected Papers*, pp. 1037–1043. On Hahn's contrary evidence and the reaction to it, see Frisch, *What Little* (note 37), p. 114, and Weart, "Discovery," p. 110.

57. The circumstances of Meitner's escape from Germany is recounted in Alan D. Beyerchen, *Scientists under Hitler: Politics and the Physics Community in the Third Reich* (New Haven and London: Yale University Press, 1977), pp. 46–47. Her invitation to Sweden is reported in Frisch, *What Little* (note 37), pp. 113–114. The relevant passages of Hahn's letter, dated 21 Dec 1938, is quoted in Fritz Krafft, *Im Schatten der Sensation: Leben und Wirken von Fritz Strassmann* (Weinheim, etc.: Verlag Chemie, 1981), pp. 105, 265. Hahn's finding was published as O. Hahn and F. Strassmann, "Über den Nachweis und das Verhalten der bei der Bestrahlung des Urans mittels Neutronen entstehenden Erdalkalimetalle" in *Die Naturwissenschaften 27* (6 Jan 1939), 11–15; paper received by journal 22 Dec 1938.

58. Meitner and Frisch's discussion and conclusion is described in Frisch, *What Little* (note 37), pp. 114–116. The quotation is from Roger H. Stuewer, "Niels Bohr and Nuclear Physics" in A. P. French and P. J. Kennedy (eds.), *Niels Bohr: A Centenary Volume* (Cambridge, Massachusetts and London, England: Harvard University Press, 1985), pp. 197–220, on p. 211.

59. On the origin of the liquid drop model and Bohr's change of conception, see Stuewer, "Bohr and Nuclear Physics" (note 58), pp. 199, 209–210, 212, where Bohr is quoted on p. 209. Stuewer's forthcoming article, "The Origins of the Liquid-Drop Model of the Nucleus," to appear in the *Proceedings* of a conference on "50 Years of Nuclear Fission" (West Berlin, 30–31 Mar 1989, describes the subject in more detail.

60. Roger H. Stuewer, "Bringing the News of Fission to America" in *Physics Today 38* (Oct 1985), 48–56, pp. 50–52, and Peierls, "Introduction" (note 40), pp. 52–55, which contain corrections of Frisch, *What Little* (note 37), pp. 116–117, and of Frisch, "The Interest Is Focussing on the Atomic Nucleus" in Stefan Rozental (ed.), *Niels Bohr: His life and Work as Seen by His Friends and Colleagues* (Amsterdam: North-Holland Publishing Company, 1968), pp. 137–148, on pp. 145–147.

61. This incident has been reported in several places. For a recent account quoting from the relevant archival material, see Peierls, "Introduction" (note 40), pp. 55–64. Frisch to Bohr 15/18 Mar 1939, BSC (19.3) (note 33) is reprinted in *Niels Bohr Collected Works: Volume 9* (note 39), p. 565. The proposed explanation was published as Lise Meitner and O. R. Frisch, "Disintegration of Uranium by Neutrons: A New Type of Nuclear Reaction" in *Nature 143* (11 Feb 1939), 239–240; paper dated 16 Jan 1939. The experimental verification was published as O. R. Frisch, "Physical Evidence for the Division of Heavy Nuclei under Neutron Bombardment" in *Nature 143* (18 Feb 1939), 276; paper dated 17 Jan 1939. The quotation is from Weiner's Levi interview 28 Oct 1971 (note 14), p. 73.

62. Arnold to me 24 Aug 1980 (note 28); see the slightly different account in Frisch, *What Little* (note 37), p. 117.

63. Niels Bohr and John Archibald Wheeler, "The Mechanism of Nuclear Fission" in *Physical Review 56* (1 Sep 1939), 426–450, paper received by journal 28 Jun 1939; reprinted in *Niels Bohr Collected Works: Volume 9* (note 39), pp. 365–450. Wheeler brings his collaboration with Bohr into a broader context in John Archibald Wheeler, "Some Men and Moments in the History of Nuclear Physics: The Interplay of Colleagues and Motivations" in Stuewer (ed.), *Retrospect* (note 47), pp. 217–282, where Wheeler describes his experience of collaborating with Bohr in Copenhagen on pp. 238–241, and the Princeton collaboration on pp. 272–282. For a complementary account, see *idem.*, "Niels Bohr and Nuclear Physics" in *Physics Today 16* (Oct 1963), 36–45. See also my interview with Wheeler in Princeton on 4 May 1988, deposited in the Niels Bohr Library of the American Institute of Physics in New York City.

64. On the method used in the first verification of fission, see Frisch, "Physical Evidence" (note 61), p. 276. The next publication reported use of the high-voltage equipment: Lise Meitner and O. R. Frisch, "Products of the Fission of the Uranium Nucleus" in *Nature 143* (18 Mar 1939), 471–472, paper dated 6 Mar 1939. The use of the high-voltage equipment for early fission studies is described in Bjerge *et al.*, "High Tension Apparatus" (note 53), pp. 4–5.

65. A list of publications from the institute is found in *Afhandlinger* (note 25). In particular, the experimental article coauthored by Bohr was N. Bohr, J. K. Bøggild, K. J. Brostrøm, and T. Lauritsen, "Velocity-Range Relation for Fission Fragments" in *Physical Review 58* (1 Nov 1940), 839–840, paper received by journal 3 Sep 1940; reprinted in Erik Rüdinger (gen. ed.), *Niels Bohr Collected Works, Volume 8:* Jens Thorsen (ed.), *The Penetration of Charged Particles through Matter (1912–1954)* (Amsterdam, etc.: North-Holland Physics Publishing, 1987), pp. 325–326. The first publication reporting the use of the cyclotron to study fission problems was I. J. Jacobsen and N. O. Lassen, "Deuteron Induced Fission in Uranium and Thorium" in *Physical Review 58* (15 Nov 1940),

867–868; paper received by journal 28 Jun 1939. See J. C. Jacobsen, "Construction of a Cyclotron" in *Det Kongelige Danske Videnskabernes Selskab. Mathematisk-fysiske Meddelelser 19* (2, 1941), 32 pp.

66. The number of publications is taken from *Afhandlinger* (note 25), the number of visitors from "Visitors" (note 38). Lauritsen's departure is described in the Lauritsen interview (note 55), pp. 19–20. On Bohr's escape, see, for example, Stefan Rozental, "The Forties and the Fifties" in *idem.* (ed.), *Niels Bohr* (note 60), pp. 149–190, on pp. 166–168, which also describes the occupation of the institute on pp. 171–172. On science in wartime Germany, see Mark Walker, *German National Socialism and the Quest for Nuclear Power 1939–1949* (Cambridge: Cambridge University Press, 1989).

67. Bohr's return and the subsequent development of his institute is briefly described in Rozental, "Forties and Fifties" (note 66), pp. 173ff. Basic information on the work at the institute after the war can be gleaned from the annual bibliographies in *Afhandlinger* (note 25). On the continuation of aborted projects, see Peierls, "Introduction" (note 40), pp. 50–52.

68. In conversations with me at the Niels Bohr Institute on 28 Oct 1980 and 12 May 1981, Hilde Levi and N. O. Lassen related the physicists' experience of experimental biology. In the former conversation, Levi also made the point about the perceived security risk with isotopes. This question is treated in general terms in Richard G. Hewlett and Francis Duncan, *Atomic Shield, 1947/1952: A History of the United States Atomic Energy Commission, Volume 2* (University Park and London: Pennsylvania State University Press, 1969), pp. 81–83, 97–98, 109–110. On Hevesy's move to Stockholm and the subsequent development of experimental biology in Copenhagen, see Levi, *Hevesy* (note 15), pp. 103–107. On the Rockefeller Foundation's support of Copenhagen experimental biology after the first five years, see, for example, the memorandum from G. R. P[omerat] to C. I. B[arnard] 10 Nov 1948, General Correspondence, Record Group 2, RAC (note 1).

CONCLUSION

1. The difficulty of tracing the origins of Bohr's idea is described in Peierls, "Introduction" in Erik Rüdinger (gen. ed.), *Niels Bohr Collected Works, Volume 9: Rudolf Peierls* (ed.), *Nuclear Physics (1929–1952)* (Amsterdam, etc.: North-Holland Physics Publishing, 1986), pp. 2–83, on pp. 16–21. Peierls, however, like others who have dealt with this problem historically, does not introduce the context of Bohr's concerted effort to redirect research at the institute.

2. The quotation is from Charles Weiner's interview with Hilde Levi 28 Oct 1971, p. 66 of original transcript; this interview is deposited at the Center for History of Physics, the American Institute of Physics, New York.

3. The transmission of the informal conferences to the United States, for example, is recounted in Thomas David Cornell, *Merle A. Tuve and His Program of Nuclear Studies at the Department of Terrestrial Magnetism: The Early Career of a Modern American Physicist*, PhD dissertation at the Johns Hopkins University 1986 (microfilms international 8609316), pp. 445–446.

Index

Index

Printed in the United States
By Bookmasters